# The 360° Gaze

# The 360° Gaze

Immersions in Media, Society, and Culture

Christian Stiegler

The MIT Press
Cambridge, Massachusetts
London, England

© 2021 Massachusetts Institute of Technology

All rights reserved. No part of this book may be reproduced in any form by any electronic or mechanical means (including photocopying, recording, or information storage and retrieval) without permission in writing from the publisher.

This book was set in Stone Serif and Stone Sans by Westchester Publishing Services. Printed and bound in the United States of America.

Library of Congress Cataloging-in-Publication Data

Names: Stiegler, Christian, 1981– author.
Title: The 360° gaze : immersions in media, society, and culture / Christian Stiegler.
Other titles: Three hundred and sixty degree gaze
Description: Cambridge, Massachusetts : The MIT Press, [2021] |
    Includes bibliographical references and index.
Identifiers: LCCN 2020025426 | ISBN 9780262045667 (hardcover)
Subjects: LCSH: Virtual reality--Social aspects. | Virtual reality in literature. |
    Motion pictures--Plots, themes, etc. | Video games.
Classification: LCC HM851 .S7485 2021 | DDC 303.48/33--dc23
LC record available at https://lccn.loc.gov/2020025426

10  9  8  7  6  5  4  3  2  1

# Contents

Acknowledgments    vii

**Ready Player One: Immersion, Postmodernism, and Liquidity**    1

**1    Rabbit Holes**    13

Virtual Reality: The Beginning Is the End Is the Beginning    13
The Politics of Immersion    35

**2    Beyond the Mediation**    55

Liquid Spaces    55
The 360° Gaze    65

**3    Intermedia Immersions**    103

Texts as Worlds    103
Performing Arts and Sound    115
*Moments of Immersion: James Seager*    130
*Moments of Immersion: Ragnar Kjartansson*    132
The Extension of the Screen    133
World(s) of Warcraft    144
*Moments of Immersion: Jordan Tannahill*    154
*Moments of Immersion: Lynette Wallworth*    157

**4    Sociocultural Immersions**    161

Un/masking the Self    161
*Moments of Immersion: Katie Kelly*    175
Immersive Parasocial    177
Binge-Watching and Algorithmic Flow    189

Posthuman Hybridity and Artificial Intelligence   202
*Moments of Immersion: Ed Atkins*   219

**5   In Limbo**   221

Virtual Reality: The End Is the Beginning Is the End   221

Notes   233
References   275
Index   297

# Acknowledgments

Immersion always begins with the *self*.

At the age of 15, I went through a particularly difficult period in my life. Like many teenagers, I felt alone, unloved, and never to be kissed. Nothing seemed to really fit: the school I was in, the clothes I wore, the girls I dated. In the years of my teenage vulnerabilities I began to isolate myself from the world around me, and I lost myself in music. Quite a lot of music, actually. Endless hours, every day. I spent most of my time hiding in the attic of my parents' house, eating cherries and listening to records by Patti Smith, Bob Dylan, David Bowie, Peter Gabriel, Elliott Smith, and Bruce ("The Boss") Springsteen.

At that time, I discovered R.E.M. The four-piece band from Athens, Georgia, got me at my most vulnerable. And boy, did I need them. Not only did they become the nucleus of my indie-rock and pop-culture universe, but I would even go so far as to admit that they have significantly shaped my personality. I was fascinated by their modesty, their charismatic quirks, and the way they handled their fame. They set the template for many guitar-based bands to follow, from Nirvana to Radiohead, Arcade Fire, and The National.

Over the course of two decades, I think it is safe to say that I became a hardcore fan. I have seen them 17 times in concerts all around the world, interviewed them on three occasions in my role as a music journalist, collected over two hundred band-related items—in what I call my "R.E.M. shrine"—and met several other fans over the years who became dear friends of mine. Whenever I was in love, I always made my partner an R.E.M. mixtape—not to convert her to my devotion, but to give her something that meant a great deal to me. Whenever I had a job interview and needed to introduce myself, the last sentence has always been: "I am also a big fan of R.E.M." That statement may not have seemed particularly revealing, but in fact it was. That I was in love with this particular band probably said more about my personality than any CV or reference letter ever could. In 2011, when R.E.M. called it a day, the amicable way the three remaining members Peter Buck, Mike Mills, and Michael Stipe walked away from their careers was as inspiring to me as everything else they did.

What happened in the years after the breakup I often like to call my "post-immersion" phase. It might have been mainly caused by the mid-life crisis of an aging 30-something. But a feeling of nostalgia and longing for the past made me foolishly attempt to recreate the feeling I once had when I discovered them and bring back the memories I valued so much. I have met the band members, but I never really *knew* them. I was merely granted access to mediated representations of them in concerts, films, interviews, and, of course, their records. But those mediations were so *all-surrounding*, so *deep*, so *immediate*, and so *real* that they had a profound impact on my identity, my selfhood, and my understanding of the world.

R.E.M. have always been *part of my reality*. In fact, even after the band's breakup I have never stopped *feeling immersed* in their music.

For over a decade now, I have been working with immersive technologies, both as a researcher and a storyteller. I have never stopped being a music fan, but I do not visit the attic as frequently as I did when I was younger. But I am aware that technologies such as virtual, augmented, and mixed reality and different modes of digitally induced immersion are able to influence and shape audiences the way music has shaped me. And this is not unique to immersive technologies, but can be witnessed in a range of mediated experiences in postmodern culture. This book is about those experiences.

I should like to express my thanks to Doug Sery, my fantastic editor at the MIT Press, who always believed in me and supported this project with patience and advice. For their invaluable help and assistance in producing this book, I would like to thank Paul Moody, Meredith Jones, Jingrong Tong, and my dear friends at ARTE and Berliner Festspiele, in particular Kay Meseberg and Thomas Oberender.

My special thanks are due to the artists who inspired this book and contributed to it with innovative thoughts and discussions. First and foremost: Björk, Ed Atkins, Lynette Wallworth, Jordan Tannahill, James Seager, Ragnar Kjartansson, Katie Kelly, Claire Hentschker, David Vintiner, Scott McPherson, Iris McPherson, Mary Hickson and the 37d03d community,

# Acknowledgments

Balmorhea, The National, Echo Collective, Dustin O'Halloran, Phoebe Bridgers, Henry Stuart, Barbara Lippe, and all the creative thinkers I have met over the years at several XR events.

I also want to express my gratitude to Patti Smith for letting me use a line from her inspiring record *Horses*, and especially to Father John Misty, who allowed me to use the lyrics to his song *Total Entertainment Forever* as a guiding principle for the chapters in this book.

For her creativity and kindheartedness I am forever grateful to Ines. *Trusty and true.*

Finally, all my love to my mother for keeping up with me all these years.

# Ready Player One:
# Immersion, Postmodernism, and Liquidity

The British anthology series *Black Mirror* has a long tradition of portraying the curious and anxious intersections, mutations, and transmigrations of modern technology. It draws our attention to immersive mediated experiences by setting up a dystopian future that alienates its viewers as readily as it terrifies them. If there is anything like an overall theme across all seasons, it might be an interest in the disconnections of the connected world. It is a crucial issue, as the question of how we use our technologies, or rather how we are used and abused by them, should be the core of any debate around digital societies, emerging technologies, and immersive media. All of the portrayed technological developments in the show, created by Charlie Brooker, seem to be part of a natural, logical, and linear evolution within human society. The rules and (very often lack of) regulations appear plausible to viewers.

For instance, in the episode "Nosedive" (S03E01, 2016), the audience gets transported to an immediate future in which interpersonal encounters conclude with the participants give each other ratings on their phones. If you get a free muffin with your caffè mocha: five stars for the barista, click. If you acknowledge the value of your colleague's pictures of his or her kid's birthday party: five stars for you, click. An overall high rating of an individual has immediate consequences for his or her overall performance in life. Five out of five stars might get you a better deal on insurance, housing, and cars—and ultimately has an effect on which people you might be able to befriend and date. Just as in high school, nobody wants to hang out with a loser.

The resemblances to Facebook and other social media platforms are too obvious to ignore. These *immersions* and *intersections* between mediated and physical performances, and how they affect both concepts of reality, are only a small step away from what Helmond and Gorlitz describe as the "Like Economy."[1] According to them, social media expand over their platforms' boundaries to establish an infrastructure of both decentralized and recentralized data flows. This immense amount of data emerges from the corresponding layers of performance, infrastructure, and social and economic capital. In other words: how (often) and what exactly we are rating will define our social capital and the relations of proximity (how we influence others) in a particular digital ecosystem. All of that ultimately impacts Facebook's economic growth. For Donna Haraway, this kind of social reality is a form of hybridization, "our most important political construction, a world-changing fiction."[2] It creates an infrastructure of invisible and instant participation in which users ultimately become exploited workers[3] within the multilayered fabric of the social web. The ratings in "Nosedive" are the most dominant social currency. Get above 4.5 stars and the world is yours. Drop below 3.5 and you are a pariah in social-network Siberia for whom it is almost impossible to maintain a veneer of niceness to boost your rating score again. In any case, the protagonists *get immersed* into digital spaces and perceive their identities according to the rules of digital platforms. In this book I will argue that digital ecosystems and the logics of ratings and algorithms create powerful immersive experiences. They play a crucial role in the perception of reality.

*Black Mirror* loves to be ahead of the game, and the show also touches on the hype around immersive technologies. In the episode "Playtest" (S03E02, 2016), a young American who is traveling abroad in order to escape some difficulties at home finds himself trapped in an elaborate mixed reality simulation as a test subject for a game company. The company offers him access to an engaging horror game experience through an implant at the back of his head, injected by a headband-like device. This "cybernetic organism, a hybrid of machine and organism"[4] merges with the functionality of the human senses, but extends, transforms, and betrays them. What in the episode is described as "layers on top of reality" turns out to be a mixed reality nightmare. The game testing involves escalating terrors in a haunted cottage (including monstrous spiders) and sets up a system that methodologically digs deeper into the player's psyche and the visualization of his own worst fears. The immersive experience builds up to the final devastating discovery that he is trapped in the immersive environment until he draws his last breath. Death is the only and ultimate escape hatch in this mixed reality scenario. The plot's stinging criticism points to a dystopia in which the technological progress binds and blinds its participants, while the virtual world uses and abuses its visitors.

But while "Nosedive" and "Playtest" both elaborate on the dangers of specific immersive mediated experiences, another of *Black Mirror*'s episodes also captures the social and cultural implications of immersion: "San Junipero" (S03E04, 2016).

One of the characters in "San Junipero" describes the ultimate virtual reality experience as some sort of "immersive nostalgia therapy." Instead of facing the unknown aftermath of their own deaths, participants can have their consciousnesses uploaded to a virtual environment that allows them to live happily forever in the same perfect spot and free of all the maladies of their physical existences. The experience is designed as an almost endless loop of nostalgic settings, and it makes users forget about the world outside and their own mortality. They can choose between different nostalgic decades. The makeover of the 1980s, 1990s, and 2000s is detailed and enriched with the representation of cultural artifacts, from distinguishing soundtracks to films, arcade videogames, and fashion accessories. In the end, as Belinda Carlisle's "Heaven Is a Place on Earth" (1987) fizzes on the radio, the portrayal of an almost too-good-to-be-true romance is carried out in the illusion of the VR experience: a lesbian couple is driving into the sunset.

The two characters decide to live their lives and their love in a virtual, almost lucid dreamlike mediated environment, designed out of the fragments of an '80s nostalgia template. They reject having to face society's dismissal of their love and the consequences of their own deaths. The scenario resembles the plot of *Vanilla Sky* (2001), Cameron Crowe's pop culture remake of Alejandro Amenábar's *Abre los Ojos* (1997). The main character, played by Tom Cruise, decides, after a car accident that scarred his face horribly, to spend the rest of his life in a virtual lucid dream. His ideal world is composed of pop-culture references tailored to his desires in which he is also able to restore his scarred face completely. His self-chosen references include a painting by Monet (*The Seine at Argenteuil*, 1873), films such as *To Kill a Mockingbird* (1962), and album covers like *The Freewheelin' Bob Dylan* (1963).[5]

"San Junipero" points out how nostalgia is connected to immersion. It is not a singular mediated experience but rather a range of different references from film to music that activate the feeling of immersion. Nostalgia plays a significant role in immersing audiences in mediated realities created out of memories, past activities, events, and experiences. Yearning for what is long gone (or has never been) is the ultimate feeling of immersion, one that is able to grow so overwhelmingly powerful that we want to bring it back, again and again. We are aware that our identities are influenced by cultural artifacts—the films we watch, the music we listen to, the art we consume. But we need to further investigate the ways in which feeling immersed in nostalgic mediations affects our identities, especially if emerging technologies are involved in the process. In "San Junipero," nostalgia is the safe comfort of a mediated past to overcome the fear of death in the present. It is a glorified version of the past that we all know too well from films, songs, and images. It does not even matter that this glorification is not based on our own physical experience. The mediation is real enough.[6]

Immersion is a phenomenon that entered the cultural bloodstream of postmodern societies a long time ago. Over the past decade, various immersive experiences like the ones portrayed in *Black Mirror* and other television shows, such as *Westworld, Humans, Electric Dreams, Years and Years, Upload,* and *Devs,* have created mediated realities that penetrate deeply into the mechanics of our everyday life, affecting people's informal interactions, as well as institutional structures and professional routines. Some of these mediated realities are so compelling, so engaging, and so seductive that we spend a majority of our time being surrounded by them. For instance, we have changed the rules and conditions of social interaction by setting up digital avatars on social media and by eliminating the dichotomy between on- and offline realities. As a result, most of us feel at least partly immersed now in digital spaces. Our profiles on Facebook are more than just databases for personal images and videos. They *represent* our identity in the digital world. If you do not have a Facebook profile, you simply *do not exist* in that specific concept of reality. On the other hand, though, if we spent most of our time living through our online identities we might end up in the ratings hell of *Black Mirror*'s "Nosedive" episode.

The phenomenon of immersion is central to understanding mass media in the twenty-first century. The rhetoric of immersion is a commonplace for immersive technologies, but it is actually a grounding principle across a spectrum of mass media and digital-media practices. We are engaged in all sorts of immersive experiences through the texts we read, the films and shows we watch, the games we play, and the images we create. Sometimes they simply augment our reality. At other times they offer us a new one.

In addition, we perform immersive practices in several phenomena of postmodern culture, such as creating profiles on social media, binge-watching, fandom activities, and nostalgia experiences. In fact, the current hype around

# Ready Player One: Immersion, Postmodernism, and Liquidity 5

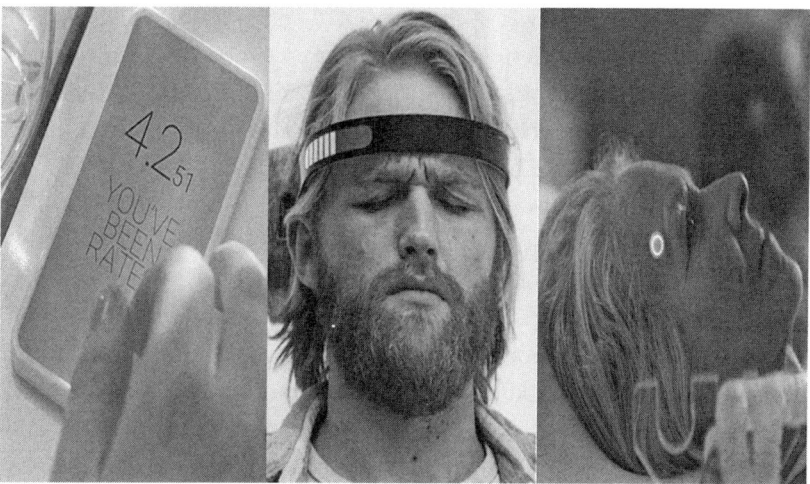

**Figure 0.1**
*Black Mirror's* "Nosedive," "Playtest," and "San Junipero": immersive experiences with the help of various digital technologies. Reprinted with the permission of Netflix. © 2016 Netflix.

human-machine interactions, artificial intelligence (AI), and extended reality (XR) like virtual reality (VR), augmented reality (AR), and mixed reality (MR) needs to be seen as the culmination of a long history of immersive experiences in media, ranging across literature, theater, film, and games.

Immersive technologies are about to herald a major cultural and economic shift in our society. They might be as influential as the invention of television and the Internet. In fact, radical change is already under way. Devices such as VR headsets and AR-ready phones represent bundles of functions and cumulative histories of immersive experiences. They question our understanding of reality, representation, and immediacy. Imagine a world in which all it takes is to put on a headset and it is all there: friends, family, shopping, news, education, entertainment, love, and sex.[7] At the heart of this experience lies the utopian desire to be fully immersed in all-surrounding mediated environments and to fully interact with those in it, as one can do in the physical world. It is a collective desire, hardwired into the human psyche, to find other worlds outside the reality in which we live, even if those worlds mean we would be wrapped up in artificial and illusionary environments—as long as these virtual realities are better than the one in which we are living.

That is also the main topic of the Holy Grail text of virtual reality, Ernest Cline's pop-culture manifesto *Ready Player One* (2011), famously adapted into a feature film by Steven Spielberg in 2018. *Ready Player One* imagines a bleak, dark future in the year 2045 in which humanity has stopped caring about the world and its problems. Instead it has started to escape despair by plugging into an expansive VR universe called OASIS (Ontologically Anthropocentric Sensory Immersive Simulation). The VR playground is the brainchild of James Halliday, a reclusive genius who failed to connect with the people around him and instead created a virtual place in which life could be so much easier. His world is a pop-culture candyland full of cultural artifacts from his childhood, such as the DeLorean from *Back to the Future*, King Kong, ABBA, the Millennium Falcon, Freddy Krueger, Chucky, and *The Shining*. The OASIS is a playground for nostalgia similar to *Black Mirror*'s "San Junipero." It is an easy place to fall in love with, allowing visitors to forget about the physical world for a while, or even forever.

Cline's novel is an updated version of Wonderland for millennials. If any story captures the desire and temptation for immersive experiences, it is *Alice's Adventures in Wonderland* (1865) by Lewis Carroll. That book's fantasy world, which Alice enters through a rabbit hole, reverses the logic, rules, and values of the Victorian middle class. At the same time, it opens subversive layers of meaning that position the fantasy world as both a seductive and dangerous counterpart to modern society. In the context of the story, we as readers believe in this utopian world of mad hatters, invisible cats, and talking rabbits. *We take it for real.* The story is about escapism, a motif that can also be found in *Ready Player One* and in several immersive experiences in postmodern culture. Alice *wants* to be in Wonderland, even though it is a potentially dangerous place with a furious Queen who is quick to declare death sentences at the slightest offense. Wonderland resembles Alice's physical reality, yet it is so much more. Beyond the rabbit hole, Alice creates meaning through friendships and by embracing her heroic role in the new environment. Her physical world, in which she is an outcast in Victorian society, expects nothing but good manners from young girls. *Falling down the rabbit hole* is a metaphor for immersion in digital media and a world beyond the physical that is etched so deeply into our own existence that we feel the need to return to it again and again. Carroll's text is the archetype for feeling immersed in a world beyond physical reality, and many of its terms and phrases are used in the terminology of popular culture: "wonderland," "rabbit hole," "to follow the white rabbit."

**Figure 0.2**
*Alice's Adventures in Wonderland* and *Ready Player One*: falling down the rabbit hole and putting on the VR headset. Reprinted with the permission of Jaap Buitendijk/ Warner Bros. Pictures. © 2020 Arthur Rackham/CC BY and © 2018 by Jaap Buitendijk/ Warner Bros. Pictures.

Postmodern culture and immersive experiences have many things in common. Both are crucially defined by feelings of uncertainty, dissolution, frictions, fluctuations, and a changing *perception of reality*. They mark a totality of experiences that surrounds and absorbs us and ultimately eliminates distance and security mechanisms. By being completely exposed to these experiences, we find ourselves in what I would like to call *liquid spaces*. The feeling of immersion evolves in these spaces. They are determined by processes of fluidity, hybridity, crossing borders, and liminal thresholds. Once we have crossed the line into them, we feel very often lost and vulnerable. They are tempting and exciting, but it often feels like we have to give up control to fully enjoy them.

But immersive experiences are not merely a phenomenon of postmodern culture. Ultimately, they are a reflection of the postmodern as well. The social, political, and economic disruptions of contemporary times leave a vast majority of people in a state of uncertainty. For instance, in 2020, the scale of the global coronavirus (COVID-19) crisis not only kept us contained in our homes, it also reshaped our relationships to the outside world. The spread of the virus has caused mass gatherings such as conferences, festivals, concerts, and sporting events to be called off or postponed. With the term "social distancing," we were urged to put space between ourselves and other people: staying at home, avoiding crowds, and refraining from touching one another. However, we did not so much practice social distancing as we implemented measures of *physical distancing* in our lives. We did not suddenly stop keeping in touch with others, but instead moved our social activities almost entirely to virtual settings. The prevalence of video conferencing in both professional and personal life has increased out of necessity.[8] With the help of digital technologies, people were able to attend birthday parties, weddings, movie nights, book clubs, dance raves, meditation sessions, yoga classes, and religious services, including funerals. After schools and universities were forced to close, classes were moved entirely to remote learning and virtual classrooms. In addition, concerts, DJ sets, theater and dance performances, and business events were streamed in real time on platforms such as Instagram, YouTube, Facebook, and Twitch for thousands of participants.

Existential crises such as the coronavirus pandemic have always made people search for alternative realities. In postmodern culture, these realities are very often digital and immersive. Edward Castronova even once anticipated a "digital exodus" in which mediated and virtual worlds would lead to a mass exodus from the physical world that would ultimately transform both spaces forever, the physical and the virtual.[9] As a result of the coronavirus outbreak, Deborah Tannen notes that "[t]he comfort of being in the presence of others might be replaced by a greater comfort with absence, especially with those we don't know intimately. Instead of asking, 'Is there a reason to do this online?' we'll be asking, 'Is there any good reason to do this in person?'—and might need to be reminded and convinced that there is."[10]

Immersion is a *state of liquidity* in which the frames of the mediation dissolve and disappear and ultimately merge with our physical reality. We are *falling* down the rabbit hole, getting *lost* in a book, getting *absorbed* in social media and streaming platforms. But what happens to our existence and our idea of selfhood as we cross the border to another world, be it social media, a fictional world, or an extended reality? What if we ultimately question the status and importance of physical reality, favoring mediated realities? What happens, exactly, in these liquid spaces?

Immersion is a concept that arrived in everyday language with a built-in meaning. Especially in the English language, the phrase *to feel immersed* is quite often used without thinking through what it entails in different contexts. In recent years, the academic discussions of immersion and immersive media have been focused heavily on a technological (or technocratic) perspective. This book aims to contribute to the conversation with a humanities and social sciences perspective that addresses the human implications of pervasive and immersive media, and highlights technological advances as influences on our understanding of the world. The lenses of the humanities and the social sciences, which are fairly established in the discussion about digital culture and social media, have been widely overlooked so far in regard to immersion. They raise a number of fundamental questions about the risks and opportunities associated with being fully absorbed in mediated realities, and our understanding of individuals, communities, and societies as part of all-surrounding experiences. If we change the point of view, we are able to reframe mass-media experiences through this specific theoretical trope and critically reflect on the increasing influence of immersion on different aspects of our daily lives.

The character of this book is strongly interdisciplinary and ranges across media studies, sociology, performance studies, film studies, business and economics, cognitive science, and games studies. This project aims to open up a knowledge base that nontechnical people can use to empower themselves to better understand immersive experiences. Over the course of this book, I will discuss a variety of examples from a range of media practices in our lives, and I will explain them with the concept of the 360° gaze. I will provide an interdisciplinary framework to explain the invisible logics and tactics of *immersion* and *liquid spaces* in media, society, and culture. The 360° gaze is my attempt to look at the specifics of immersive experiences from a humanities and social sciences perspective. The framework tackles the fluidity of immersion in postmodern mass media, and uncovers the dominant metaphors and prevailing ideologies within everyday mediated practices. It does so by looking at their specific *semiotic codes and social activities, performativity, processes and practices*, and by considering their implications on *psychology and reception*, and their role in *cultural industries and digital ecosystems*. The 360° gaze refers to every mediated experience that creates a *mediated self* and introduces audiences to *different modes of reality*. The framework reveals the connections between immersive settings in literature, theater, film, television, and other media, as well as the everyday immersive acts. It positions rapid developments in the field of emerging technologies, such as XR, AI, and robotics, alongside other immersive experiences, both analog and digital. By proposing a novel, interdisciplinary theory, which can be discussed, applied, reframed, extended, and even rejected, this book wants to provide a clear argumentative focus to fully comprehend the state of immersion in a variety of mediated realities.

The first and final chapters of this book are formed like a bracket. The book begins by looking at the current hype around immersive technologies, embraced by several media and cultural industries and met with high expectations from audiences. It will further outline the dominant power mechanisms of digital industries and their tactics to make immersive technologies and *empathy machines* the next big thing for the future. The last chapter, on the other hand, reflects on the dangers and ethical implications of these developments, and ultimately imagines a possible *end of immersion*—a state in which powerful individuals and institutions with commercial, governmental, political, or religious interests use immersive experiences to abuse audiences and manipulate their behavior.[11] This chapter is exclusively dedicated to the future of emancipation and counterstrategies to develop a critical dimension of immersion.

In between, I will assess several immersive experiences to make clear that they are not singular phenomena, but deeply built into the social, political, and economic mechanics of postmodern culture. In two corresponding chapters—"Intermedia Immersions" and "Sociocultural Immersions"—the framework of the 360° gaze is used as a compass to navigate through immersive experiences in literature, performing arts, music, cinema, television, digital games, and immersive technologies.[12] "Intermedia Immersions" tackles different immersive phenomena in media history, for instance the semiotic codes in literature, the performative practices in immersive theater, and the role of digital screens and world-building activities in games and VR. In "Sociocultural Immersions," it will become evident how immersion deeply affects our understanding of social and cultural conditions, while not being limited to a singular mediated experience. For instance, the mediation of the self, binge-watching, nostalgic experiences, fandom, and forms of human-machine hybridization and artificial intelligence stretch beyond the limits of one medium and have a profound impact on various aspects of our lives. Even though I will look at these phenomena in separate chapters, it will become clear that in the age of convergent media there are several overlaps. For instance, immersive experiences in screen cultures are often also examples of human-machine hybridization, and vice versa. This will illuminate the similarities of these experiences and uncover the ideologies behind the configurations of the physical and the mediated.

By defining the politics of immersion and the concepts of reality, presence, and immersion, my aim is to contribute a new thought to the discussion.

Over thirty years of research in human-computer interaction and digital games have provided significant insights in the topic, while often using the terms "immersion" and "presence" interchangeably. That is a valid viewpoint in regard to specific mediated experiences (such as games), but is often insufficient when adapting it to other areas, and particularly when addressing the human implications of immersive experiences. I argue from a humanities and social sciences perspective that presence and immersion should be discussed as two different forms of engagement. Immersion is a long-lasting state of absorption in which the frames of mediation dissolve and physical and mediated experiences merge with each other. This phenomenon is not limited to the consumption of mediated content, and can be experienced long after the mediation has ended. I will outline the ways in which immersion is activated by mediated realities, and how powerful it is as an invisible and very often unconscious feeling of engagement.

Immersion goes *beyond the mediation*.

To underline the interdisciplinary approach of this book, several artists, producers, and authors whose works are related to immersion contribute their thoughts in brief exclusive interviews. In cooperation with the Franco-German television network ARTE[13] and the three-year program "Immersion: Analogue Arts in the Digital Age" at the Berliner Festspiele in Berlin (2016–2019),[14] these *moments of immersion* offer artistic and philosophical insights to complement the ideas and arguments of the book, and establish a rich exchange between theory and practice.

My premise is that immersion is a key term in postmodern culture, and a necessary lens through which to understand concepts of reality and mass media in the twenty-first century. By the time you read this, rapid developments in digital and immersive technologies will have introduced more new hardware and software, various XR-based platforms and services, and several popular digital trends. You might even be the owner of a VR headset by now. The proposed framework in this book is transferable, even though future trends can only be anticipated for now. That said, my proposal is bound to a specific perspective and by no means intended to capture every aspect of mass media history. But it will reframe several contemporary media and cultural practices. And while it would be foolish to believe that definitive answers could be found here, my goal is to craft a path to rethink immersion in mediated experiences from the past to prepare us for what is ahead in the future.

# 1  Rabbit Holes

**Virtual Reality: The Beginning Is the End Is the Beginning**

> I haven't fucked much with the past, but I've fucked plenty with the future
> —Patti Smith, "Babelogue," *Horses*, EMI-Electrola, 1975

In medias res: Barcelona, February 2016. Mark Zuckerberg has a slightly menacing grin on his face as he walks past a sea of more than 5,000 attendees at the Mobile World Congress. Golden light shines down on the Facebook CEO. Zuckerberg passes his subjects, cast in gloom, toward a stage where he will accept their adulation. He is the star of the event. But no one in the audience notices the tech leader yet. The attendees are not able to see the Facebook boss as he breezes past them. Instead, they are all wearing VR headsets. They are plugged into their machines, gawking into dead air, drifting off into some mediated space, blinded to reality by the screens in front of their eyes.

It is Zuckerberg's version of reality they see instead.

The striking nature of this haunting image caused a subsequent debate online, in which the scene was compared to George Orwell's *1984* and Aldous Huxley's *Brave New World*. The image provokes fears of a dystopian and nightmarish future in which those who control technology ultimately control everyone else with it. In this future, mass audiences are degraded to be the slaves of the modern digital era and forced to play their part in a preprogrammed world built around the objectives of a big corporation. They are trapped inside technologies and mediated experiences that promise to fulfill their pleasures, hopes, and dreams, while these same technologies undo their capacity to think on their own. Neil Postman once called this phenomenon an "inflicting pleasure" and "man's almost infinite appetite for distractions," nurturing the fear that the technologies "we love will ruin us."[1]

But the picture visualizes more than just "the surrender of culture to technology"[2] and how we are slowly becoming the "tools of our tools."[3] It also displays the "governmentality" (a concept formulated by French philosopher Michel Foucault to describe the governing of people's conduct) of immersive experiences by multinational corporations. For nearly two decades, Zuckerberg has been the face of social media, transforming the conditions of the public sphere by coding "social behavior into algorithms and interfaces—technologies that end up at the heart of people's everyday routines."[4] The same person is now trying to become just like *Ready Player One*'s mastermind James Halliday, the leader of a new age of sociality and connectivity through immersive technologies. For many, this idea would be a paradox, paving the way for an antisocial future in which we would interact through headsets and avatars in alternative realities, while we would be "in a kind of iron maiden that encloses [us] on the spot."[5] It is a singular vision of access and availability, no ethics attached. Our capacity to dream is the ultimate human resource for profit-oriented multinational conglomerates as humans are about to get lost in their headsets and are too distracted and preoccupied to notice. That, of course, is merely the dystopian view of it.

**Figure 1.1**
Facebook CEO Mark Zuckerberg at the Mobile World Congress in Barcelona, 2016. Reprinted with the permission of Facebook. © 2016 Facebook.

On October 6, 2016, the Facebook CEO gave a first glimpse of his vision for social networking through VR technology. When he took the stage at Oculus Connect 3 in San Jose, California, to present a demo version of the company's ideas for VR, the announcement not only included the launch of Facebook's 360° Spatial Workstation, a software for producing spatial audio for 360° videos and cinematic VR. It also showcased the kickoff event for Facebook Spaces,[6] a VR chat environment designed for Oculus Rift, which enabled users to walk around in virtual spaces and connect with others. Although the platform was still untitled back then, the video of the "social VR" presentation had been watched over eight million times as part of Zuckerberg's original post on Facebook: "Here's the crazy virtual reality demo I did live on stage at Oculus Connect today. The idea is that virtual reality puts people first. It's all about who you're with. Once you're in there, you can do anything you want together—travel to Mars, play games, fight with swords, watch movies or teleport home to see your family. You have an environment where you can experience anything."[7]

Zuckerberg's statement contained the ultimate promise. It resonated with the preoccupations of contemporary media, the transparent presentation of the real, and the immediacy of the mediation. To "experience anything" opens a whole lot of different options without being too concrete. It is almost like claiming you have invented the best product in the world without really letting anybody know what they can do with it. That way, everyone's secret wishes, hopes, and dreams fit right in. It is like dangling a carrot in front of the audience. Facebook promised to offer the ideal environment to become *real in VR*. Zuckerberg's demo imagined Facebook's social media platform as a three-dimensional space in which users could move and interact, feel involved and engaged with others. Users' profiles on Facebook Spaces would get enhanced by the creation of a three-dimensional digital avatar to fully interact with others in virtual locations. The social VR app was about to add a spatial dimension to the daily communication routines in digital spaces.

But you would not want to use VR to meet your friends at some everyday location. You would want to experience the extraordinary, your own kind of "wonderland." After putting on the Oculus headset, Facebook Spaces "transported" users to unusual places, which would be difficult or almost impossible to visit in physical reality: a tropical island, a rainforest, the bottom of the sea, the surface of a faraway planet, or Mark Zuckerberg's office. In the demo, Zuckerberg and his colleagues are transporting themselves to his luxurious home to have a video chat with his wife, Priscilla, and watch over their dog, Beast. These locations are all *rabbit holes*, a metaphor borrowed from Carroll's *Alice's Adventures in Wonderland*—a gateway into another reality, a different environment, a new world.[8]

Facebook Spaces felt like the next logical step in full media convergence and immediacy. Everything becomes part of virtual spaces in which users can communicate, watch films, take pictures, play multiple-player online games, draw virtual 3D objects in the air with the dedicated VR painting and animation tool Quill, have real-time video calls with Facebook's own messenger voice over IP service, and even livestream their VR exploits directly to Facebook Live. In addition, Spaces allowed users to view 360° content by becoming surrounded by it, enhance videos to the size of cinema screens, and watch them together with friends in virtual cinemas. Spatial audio was added to give the impression of location-based conversations and sounds. Users could celebrate virtual events such as birthday parties, take pictures with virtual selfie sticks and share them directly on their Facebook timelines, Instagram, or WhatsApp. Through controllers (Oculus Touch) and other haptic extensions of the virtual environment, users could connect the physical with the virtual and increase the level of interactivity and connectivity within their experiences.[9] Spaces had been designed as shared experiences with photo- and videospheres of all kinds mixed together. It felt as real as it could at this point in time, when the demand for VR technology to become a social platform was about to be explored. In any case, it felt like a world away from simply posting on social media.

At that point, Facebook had already invested heavily in its VR play. In 2014, it acquired the startup company Oculus and its VR head-mounted display (HMD) technology Rift and announced plans to invest an additional $3 billion over the next ten years,[10] including a standalone low-end headset (Oculus Go, released in 2018, but discontinued in 2020), a wireless high-end headset named Oculus Quest (released in 2019, and followed by the Quest 2 in 2020), original VR content, its own volumetric VR cameras, an open-source JavaScript library to create cross-platform WebVR experiences (React 360), and a dedicated section for educational content in the Oculus Store (funded with $10 billion). Oculus had also published two social VR apps: Rooms and Parties. Parties allowed users to chat in a virtual space; Rooms put them in a shared space to watch videos, play games, and perform various other group activities. Both apps were fully interactive versions of

Spaces and were connected to Oculus's room-scaling software, which tracks users' movements in the physical world and translates them to the virtual environment. Another social VR project, in partnership with the VR broadcasting company NextVR, named Venues, had been announced at Oculus Connect 4 in 2017, a way of viewing massive live events such as sports and concerts in VR with multiple virtual participants.[11] Zuckerberg was convinced that immersive technologies would be the next big thing for entertainment, connectivity, and social networking in digital ecosystems such as Facebook: "[He] claims that the future of virtual reality will be social, stating that four of the top mobile applications in the world today are social, including Facebook and Instagram."[12]

A few years after Zuckerberg's announcement at Oculus Connect 3, the future of social VR and Facebook's strategy for immersive technologies looks slightly different. As of October 25, 2019, Facebook Spaces and Oculus Rooms were shut down to "make way for Facebook Horizon,"[13] the company's next try at a social VR platform in 2020. Despite high hopes, Spaces in particular has not been able to achieve enough traction to convince the masses to buy in. The company's newest platform Horizon wants to combine features of both Spaces and Rooms but plans to focus more on the idea of an "ever-expanding world"[14] instead of multiple separated meeting places. What Marie-Laure Ryan described as "digital wonderland" almost twenty years ago seems like an accurate description of what Facebook imagines Horizon to become: "[a] computer-generated three-dimensional landscape in which we would experience an expansion of our physical and sensory powers; leave our bodies and see ourselves from the outside; adopt new identities; apprehend immaterial objects through many senses; including touch; become able to modify the environment through either verbal commands or physical gestures; and see creative thoughts instantly realized without going through the process of having them physically materialized."[15]

Despite the setback with Spaces, Facebook remains committed to the idea of social VR. With the social VR apps Horizon, Parties, and Venues, Facebook wants to extend its digital ecosystem—alongside other communication tools like WhatsApp and Instagram—so that users spend most of their time on Facebook's digital platforms. In fact, in 2019, Facebook decided to put its corporate logo on both WhatsApp and Instagram to ensure that users know about its corporate parent.[16] As of October 2020, new Oculus owners are even required to have a Facebook account to use the VR service.[17] Immersive

**Figure 1.2**
Facebook Horizon: Facebook's newest try at a social VR app. Reprinted with the permission of Facebook. © 2020 Oculus.

experiences are part of Facebook's strategy to expand its digital ecosystem to full sensory technology and to increase revenue that way. Facebook claims its users already spend more than 50 minutes a day across Facebook's suite of apps globally, a number that grows if you include communication on the instant messenger service WhatsApp. According to Facebook's full-year results, monthly active users (MAU) were 2.5 billion in 2019, which is an increase of 8 percent year over year.[18] Mobile advertising revenue represented approximately 94 percent of the advertising revenue, a figure that, along with ad impressions and average price per ad, is about to increase with the market growth of mobile VR technology and eye-tracking techniques for studying visual audience behavior.

With Facebook's acquisition of Oculus and the launch of multiple new platforms to access immersive content, VR is already very much part of our cultural landscape. For many, it is even the new frontier of mass media. There is a new wave of immersive technologies in our society, be it in the military, education (with companies such as VirBela and Engage), health care, tourism, sports, architecture, automotive industries, entertainment, or pornography. They are creating experiences using physical-world content, purely synthetic content, or a hybrid of both. It seems whatever can be done in the physical world could potentially work better with the help of immersive technologies.

Currently, the market for immersive technologies remains fragmented and is focusing on three related but quite different segments of extended reality: virtual reality, augmented reality, and mixed reality. The use of these terms can be quite confusing, especially since several companies try to redefine them for their own purposes. AR and MR both use the physical surroundings of their participants and insert digitally created elements into those surroundings. The distinction between AR and MR has not been clearly drawn yet, as the terms are often used interchangeably. In such cases, one term winds up being favored over the other, and right now the term AR is winning. There is a significant difference between those terms, though, that is worth addressing. AR, famously responsible for the hype around *Pokémon Go*, Snapchat, and TikTok, creates an overlay of digital elements *on top of* a physical environment. AR presents digital information *as if it exists* in the physical world. Those elements can only be seen through a mobile device (smartphone, tablet, laptop), thanks to the integrated camera that shows the physical environment in real time and is capable of sensing the world around it. However, there is no way to interact with the digital elements, because they are not anchored to or part of the physical world. They are like pre-recorded footage, which starts to play by getting activated, e.g., by the shape of physical-world objects. In that way, AR is a powerful tool for reimagining the physical space. For instance, Swedish furniture company IKEA uses AR technology in cooperation with Apple to let costumers imagine how a specific piece of furniture would look like in their apartment.[19] With the failed Google Glass project,[20] several other companies tried to redefine AR for their benefit, although Google never used the term "augmented reality" for the product. Instead, Glass was one of the earliest "smartglasses": a wearable device, like a smartwatch, that was capable of presenting information to the user using a transparent display. This information could have been a text message, the wearer's heart rate, or an incoming phone call, among other possibilities. Apple, in particular, currently invests strongly in AR technology and already owns the world's largest AR platform. In 2020, Apple shared plans to work on its own "Apple Glass" project similar to Google Glass, which potentially then could be connected to other Apple devices such as the iPhone, AirPods for audio, and the Apple Watch. In another example, the alternative rock band Pearl Jam has been employing AR to promote its album *Gigaton* (2020). The experience asks audiences to point their smartphones to the actual moon, which then activates onscreen animations and the preview of a new song.[21]

MR is similar to AR, but positions the digital elements *in* the physical environment in the form of a hologram.[22] It is not just a layer on top of it. Instead, the digital components can be changed, adapted, and interacted with. The digital overlay is anchored to or part of the physical surroundings, and both worlds react to each other in real time. The merging of both realities can only be seen through MR headsets and special glasses such as Microsoft's HoloLens or Magic Leap's One. Having witnessed the problems with Google Glass, Microsoft is reluctant to release a consumer version of its HoloLens. Currently, it is mainly aimed at businesses and ventures to improve manufacturing processes and internal communication. Microsoft announced that a mass audience version is planned "once the technology matures,"[23] while the startup Magic Leap initially planned to position MR also as an entertainment technology.

VR, however, is the ultimate immersive experience. VR technology creates a digital world, which is visually separated from the physical one. All sensorial and perceptual levels are focused on the aesthetic illusion, "a lucid dream . . . difficult to distinguish from actual experiences until the laws of physics are suspended,"[24] and manifested by headsets, tracking systems, and controllers. Headsets become like blinders—what is around might only distract. This is where the industry gets excited. Major digital companies are investing in VR technology, and their influence spans across digital communication, information, entertainment, sport, games, journalism, film, and television. According to PricewaterhouseCoopers (PwC), VR will be a $15 billion global market in the year 2021, compared to $869 million in 2016.[25] ResearchAndMarkets goes even further and anticipates the global VR market will reach $212 billion by 2026.[26] For companies such as Facebook, Google, Sony, and HTC, VR promises to open the door to new business opportunities. On the hardware side, high-end market competitors Oculus, HTC, and Sony are linking their own individual headsets to premium-content products (such as games and interactive VR experiences) to compete with cheaper cardboard and mid-range plastic solutions for mobile phones (Samsung, Google) and their focus on 360° panoramic videos and browser-based experiences called WebVR. In addition, there is a new market for equipment like 360° cameras for amateurs and professionals such as the Samsung Gear 360, Insta360 Pro, GoPro Odyssey and GoPro Omni, Ricoh Theta, and Jaunt One. On the content side, the language of VR storytelling is about to get explored—from 360° panoramic videos (cinematic VR) of physical-world scenes to highly interactive computer-generated content, and a combination of both—with widely recognized VR productions such as the *Mr. Robot VR* experience (2016), the Academy Award nominated short *Pearl* (2017), Alejandro González Iñárritu's VR installation *Carne y Arena* (2017), and ILMxLAB's Darth Vader VR Experience *Vader Immortal: A Star Wars VR Series* (2019). VR now fulfills the long promise of being fully surrounded by mediated narratives such as the *Star Wars* or *Harry Potter* story universes, but it also forces content producers to rethink dramaturgical structures, the role of audiences and their ability to move and interact in all-surrounding story worlds.[27] In 2019, more than one hundred VR games, such as *Beat Saber*, exceeded $1 million in revenue.[28] In addition, PwC predicts revenues of VR video (cinematic experiences) will exceed the revenues of both VR gaming and VR apps until 2021.[29]

Facebook and other multinational companies promise their costumers a world beyond imagination. Samsung, for instance, supported the marketing campaign for its Galaxy S8 smartphone and the Gear VR headset with the tagline "We make what can't be made, so you can do what can't be done" and the hashtag *#DoWhatYouCant*. In one of the campaign's TV spots (March 2017), audiences saw an ostrich stumbling off the African plain onto the porch of a house, carelessly picking up crumbs off the breakfast table, as well as accidentally putting on a Gear VR headset. While wearing the headset, the ostrich sees the world through a VR flight simulator. To the sound of Elton John's "Rocket Man," the bird suddenly envisions what it might be like to fly and takes off into the air for real. The bird knows as well as we do that ostriches are not able to fly in physical reality. But Samsung argues that its VR technology can make even the impossible happen. Only three months after its launch, the spot "Ostrich" reached more than 24 million views on YouTube. Another Samsung spot, called "The New Normal" (more than 11 million views on YouTube), takes a school class back to the Jurassic and Cretaceous periods to introduce the students to dinosaurs and the process of evolution. It envisions the future of VR in education: imagine you would be able to witness the signing of the American Declaration of Independence, the invention of the telephone, or Einstein developing the theory of relativity. VR productions could become a powerful didactic tool, since learning from experience has a significant impact on remembering subject matter. It might become "the new normal" to learn about the world through virtual experiences.

With the development of global platforms, universal narratives, and affordable technology solutions, market reach for immersive technologies will increase in the foreseeable future. But this expansion also poses fundamental questions concerning how immersive technologies with emerging virtual communities, new democracies, and new economies will shape the future of digital societies. Tech pioneer Ray Kurzweil expects that within the next decade humans will be part of "full-immersion visual-auditory environments, with images written directly onto our retinas by our eyeglasses and contact lenses."[30] *Wired* editor Kevin Kelly imagines so-called "mirrorworlds," in which "every place and thing in the real world—every street, lamppost, building, and room—will have its full-size digital twin in the mirrorworld."[31] Philippe Fuchs argues that the purpose of virtual reality is to create "a simulation of certain aspects of the real world . . . a symbolic world . . . an imaginary world."[32] This kind of "pseudo-natural-immersion"[33] could also be discussed from a pessimistic and dystopian angle with warnings against the digital divide, information glut, and extensive surveillance. Both outlooks are rooted in concerns related to new-media technologies such as hypertext, cyberspace, and VR, which determine the shift from the material to the immaterial, from atoms into bits[34] and matter into mind.[35] In the spirit of Marshall McLuhan, VR controls and shapes the forms of human interaction, and therefore should be considered as a new medium.[36] It incorporates existing digital electronic and communication technologies, while acting as an extension of the human senses and as an account of the cultural trajectory in cyberspace and contemporary digital culture. This new medium considers the social relations and contexts already factored into the technology.

> The market for extended reality (XR) technology currently consists of three segments: *augmented reality* (AR), *mixed reality* (MR), and *virtual reality* (VR). AR and MR create digital overlays in physical environments with different degrees of interaction. VR technology is able to create complete digital worlds that are visually separated from the physical surrounding.

Already in the early years of postmodern media, people were intrigued by the idea of finding a digital wonderland with the help of VR technology. One of them was American computer scientist and engineer Ivan Sutherland. His conference paper "The Ultimate Display" (1965) and his work on the first HMD are considered to be the beginning of the era of immersive technologies. Sutherland wanted to invent a highly sophisticated, complex, invasive, and technological illusion-producing machine, inspired by science fiction stories like Stanley G. Weinbaum's "Pygmalion's Spectacles" (1935). Weinbaum's story included the idea of goggles that let the viewer experience the freedom of fictional worlds. The inventor in the story, a professor, has a clear vision for his tool—to "make real a dream." He wants to invent "a movie that gives one sight and sound. Suppose now I add taste, smell, even touch, if your interest is taken by the story. Suppose I make it so that you are in the story, you speak to the shadows, and the shadows reply, and instead of being on a screen, the story is all about you, and you are in it. Would that be to make real a dream?"[37] The materiality of virtual realities as illusionary experiences lies in the heart of immersive technologies, as Sutherland once put it: "A display connected to a digital computer gives us a chance to gain familiarity with concepts not realizable in the physical world. It is a looking glass into a mathematical wonderland."[38]

With that vision in mind, Sutherland continued to work on ideas for headsets. The Telesphere Mask (patented 1960) became the influence for Headsight (1961), the first precursor of the HMD as we know it today. Funded by military money from the Advanced Research Projects Agency (ARPA), Sutherland created the first computer-aided HMD in 1968, which was able to track user's head movements through internal sensors. Unlike other visual display paradigms,[39] HMDs were not designed to be stationary, but to move in conjunction with the user's head. A wider range of tracking systems allowed the system to measure the joint angles of the linkages to determine the positions of the two miniature monitors placed directly in front of the eyes to create binocular 3D vision. Unfortunately, a prominent drawback of any HMD is that any misfit in the tracking and image-generating systems can cause visual confusion and a misalignment between the body's visual system and its vestibular system. The symptoms are often referred to as "virtual reality sickness."[40] Sutherland's helmet design was so heavy and bulky that it needed to be suspended from the ceiling. The setup inspired its name: The Sword of Damocles. Although its graphics, appearance, and interface were primitive and simple, the basic design of placing screens in front of the eyes remains the same to this day.

More important than the design were Sutherland's visionary predictions at the time. Not only did he envision processes of measuring gaze and movement (eye-tracking, gestural interfaces, body movement). He was also confident that one day computers would be able to transform abstract mathematical constructions into immersive, meaningful environments in which people could move and interact. Decades before massively multi-player online role-playing games (MMORPG) such as *World of Warcraft* in which participants "lose themselves" in digitally created worlds, Sutherland was convinced that one day virtual worlds would be coexisting with each other, each of them with its own rules and possibilities. He was confident that parallel realities, created by computers, could be as powerful and meaningful as any experience in the physical world.

> There is no reason why the objects displayed by a computer have to follow the ordinary rules of physical reality with which we are familiar. The kinesthetic display might be used to simulate the motions of a negative mass. The user of one of today's visual displays can easily make solid objects transparent—he can "see through matter!" Concepts which never before had any visual representation can be shown. . . . By working with such displays of mathematical phenomena we can learn to know them as well as we know our own natural world. Such knowledge is the major promise of computer displays. . . . With appropriate programming such a display could literally be the Wonderland into which Alice walked.[41]

Sutherland's predictions were quite spot on at the time. He recognized the power of parallel worlds, even though he glorified the idea of the wonderland a bit too much. Alice certainly did not walk down the rabbit hole. At one point she simply fell, with nothing to hold onto or grasp. Her fall into another world is like a unidirectional dive into a containing vessel and "a split between the physical 'here' and the virtual 'there.'"[42] Just like Alice, users give up control while falling into immersive environments without having full knowledge of their rules and limitations. This is the ultimate fear we have in relation to emerging and pervasive technologies. We try to process our concerns and fears in stories such as the ones in *Black Mirror*. Sutherland's name for his HMD, Sword of Damocles, also casts a menacing shadow, alluding to an imminent and ever-present peril. It is as if he wanted to send out a warning that even if immersive technologies were an opportunity to change the world, there was no guarantee we would not be changing it for the worse.

The next two decades felt like a rollercoaster. After Sutherland's groundbreaking work, VR almost disappeared completely from the scene. It was computer scientist and visual artist Jaron Lanier, founder of the first VR company VPL Research, who brought it back into the mainstream in 1983 by using the term "virtual reality" to describe immersive digital worlds. His first invention was a commercially available data glove (1984), which could be used for stepping into computer-generated settings. Together with his colleague Tom Zimmerman, he designed a range of VR haptics like glasses, gloves, and an inexpensive HMD. None of them were ever fully embraced by a mass audience. But his creative and aesthetic philosophies were refreshing within a technology that was too often compared to traditional computer simulations.[43] In Lanier, the technology found its first "VR evangelist": a person who established a critical mass of support for VR. Where others saw a pure technological evolution, Lanier recognized the aesthetic potentials, the communicative dimensions, and even the entertainment possibilities. Lanier dreamt of a superior medium for human expression. As so often before, the idea of metamorphosis, akin to Alice's in Wonderland, became part of Lanier's vision: "you can visit the world of the dinosaur, then become a Tyrannosaurus. Not only can you see DNA, you can experience what it's like to be a molecule."[44]

However, the 1990s saw a major setback again. At this point, VR was probably too many things at once: a philosophy for tech geeks, a sociospatial practice, and a utopian technology that would turn us into digital serfs. It was seen as nothing but hype, never meant to come true. Only Hollywood was still hooked on the new medium. The impression audiences had of virtual environments was mostly inspired by science fiction films that portrayed VR technology as a threat like *The Lawnmower Man* (1992), based on a Stephen King story, *Virtuosity* (1995), and *The Matrix* (1999). These films suggested that VR would be a danger to the whole fabric of society, because it could erase the difference between reality and the medium. Although *The Lawnmower Man* (and before that *Tron*, 1982; *Blade Runner*, 1982; *Brainstormer*, 1983; *Total Recall*, 1990; *Arcade*, 1993; and *Disclosure*, 1994) portrayed virtual worlds as accessible through technology that pulled users inside an imaginary adversary, it was the *Matrix* series that made characters (and audiences) believe the virtual world was, in fact, the real one. The aesthetic promise of the films positioned their protagonist, Neo, in a clean and transcendent simulated world and mixed it with martial-arts aesthetics and postmodern theories. *The Matrix* succeeded in presenting media theory to a wider audience in explicitly linking it to emergent digital technologies. For the cognoscenti, it name-checked and illustrated the number one theorist of simulation, Jean Baudrillard, and his book *Simulacra and Simulation*,⁴⁵ and revived several terms and metaphors used for immersion, from the terminology of *Alice's Adventures in Wonderland* to references to Dorothy's journey in *The Wizard of Oz*.

However, while the visionary promise of being surrounded by virtual worlds continued to experience a renaissance in fictional storytelling in such films as *Johnny Mnemonic* (1995), *Strange Days* (1995), and *eXistenZ* (1999), technological developments for the consumer market were not able to live up to the promise. Games market competitors Sega and Nintendo both announced VR headsets at the start of the decade. Sega planned to release wrap-around glasses with stereo sound and LCD screens, which could be plugged into the Sega Genesis console. Technical difficulties were given as the reason the device has remained forever in its prototype phase despite Sega having already developed a handful of games for it. Sega's headset was never commercially released to the public. Nintendo did not have much more luck. Its Virtual Boy, a portable console with 3D glasses, was nothing but a disaster. A lack of color graphics and software support resulted in the discontinuation of production and sales in 1995. In the '90s, it seemed like the dream of entering virtual realities was a utopia reserved for Hollywood.

Fast-forward to today, and *Ready Player One* and *Black Mirror*. After all the failures, false hopes, and disappointments, immersive technologies in the twenty-first century are in a different position. The cultural and economic landscape has changed. Facebook and Google have much more influence now than Sega and Nintendo ever had in the '90s. Cultural industries have experienced a drastic change in the last three decades, with the process of digitalization having the most immediate impact on technologies of cultural production. With the growing importance of the Internet, social media, and digital communication tools, digital media and digital production spread through all kinds of industries. The likes of Facebook, Google, Apple, Netflix, and Amazon are now major players in the cultural industries and influence our understanding of knowledge, heritage, and identities.⁴⁶ The growing importance of digital ecosystems is reshaping entire markets by relying on new digital business models, digital consumers, and digital technologies such as immersive media.

The idea of the ecosystem, historically used to refer to living organisms, is well established in the tech industries to refer to the community of platforms and devices that are connected with each other and with their users. What Henry Jenkins described as a shift from old to new consumers can be considered one of the reasons why digital companies currently invest in VR technologies: "If old consumers were assumed to be passive, the new consumers are active. If old consumers were predictable and stayed where you told them to stay, then new consumers are migratory, showing a declining loyalty to networks or media. If old consumers were isolated individuals, the new consumers are more socially connected. If the work of media consumers was once silent and invisible, the new consumers are noisy and public."[47] For Jenkins this development is part of a convergent media culture that culminates in the increasing ability of immersive technologies to satisfy our desire to be *inside the media*. In his 2017 book, Jaron Lanier even called the subversion of immersive technologies the *Dawn of the New Everything*.[48]

The potential might be there, but the consumer market for VR is still too fragmented. The highly immature market involves several underdeveloped business models and lots of experimental or low-quality content. There are currently way too many unsatisfactory headsets, platforms, and market segments, and not enough market expertise or functional business models. Creators struggle with distribution and monetization and therefore try to diversify their products to reach as many audiences as possible. The current range of products can be confusing. It offers everything from low-end cheap cardboard solutions to mid-range headsets with additional features and high-end headsets that run off game consoles, or computers with serious processing power and high-quality graphic cards (see table 1.1).

There are different VR headsets for different users, and a lot depends on what users want to experience with them. Philippe Fuchs and his team classified the functionalities of VR headsets. They divide them into four categories: observing, moving, acting, and communicating with others within the virtual world.[49] In 360° cinematic videos and WebVR experiences, audiences can only observe: a static experience like in a fishbowl, in which users are not able to interact or move. Interactive VR worlds that would allow users to move, act, and communicate—apart from the digital games sector—is still a niche product. For a fully interactive experience with positional tracking, users currently would need to overcome the obstacle of acquiring a relatively expensive high-end VR headset in addition to a powerful computer, raising the bar to thousands of US dollars. Standalone VR headsets, such as the Oculus Quest 2, are gaining popularity as they are less expensive, but they still underperform compared to the full potential of room-scale VR.[50] One option in between might be volumetric video,[51] a technique to extract three-dimensional content and place it into other live-action or CGI environments to give audiences a greater sense of depth and texture. This type of content could be watched on any VR device, even on the web.

Looking at the market for high-end VR headsets, Oculus is the company that has been most associated with the latest VR hysteria. Its headset Rift is often referred to as the first HMD of a new generation, while the Oculus Quest accounted for almost half of all VR headsets sold in 2019.[52] Funded by a crowdfunding campaign on Kickstarter and ultimately bought by Facebook, the Rift and the standalone headset Quest 2 track head movements and provide 3D imagery on stereo screens.[53] Oculus benefits from Facebook's global ecosystem. It exclusively offers social VR apps Horizon, Parties, and Venues and connects them to a room-scaling software for interactive spatial communication. But Oculus's aim is not merely the creation of 3D environments for Facebook's social media apps, but also the design of haptic social gaming experiences. In the Oculus Store, users find games like *Killing Floor Incursion*, a first-person shooter, allowing users to touch objects in the VR environment with the controller Oculus Touch. The Quest 2 features hand controllers that have six degrees of freedom, which means they can effectively track users' hand movements. In addition to that, Facebook's ecosystem invites them to socialize with other players to transform VR to a shared experience, which is a hallmark for the games industries.

HTC, one of Oculus's main competitors in the high-end sector, has an overwhelming variety of headset packages including the original Vive (discontinued in 2019), Vive Pro, Vive Pro Eye, Cosmos, and the standalone headsets Focus and Focus Plus. In early 2020, the company decided to discontinue Vive Pro and Focus in favor of newer versions, but it also unveiled three more Cosmos headsets (Elite, Play, and XR) to appeal to a range of consumers and enterprise users from the beginner to the high-end enthusiast and everyone in between. In addition, HTC revealed the prototype Vive Proton, two versions of the same headset: a standalone version with its own processing and battery power, and a "VR viewer" that needs to be powered by either a smartphone or a computer.[54] HTC also launched its own subscription-based app store called Viveport Infinity for VR games, apps, and videos. As a result of the coronavirus lockdown in 2020, the company pitched the VR app bundle Vive XR Suite that offers a variety of cloud-based VR experiences in cooperation with partners such as VRChat, VirBELA, and Museum of Other Realities, which often do not even require a headset. HTC also unveiled the meeting service Vive Events to host conferences and exhibitions in VR. However, HTC is the only major player in the field that does not control a major content distribution platform exclusively for games. Instead, to enable users to access VR games such as *Eve: Valkyrie*, *Skyrim VR*, *Beat Saber*, and *Half-Life: Alyx* of the legendary *Half-Life* series, the company cooperates with games giant Valve and its games-dedicated platform Steam. Valve has already been involved in the development of the Vive, but since 2019 also has its own high-end headset on the market, called Index.

The most important feature of most HTC VR headsets since the original Vive is the unique laser tracking system called Lighthouse. In this system, external base stations are placed at the top corners of a room to translate users' actual movements into the VR environment. The use of controllers further enhances the experience. This allows users to play games like *Star Trek: Bridge Crew* and take on the role of a captain on a Federation starship to relive adventures from their favorite film franchise. In addition, HTC collaborates with artists such as the Icelandic singer Björk to introduce the technology to nongaming audiences.

Sony's PlayStation VR (formerly Project Morpheus, named after one of the characters in *The Matrix*) wants to tackle the mass market for games. With the lowest price in the high-end sector, the PS VR is a PlayStation console accessory with a wide range of content opportunities, having sold five million units worldwide since its launch in November 2013.[55] Sony is one of the biggest global media conglomerates and benefits from the synergy effects of its product portfolio: "The idea . . . was that different parts of a corporation should relate to each other in such a manner as to provide cross-promotion and cross-selling opportunities, so that sales would exceed what was possible when divisions acted separately."[56]

The Sony Group is not only a multinational conglomerate, which offers both hardware and software products, it engages business through electronics, motions pictures, music, and financial services. By being able to provide cross-selling opportunities between entertainment content and VR technology, Sony introduces long-established intellectual property (IP) on PlayStation VR such as *Batman: Arkham VR*. In the game, players wear the Batsuit and experience how it must feel to act as the Dark Knight. Although the technology itself creates an immersive environment, it builds upon past experiences with the *Batman* universe and connects it with other immersive sociocultural experiences such as fandom.

While Sony's PlayStation VR is aimed to gamers, low-end and mid-range solutions are predominantly important for VR technology to reach a mainstream audience. Putting a smartphone into a cardboard box may sound like a joke, but it is actually the simplest way of entering a digital environment with a headset, since smartphones contain the sensors and positioning systems to accurately track movements. The idea had been around for years, but only Google's branding with the Google Cardboard made it known to a wider audience. Cardboard-compatible headsets offer limited interactivity, and most of them are suited for watching 360° video content only. In addition, lenses cannot be adjusted and devices must be held in place by the user, which after a while makes them uncomfortable to hold.

The softer fabric headset Daydream View had been Google's answer to compete with Samsung Gear VR (discontinued in 2019), Zeiss VR, and the French Homido device, all mid-range solutions to connect a smartphone with headsets that are offering additional tracking sensors, more sophisticated built-in controls, focus wheels, or even their own screens. Daydream View only runs with powerful smartphones, so Google coupled its VR service with its own branded mobile phones Pixel and Pixel XL, hoping that users would access Daydream mainly through Google's Pixel smartphone generation. The list of Daydream-ready phones also includes Samsung's Galaxy S9 and S9+, Huawei's Mate 9 Pro, and Motorola's Moto Z.[57]

In late 2019, however, Google discontinued the Daydream View due to "decreasing usage over time,"[58] and decided to shift its once sky-high ambitions for VR toward AR technology instead.[59] For the time being, the Daydream VR platform remains open for a variety of content apps containing news, entertainment, games, art, and adventures. Users can choose to hang around in the virtual Guggenheim Museum to take a close look at famous paintings, sit in a virtual newsroom to watch the news on CNN, or relax on a virtual sofa in a ski lodge to binge-watch shows on Netflix. They can walk into virtual theme parks, take meditation sessions in VR, and play games with the Daydream controller—using it as a sword to kill dragons, for instance, or as the wheel of a racecar or to play basketball or golf. Users can also access Google's services YouTube and Street View as three-dimensional experiences.

Immersive technologies are still a field of experimentation. While VR promises to offer full sensory experiences, for many audiences non-headset solutions such as AR might be the first step in getting introduced to XR

Table 1.1

| | Google Cardboard | Sony PlayStation VR | Oculus Quest 2 | HTC Cosmos (Elite), Vive Pro (Eye), and Focus Plus | HP Reverb G2 | Valve Index |
|---|---|---|---|---|---|---|
| Type | Mobile | Tethered | Mobile/tethered | Tethered/mobile | Tethered | Tethered |
| Price | $10–40 | $300 | $300 | $550–1600 | $600 | $1000 |
| Platform | Phone | PlayStation 4 and 5 consoles | Standalone and PC (with Oculus Link) | PC and standalone (Focus Plus) | PC | PC |
| Position tracking | None | Stereo camera (external) | Oculus Insight | Inside-out tracking/Lighthouse base stations (external) | Quad on-board camera | SteamVR tracking 1.0 or 2.0 (external) |
| Controller | None | DualShock 4, PS Move, PS Aim | Oculus Touch v3 | Motion controllers | Reverb G2 controllers | Valve Index controllers |

Range of exemplary VR headset solutions (as of January 2021) technologies. With the help of mobile devices, browser-based AR features (such as AR animals in Google's search engine), and apps like Snapchat, Holo, Pokémon Go, and Inkhunter, users are able to combine photographic images shot with their phone cameras with digital overlays, which are displayed onto real-world settings in real time. This does not create a different reality, but rather alters the ways physical reality might appear: digital tattoos on physical body parts, digital furniture in physical living rooms, digital monsters in physical street environments. With the help of AR, it could also soon be possible to copy and paste physical objects into digital documents.[60] The process of creating AR is less costly than VR experiences, and users have easy access through their mobile phones and do not need to acquire a headset. For instance, Apple's iPhone X had been advertised as custom-designed for the "ultimate augmented reality experience,"[61] with specially calibrated cameras that are also part of Apple's AR eyeglasses project. Also, Microsoft introduced its ambitious MR device HoloLens, which combines real-world elements with virtual holographic images already used for medical training (e.g., to practice on virtual patients). In addition, the tech giant launched its own high-end VR headset HP Reverb 2 in cooperation with HP and Valve, and works on its own platform, which blends VR and AR applications, called Microsoft Mixed Reality. Facebook, on the other hand, is preparing on its own phone-based AR technology to introduce data gathering from AR experiences activated by, e.g., movie posters.[62]

For a long time, another notable contender in this phase of experimentation has been the secretive startup Magic Leap, hyped within the tech community because its MR glasses promised to be much more comfortable to wear and are able to project virtual objects that are indistinguishable from the physical-world scenes around them. Despite facing tough competition from larger tech companies, Magic Leap's first major production was the app *Tónandi*, a collaboration with the Icelandic art-rock band Sigur Rós, which positions waveforms within the physical space of the listener (both visually and in the audio playback): potentially a new way to listen to music. However, due to the lack of investment and clear design strategy, Magic Leap stopped focusing solely on entertainment productions and also turned to more promising sectors such as health care.

The market for VR headsets targeted at consumers and enterprise users is rapidly changing as well. Both Oculus and HTC have been releasing more affordable, standalone VR headsets with the Oculus Go,[63] Oculus Quest and

Quest 2,[64] and Vive Focus (Plus),[65] respectively, given that dangling cables are not only annoying but ultimately an immersion detractor. With three new Cosmos headsets and the Vive Proton, there are four more devices in the pipeline by HTC alone. Apple is working on its own AR eyewear, and a wireless headset that combines both VR and AR technologies.[66] Despite having lost its faith in VR, Google remains committed to its 3D VR creation tools Tilt Brush and Blocks and introduced its first standalone VR headset with positional tracking called Lenovo Mirage Solo. In addition, crowdfunding campaigns for VR headsets have been highly successful since the launch of Oculus Rift, given that the community of early VR adopters loves to push interesting prospects. Pimax, a Shanghai-based VR headset manufacturer, has raised $4.2 million in a Kickstarter campaign (more than Oculus did) and nearly $15 million in its first post-crowdfunding investment round to produce Pimax 4K, 5K, and 8K, three high-resolution VR headsets with an immersive 200-degree field of view. The Oculus Quest 2 has a field of view of about 100 degrees, while the human eyes have a field of view of about 190 degrees. We can expect that headsets will get closer to that goal, and it is plausible that they will achieve it. Their accessories include, for instance, a wireless transmitter, a prescription-glasses frame, an eye-tracking module, a cooling fan, and a headband with integrated audio. Fove VR, another company in the field, is working on a device that specializes in eye-tracking. Through the development of control-based methods focusing on eye movement and eye view, the production of interactive content ideas would be become more feasible and would challenge the use of controllers. In addition, the company Feelreal developed a prototype for a sort of "smell-o-vision" add-on for VR headsets that provides users with a number of scents through a clip-in cartridge system.[67] It is safe to assume that in coming years headsets will offer better resolution, sound, and improved tracking. In a few years' time, it is probable that VR systems will be able to track hands directly without controllers.

If that is not enough, there are more ways to bring audiences to VR. Live event broadcasting might be an opportunity to sell tickets to already sold-out concerts and sports events and offer visitors the best seats in the house from their sofas. Also, VR is used for out-of-home entertainment merging both physical and virtual environments. The Void in Lindon, Utah, claims to be the first VR theme park. The VR Entertainment Center has seven 60-by-60-foot spaces, each room with its own layout, and all of them completely modular and adjustable. Visitors move through the theme park with a custom-designed headset, including sensory tracking. With the help of physical equipment (such as gun controllers) and special effects, players are able to feel present in battle scenes and war zones on other planets. As part of the experiences, participants behave as if everything around them is real and actually happening, the prerequisite for developing a feeling of presence in a mediated environment and to feel immersed in its conditions. The Void calls this form of VR-out-of-home-entertainment "hyper-reality" and teams up with the likes of Lucasfilm, ILMxLAB, and Madame Tussauds for several franchise-related pop-up experiences such as *Ghostbusters: Dimensions*[68] and *Star Wars: Secrets of The Empire*[69] in Anaheim, New York, Orlando, Toronto, Dubai, and London. Cliff Plumer, The Void's CEO, explains that the feeling of "being" a Stormtrooper is achieved through several illusionary tricks enhanced by technology: "[W]e play tricks on your brain that you actually believe—such as something called redirectable walking. You are convinced you are walking down a straight hallway when you are actually walking in a curve. The result is we can maximise a very small footprint but you think you're walking a great distance."[70] One reviewer adds: "The Void not only matches the VR environment with the physical one, it takes advantage of the fact that you cannot see your real-world surroundings during the VR ride so it can mess with your head."[71] A similar VR amusement park has been opened in Guizhou province in one of the poorest regions in southwest China. In Oriental Science Fiction Valley across 330 acres, traditional roller coasters have been replaced with nearly 40 VR rides ranging from spaceship tours to shooter games. According to a policy guideline of the Chinese Ministry of Industry and Information Technology, China is hoping to become the worldwide leader in XR technologies by 2025.[72]

Everything seems possible in the age of immediacy, and the future of XR technology is promising. Developers know exactly what their communities want, and XR will improve significantly in regard to a greater sense of presence and modes of interaction. Headset technology will become better with time, and for many XR experiences might even become obsolete. Interface designer Meredith Bricken once described the ultimate immediate experience: "You can be the mad hatter or you can be the teapot; you can move back and forth to the rhythm of a song. You can be a tiny droplet in the rain or in the river."[73] Ultimately, immersive technologies are about to expand the frontiers of our physical experiences and subvert *our perception of the "real."*

## The Politics of Immersion

Immersive experiences claim to be a *formation of reality*. So before we can continue to talk about being immersed in another world and *take it for real*, we have to ask first what reality is, and how we define the conditions of *mediated realities*.[74]

If we imagine all possible formations of reality as a linear continuum, one extreme would be the perception of *physical reality*—for many the benchmark of what they perceive as real. "Real" is ultimately what they can see, hear, touch, smell, and feel with their own physical senses. The other extreme on the continuum would be a *virtual reality*, a mediated world on its own, which blocks out physical reality entirely and lets users access their senses only with the help of technology, for instance a VR headset with motion controllers. In between, though, are various grey zones of mediated environments that establish immersive experiences inside physical reality (anything from literature, theater, and film, to games), or mix up physical with digital elements to form an alternative perception of physical spaces (*augmented reality* games like *Pokémon Go,* visual AR overlays in televised sports events, and *mixed reality* technology).[75] Mediated experiences created by the hard- and software of immersive technologies are often referred to as "X reality" (also known as "XR," "extended reality," or "cross reality"). The "X" is a variable that is intentionally not specified, as it is in flux and fluid and can be always adapted. The X might stand for virtual (V), augmented (A), or mixed (M), among many other forms of realities. In addition, the X represents the extension of the meaning of the term "reality." Because what is "real," anyway? What are the relationships between concepts and ideas,[76] the semiotic codes such as words or images that we use to express them, and the things outside our minds and their representations?

There has been an ongoing discussion in Western philosophy for over two thousand years as to whether the world is really "there" or is just constructed by our perception. In fact, postmodern theory introduced a new version of this discussion by exposing ideas as part of now-deconstructed ideologies. In postmodernist thought, pure origins and identities fall apart as we recognize a disruption of traditions and conventions through various forms of hybridity, remixes, and fusion in almost every aspect of our lives: art, entertainment, globalization, social structures, gender, race, consumer culture, mass media, and many more. Postmodern mass media created a "society of the spectacle."⁷⁷ There is no objective, independent, general unified reality, but rather a pseudo-world apart, an object of mere contemplation. Reality is mediated and everything that has actually once lived has moved into representation. When it comes to visual media, "[t]he spectacle is not a collection of images, but a social relation among people, mediated by images."⁷⁸ Whatever we perceive as origin, original, true, and real is actually a copy, replica, reproduction, simulation, simulacrum, and derivation. With that in mind, why should a "virtual reality" be any less "real" than any other concept of reality, if all reality formations consist of the same defining elements: the representation and mediation of ideas. In fact, Oliver Grau argues the term "virtual reality" is an oxymoron, "a paradox, a contradiction in terms, and it describes a space of possibility or impossibility formed by illusionary addresses to the senses."⁷⁹ Virtual realities of any kind are mediated, and both reference and representation are incorporated in them. In his 2004 filmed lecture, Slavoj Žižek even claims: "I think that a much more interesting notion, crucial to understand what goes on today, is the opposite: not virtual reality, but the reality of the virtual."⁸⁰

The idea that reality is mediated goes back to the philosophical concept of constructivist epistemology in opposition to realism. While realism believes in the existence of a reality independent of its observers, constructivism argues there is not one single universal concept of reality. Instead, there are different *modes of reality* depending on the perception and knowledge of the observer: "Realism starts from the position that it is more likely that it is reality or it is only reality which has an effect on the agent (and not the reverse); while constructivism asserts that it is more likely or only the agent that, in the act of perceiving reality, creates it."⁸¹ Already the works of René Descartes, Gottfried Wilhelm Leibniz, and Immanuel Kant reflected on the mediation of perception, and that it relies not on objects but "views, perspectives, and relations."⁸² Theories of radical constructivism (e.g., Heinz von Foerster, Ernst von Glasersfeld), but also philosophical works by Friedrich Wilhelm Joseph Schelling, Georg Wilhelm Friedrich Hegel, and Johann Gottlieb Fichte's *Foundations of the Science of Knowledge*⁸³ (1794/95) have been major influences on constructivist approaches. Stefan Weber lists a diverse range of interdisciplinary fields that are associated with modern constructivism: neurobiological, sociocultural, and media-cultural constructivism, to name just a few. Weber outlines a specific constructivist terminology consisting of terms such as "making," "constructing," "planning," "designing," "embodying," and "inventing."[84] For instance, neurobiology proposes that "reality" is merely a statement about what we are actually able to observe. We also find this terminology in the field of media-cultural constructivism, which poses the fundamental question of whether media portrays reality or actually creates it in the first place. Its distinctive characteristics are:

- There is no independent, ontological reality;
- There is no reality without an observer, who (re-)constructs it;
- There is no direct access to reality; every experience of the physical world is mediated by the human senses and complex perceptual processes;
- What can be known about the world can only be known through mediation;
- We do not only construct ideas of our self, we also construct ideas of everything around us.

If you have never been to the Moon yourself, how can you know anything about it? How do you know what is like to be up there and to walk on its surface? How it must feel to float in space? You probably do not know it from your own physical experience. Still, you might have *an idea of it*. You got the idea from books, images, films, and documentaries, among many other media sources. But those are merely representations, and for many aspects of our physical reality there is no world outside of representation available for us (at least as long as we are not able to travel to the Moon ourselves). In that sense, our minds do not have a cognitive dimension of how it must feel to be physically on the Moon. However, we might have an idea of it based on mediated representations, and they become the benchmark for our knowledge.

Weber also lists a range of modalities of reality constructions in different media, from reality TV to faction journalism.[85] I would argue that, while Weber focuses on info/entertainment formats, media-cultural constructivism can be found in many other mass-media and digital-media practices, such as social media, online dating, and immersive technologies as well. How do you know someone, if you have never met that person in physical life? How does anyone develop real feelings for someone based on profiles on a dating app? For many, the representation of the digital self on social media is already the benchmark for selfhood and reality.

Western society has always tried to capture the "real" through the use of media, from the development of photography toward the production of computer-generated images with increasing realism and visual fidelity.

Technologies such as cinema and television improved this realism in the twentieth century, but VR technology has always embodied the utopian hope to create a perfect simulation that would match the sensation of physical reality.

Looking at the history of images and their reception, it becomes apparent that images never merely represent reality, but also present themselves as what they represent. For instance, take René Magritte's famous painting *The Treachery of Images* (1929). Magritte painted a pipe, adding below the handwritten expression "Ceci n'est pas une pipe" (This is not a pipe). The painting uncovers the code of the real. Magritte did not actually paint a pipe. He painted a *representation* of a pipe in the form of an image, adding not a handwritten expression, but a *representation* of a handwritten expression in the form of an image.[86] To make his idea "real," it needed to be mediated. Magritte needed a tangible form (an image) for his intangible idea (his thought of a pipe and a handwritten expression). An idea needs to be embodied by a medium. But it can only be created and perceived under the rules and conditions of that specific medium. An idea becomes a mediated reality by the standards of the medium. Mediated realities are created and mediated by media, be it a text, image, video, sound, or VR experience. Bodies and architecture can also function as media to create mediated realities.[87]

To better understand the conditions of mediated realities, it is worth taking a trip to London. The British capital is quite a remarkable place for immersive spectacles of the real. For instance, tourists travel to the physical street address Baker Street 221b to visit the house of the fictional detective Sherlock Holmes. Holmes never existed in physical reality. That the physical house only represents the mediation of Holmes's apartment based on the texts by Arthur Conan Doyle is not really relevant for visitors. Mostly triggered by the success of modern film adaptions and TV shows such as BBC's *Sherlock*, audiences want to see how the famous detective *really* lived, walk through his living room, and put their names in his guestbook. Only a few miles away, tourists also queue for hours to attend a play in the historic Shakespeare's Globe Theatre, which is actually nothing but a modern reconstruction built in 1997. It is not real, but the mediation defines its meaning.[88]

But to find the *über-reality* experience, tourists are traveling straight to Hogwarts.

Harry Potter might be one of the defining stories and cultural artifacts of the twenty-first century. Its mediated representations can be found in several books, films, comics, websites, VR experiences, and merchandising products. But to actually physically enter the world of Harry and his friends, one needs to take a trip to the Warner Bros. Studios in Leavesden, 30 kilometers outside London. The former film studio, in which all eight films were made over the course of a decade, has been, since 2012, completely devoted to exhibiting all the sets, props, and costumes that appeared in the films. The Harry Potter Studio Tour is like a candy store for Potter fans, but it is also a gigantic immersive experience.

Right at the beginning, small groups of roughly 30 people are directed to a cinema to watch a six-minute film about the tour. Daniel Radcliffe, Emma Watson, and Rupert Grint appear in the short film not as their fictional characters from the blockbuster series but as their physical selves to welcome the visitors to the tour—a first hint of mediated realities. What could be a disruption in the flow of the experience does not really matter to the audience. Most of the visitors know Harry Potter from the films and associate the face of the wizard with the actor Daniel Radcliffe anyway. The last image of the short film is the key to the feeling of presence. The screen shows the representation of a giant, massive wooden door, which in the film series leads to the Great Hall in Hogwarts. This is where we will recognize a liminal threshold, a fraction between different concepts of realities, the one of the studio tour and the one of the *Harry Potter* world. The screen with the image of the wooden door rolls up and exposes a physical door behind it, in exactly the same colors, proportions, and size, as if the image has never disappeared. Or has never been there in the first place.

This is our rabbit hole.

In this moment, realities collide. Visitors scream out of surprise, others out of disbelief. They approach the door, touch it, smell it, and take pictures. In this very moment, the mediated reality of the books and films merges with the physical reality of the studio tour. And more importantly, the representation within the mediation becomes "realer" than everything else. Gordon Calleja explains this transformation with the constitution of the real through the virtual: "The virtual is characterized by movement and creative transformation. It is a force whose coming into being, its actualization, is never fully determined at its origin. The virtual is an event, a processual generation of an outcome through the interaction of a multiplicity of elements and contexts."[89] The simulation is completed once the physical door opens, and visitors are entering the Great Hall—or better the film set of the Great Hall. But no one really cares about such details anymore. In this moment the feeling of presence is fully developed and connects with the desire that both body and mind become totally immersed in the *Harry Potter* universe, a story that most of the visitors might have read when they were younger and that has been a crucial cultural artifact for their identity formation. For a person on the outside, it may look like an almost absurd scenario: people taking photographs of the (mediated) Diagon Alley in *Harry Potter's* studio version of London, even though they have (physical) London only a few miles away at their doorstep. But the city outside of the studio doors is not the one in which Harry bought his first magic wand. That is the one inside the studio tour.

And that is the one they choose as their reality now.

"Reality itself . . . has been confused with its own image."[90] That is what French sociologist and media theorist Jean Baudrillard once wrote. Baudrillard's theories focus on socialization that is determined by the desires for mediated realities and the visually represented—the "hyperreal"—in photography, film, television, and digital media. *Hyperreality* refers to the social belief in mediated realities that become the definition of the real over the

**Figure 1.3**
The Harry Potter Studio Tour in Leavesden, UK: visitors in front of the massive wooden door that leads to Hogwarts' Great Hall. © 2012 the author.

direct experience of the world outside of representation. It is the *über-real*, the belief that the representation is "realer than reality itself." For Baudrillard, the hyperreal is a simulacrum, a complete delusion, images that do not resemble the real anymore. The word "simulacrum" can be traced back to Plato's dialogues, where it was used to refer to phantasm and semblance.[91] It is the reflection of a basic reality, but it masks and perverts it by having no relation to it anymore. Hogwarts resembles physical castles with towers and battlements, but none of those really exist outside of mediation. Referring to Jorge Luis Borges, Baudrillard mentions a fable about a map that is so detailed that it covers the territory it represents. Baudrillard argues that in contemporary culture this relationship between the real and the represented is threatened as the symbolic order overlays our experiences. If you have ever checked out the photographic representation of a location on Google Street View before actually arriving there, you will know what he means. Gradually Google is amassing a photographic reflection of our world. The digital map will influence your perception of the location long before you are even able to physically experience it. The mediated version of the real is so omnipresent that it is bolstering the notion of a prior reality.

In fact, physical reality is absent in the simulacrum. The environment is a mediation. Every agency involved is also a mediation. All of them are part of an enactment, a performance as part of the mediation. For instance, the symptoms of a cold can be simulated and faked: sore throat, sneezing, coughing, and fever. But how does the act, the performance, the simulation of the symptoms, differ from the real sickness? The answer is, from the outside they do not. For all it concerns, anyone performing the role of someone being ill might be actually ill, but we simply do not know based on the information we get. However, we might classify the mediation as real because we want to believe in it as much as Alice believes in the existence of a mysterious fantasy world down the rabbit hole. The simulation of London in The Harry Potter Studio Tour is a simulacrum, a hyperreality. As soon as visitors enter it and feel present in the simulated environment, they accept the visually represented as the definition of the real for a particular period of time. You might wonder why they do not know better. How could anyone's mindset switch between different modes of reality that easily within a relatively short period of time? How could anyone possibly forget they have bought a ticket for the tour in advance, printed it out, took a bus to the studio, and waited in a queue with others inside a ticket hall, only to forget about all that after entering the door to the Great Hall? They have enough facts to know that this is not "real." The point is, they have not forgotten, but what I will later define as the *360° gaze*—the interrelations between semiotic codes, performative practices, psychological dimensions, and cultural industries—is simply more overwhelming in that moment. Similar to Baudrillard's famous analysis of Disneyland, the experience escapes the internal relationship to reality, while at the same time it marks reality as a myth:

> Disneyland is a perfect model of all the entangled orders of simulacra. It is first of all a play of illusions and phantasms: the Pirates, the Frontier, the Future World, etc. This imaginary world is supposed to ensure the success of the operation. But what attracts the crowds the most is without a doubt the social microcosm, the religious, miniaturized pleasure of real America, of its constraints and joys. . . . The imaginary of Disneyland is neither true nor false, it is a deterrence machine set up in order to rejuvenate the fiction of the real in the opposite camp. Whence the debility of this imaginary, its infantile degeneration. This world wants to be childish in order to make us believe that the adults are elsewhere, in the "real" world, and to conceal the fact that true childishness is everywhere—that it is that of the adults themselves who come here to act the child in order to foster illusions as to their real childishness.[92]

What Baudrillard calls the "fiction of the real" is the staging of the illusion in both modes of reality. By setting up an almost timeless environment without referring to any particular historical era, Disneyland becomes nostalgic, dreamy, and traditional. Werner Wolf emphasizes the importance of representation in an aesthetic illusion: "This 'projection' takes place in the mind of the recipient, yet in the state of aesthetic illusion, the mind's activity is not free-floating but 'guided' by the illusionist work or representation . . . and both recipient and representation are influenced in turn by various contexts."[93] From the architecture to the music and the characters, they all are a "play of illusions and phantasms" to escape the real. But at the same time, they confirm reality by being based on difference and by distinguishing themselves as copies of it—a superior, more desirable version of the real. For American visitors, Disneyland becomes a nostalgic, dreamy blur, referring back to a romanticized version of the past, but also to an idealized version of the future (Tomorrowland). Positioned right in the middle of this hyperreality is Sleeping Beauty's castle, a simulacrum itself, and a reduplication of the Neuschwanstein Castle in southwest Bavaria, Germany. In fact, many tourists travel to the castle in Germany not to learn about Neuschwanstein's history but to visit Cinderella's castle in reality. Baudrillard adds: "Whence the characteristic hysteria of our times: that of the production and reproduction of the real. The other production, that of values and commodities, that of the belle epoque of political economy, has for a long time had no specific meaning. What every society looks for in continuing to produce, and to overproduce, is to restore the real that escapes it. That is why today this 'material' production is that of the hyperreal itself."[94]

Human experience has been (trans)formed by the use of mediation. Our brains select, amplify, and reduce experiences in various ways. The human mind is highly selective while defining reality. We tend to look at old photographs and videos to trigger our memories of the past, while at the same time organize, select, and focus our process of remembering. We are only able to remember certain things, to consciously and unconsciously highlight some, but forget about others. Very often we do all of that to the extent that we glorify our own past, remember things completely differently compared to how they actually happened, or reimagine events that we could not possibly remember at all (for example, events from our early childhood or times when we were not even physically present). That process might get activated by looking at a photograph, a mediated representation of a past event. Langdon Winner calls this the "reverse adaption," or "the adjustment of human ends to match the character of the available means."[95] So while the human mind is locked into its own limitations, mediation through physical tools (technology or the simple use of a pen) defines the perception of our reality. This continuous process of creative production and the fluidity of individual perception are operating under the rules and regulations of that specific medium. We do not have direct access to our memories, and there are "no guaranteed techniques . . . to discriminate between real and false memories."[96] We are only able to write about the past, take photos and films of it, or try to digitally recreate it. But by doing so, these mediations become their own formation of reality.

In postmodernism, the mediation is the benchmark for the perception of the real.

> Mediated realities are *formations of reality created through the use of media*. They do not have to refer to a form of physical reality. They produce meaning through the representation and mediation of ideas.

In his groundbreaking texts *Kommunikologie*[97] and "The Codified World," Vilém Flusser argues that our world is mediated through semiotic codes. He demonstrates how humanity is losing the capacity for immediate communication, as throughout history we are constantly trying to grasp a glimpse of reality with the help of mediation—from the first use of cave paintings to tell the story of hunting scenes (e.g., in the cave of Lascaux) to the invention of writing, photography, film, and images created by other technologies (he calls them "techno-images"). Techno-images are universally used as ideograms and icons, which can be found in almost every aspect of modern communication—for instance, consider the use of emoji in online communication. While emoticons are typographic displays of facial expressions, an emoji is a picture representing a text. The smiley face is an image that actually incorporates the meaning of a text. It mediates an idea, a feeling, an action: "I am happy" or "I am smiling at you." As Vlusser puts it, "the 'belief in texts'—in explanations, in theories, in ideologies—is lost, because texts are now recognized as 'mediations,' just as images were once upon a time."[98] This argument is essential for immersive mediations and the alienation through technology. We are using simplified, universal codes to communicate the complexity of meaning, including our beliefs, hopes, thoughts, and feelings. While Flusser obviously means the transition from texts to images that become texts, there is a wide range of semiotic variations in digital communication practices. For instance, think of the use of specific filter options to modify an image on Instagram. Think of dog-ears or big eyes-filters to *un/mask* faces/reality on Snapchat. Think of the use of hashtags. Think of the ridiculous overuse of emoji on instant-messaging apps (anything from "ninja" to "naked avocado" emoji). Postmodern societies have developed a perversion for techno-images. However, to feel immersed in mediated realities, these images are essential parts of the conventions of modern communication. Just think about the moment when an emoji is missing in a reply from your loved one on WhatsApp, and how that makes you feel.

Since we now have a better understanding of the modes of reality, there are two other phenomena that we need to look at. The sensation of "being there" in mediated realities is often described with two terms: *presence* and *immersion*.

In most discourses—and even in everyday language—both terms are often applied interchangeably, while at other times they have conflicting meanings. In his groundbreaking work on engagement in digital games,

Gordon Calleja notes: "Although there has been consensus that the experience of presence or immersion is important, there has been confusion over precisely what the terms mean."[99] He adds: "Of the two terms, *immersion* is particularly awkward because it has also been applied to the experience of non-ergodic media such as painting . . . , literature . . . , and cinema . . . , all of which provide forms of engagement that are qualitatively different from those of game environments."[100] However, in the edited collection with the misleading title *Immersed in Media*, the authors also consider entertainment media outside of digital games, such as film and television, but they mostly use the term telepresence to describe them.[101]

Early investigations in the terminology as part of human-computer interaction and digital games emerged from the fields of computational science and cognitive psychology.[102] It can get quite difficult to distinguish between the different approaches, definitions, and descriptions. For instance, Mel Slater and Sylvia Wilbur argue that *presence* is "a state of consciousness, the (psychological) sense of being in the virtual environment."[103] Bob G. Witmer and Michael J. Singer, on the other hand, think of *immersion* as "a psychological state characterized by perceiving oneself to be enveloped by, included in, and interacting with an environment that provides a continuous stream of stimuli and experiences."[104] In an attempt to focus solemnly on the understanding of immersion, Emily Brown and Paul Cairns conducted a grounded theory study in which they asked seven players about their experiences playing computer games. The participants described the degree of involvement by referring to the feeling of presence ("You feel like you're there").[105] The authors, however, merely mix the two terms once again by claiming: "Total immersion is presence."[106] Presence and immersion refer to similar physical and psychological sensations, but we have to make a distinction between them to understand how they influence mediated realities. I argue from a humanities and social sciences perspective that presence and immersion are not the same, and should not be used to describe the same phenomena.

What is "presence"? What does it mean to "feel present" in a mediated environment, communication, narrative, or performance? From an etymological point of view, "presence" indicates what is *prea* and *sens* (Latin), so in other words "before I am" or "in front/in view of me." The phenomenological idea of consciousness—"being there" in a mediated space—originates from a paper by Marvin Minsky titled "Telepresence" (1980) (*tele* means distant, *presence* the state of being there). Minsky argued that a sense of presence could establish itself through the use of transducers such as video cameras and microphones, which are becoming substitutes—or in the McLuhanian understanding "extensions"[107]—of the human body. The user is only able to see and hear with the aid of remote sensing devices, which offer the ability to directly interact with a mediated environment. While users experience the world through remote devices, they build a conscious belief in the world presented to them and accept it as their definitive surrounding for the duration of the mediated transmission.

Let us use an example. Imagine you are sitting in an astonishing orchestra hall—for instance, the Royal Albert Hall in London or the Beacon Theatre in New York City. You have done your research. You know exactly which seat you would need to pick to get the best sound experience. Probably near the stage, front row of the middle section. There is no way you would like to sit directly next to the stage, unless sheltering from Mahler or Bruckner at their most bombastic. You need to consider the location to get the best listening experience. You are looking for ambiance that is subtle but dynamic to the overall sound. So while you are getting the depth of the sound waving past your ears, you will get it bouncing back from behind as well, giving you a good listening experience. This is your perfect spot. The orchestra begins to play, and you are closing your eyes. This is a unique, "first-order" experience in the physical world, and even though it is still mediated to a degree (by your physical senses, the architecture of the orchestra hall, which has impact on the place you sit in), it is the best you can get. Or is it?

According to Wijnand IJsselsteijn and Giuseppe Riva, presence can be defined as the experience of feeling present in a three-dimensional environment that is created through media technology.[108] In their understanding, the feeling of being inside an orchestra hall could be recreated with the help of technology and might be potentially even better than sitting in the physical concert hall.

Now imagine sitting on the floor of your living room, with headphones on, listening to a recording of the exact same concert you have witnessed in the venue (or maybe never attended at all). Could it come remotely close to the physical experience? Mel Slater is convinced of it: "Suppose you shut your eyes and try out someone's quadrophonic sound system which is playing some music. 'Wow!' you say 'that's just like being in the theatre where the orchestra is playing.' That statement is a sign of presence."[109] You are "transported" to another place and develop a "sense of being there," while technology helps you to take your mind somewhere else. It is "the pleasurable experience of being transported to an elaborately simulated place" and the "sensation of being surrounded by a completely other reality, as different as water is from air, that takes over all of our attention, our whole perceptual apparatus."[110] In other words, you forget about sitting on the floor of your apartment, but rather feel entirely present in the concert hall through the mediation of the music. With that in mind, it would be possible to give you the same kind of sonic experience on headphones, which you otherwise would only get while sitting at the best spot in the concert hall. You are now able to have that experience at a location of your choice, and for 99.9 percent of all people it will be the better listening experience, especially for those who were not be able to sit at the best seat at the physical concert or were not able to attend it at all.

Lev Manovich notes how this experience alters the perception of physical reality: "I don't have to be physically present in a location to affect reality at this location."[111] At the same time, listening to a recording of a concert creates a mediated version of our self in that space, in the etymological sense of the one "in front/in view of me." This *mediated self* is the one we are imagining while listening to the recording. It might very well be the illusion of our mediated self that is sitting in a concert hall, but very often it is not. Think about the thoughts you are having while listening to music. They might be a memory of your past, a vision of the future, or a fictionalized design of a physical or imaginary encounter. Your mediated self might be dreaming away on an island, or looking out at the ocean, while listening to the recording on your headphones. Whatever it is, it becomes part of the mediated experience in which you are feeling present.

Minsky's ideas of telepresence lead to a long tradition of over 30 years of research on discussing presence in highly recognized academic journals such as *Presence*, trying to capture and measure user experiences with media technology. Research on presence is a growing interdisciplinary field and bridges computational science and psychology, human-computer interaction, neuroscience, consciousness research, hardware and software engineering, and sociology. Such an intense discourse produces a wide range of different perspectives as well as considerable disagreements about how presence should be defined. Most of these differences are ontological, not necessarily terminological. In addition, for over two decades these definitions have been made in respect to digital games,[112] while neglecting many other mediated experiences in digital and analog media. To distinguish it from immersion, the most common definition of presence relates to the *use of technology* and the *levels of consciousness*. The International Society for Presence Research has the following definition on its website: "Presence . . . is a psychological state or subjective perception in which even though part or all of an individual's current experience is generated by and/or filtered through human-made technology, part or all of the individual's perception fails to accurately acknowledge the role of the technology in the experience." And later: "Except in the most extreme cases, the individual can indicate correctly that s/he is using the technology, but at *some level* and to *some degree*, her/his perceptions overlook that knowledge and objects, events, entities, and environments are perceived as if the technology was not involved in the experience."[113] Calleja specifies: "One such experience is the potential to have a sense of inhabiting the simulated spaces they offer, not just through the use of player's imaginative faculty, but also through the cybernetic circuit between player and machine."[114]

We have learned that presence is a multidimensional concept, and there are different variations of it. For instance, "spatial presence" means users fail to acknowledge the role of technology (also known as "physical presence," "a sense of physical space," "perceptual immersion," "transportation," and "sense of being there"). On the other hand, "sensory presence" makes them notice the physical location and the role of technology (also known as "perceptual realism," "naturalness," "ecological validity," and "tactile engagement").[115] For instance, VR experiences offer the possibility to move and interact within a virtual environment, however, users might not forget about the role of technology yet as most high-end headsets are bulky and uncomfortable to wear. Digital games accessible through smartphones, on the other hand, might make you forget that you are actually holding a phone, and you certainly do not think about the headphones while listening to music. The International Society for Presence Research lists several other presence variations, such as "social realism," "engagement," "social presence," "co-presence," and "medium as a social actor."[116] All of them are relevant distinctions, but for our purpose we want to focus on the interaction with technologies.

Have you ever listened to music during a commute? Maybe you listened to some good old Dylan or Stones records while watching the cars and trees go by? This is more than just escaping the boredom of traveling. It is a sign of presence as you create your own augmented reality while listening to music. The mediation of music distorts your perception of physical reality. Presence occurs during an encounter with technology, but not before, and certainly not after it. But we still do not know why and how we are actually developing a feeling of presence. Giuseppe Riva and John A. Waterworth[117] draw from findings of cognitive science to locate a broader understanding of the phenomenon and to look at the levels of consciousness and the actions performed within a mediated environment. They argue that presence changes as part of the evolutionary process. Humans are able to change and adapt their abilities to feel present in mediated environments—for instance, children are able to adapt to the mechanics of gaming controllers much more easily than adults, who have never played any video games at all or are not up to speed with the newest technology. Therefore, children are more likely to develop a sense of presence in the mediation of the video game. Presence is an intuitive process and not something that is institutionally learned. We grow up with it and adapt it in relation to our knowledge of media literacy.

Therefore, Riva and Waterworth suggest framing presence not only in regard to technology, but also in relation to selfhood, and more precisely, to the idea of the self as a mediation of the individual. For example, while playing an interactive game such as *Virtual Tennis* on the Nintendo Wii, our self is able to experience three layers of presence:[118]

- *Proto-presence*: "the intuitive perception of successfully differentiating the self from the external world through action." While we are playing *Virtual Tennis* in our living room, we see the mediated tennis court on our TV screen. Our opponent is hitting a strong forehand. The ball is about to land on "our" side, which is actually still on the screen and not in our physical living room (we are not expecting to see the ball jump up next to our feet). This is when we are able to define our mediated self (the player in the game) in relation to the external world (the tennis court and our opponent) through action (the ball that we are about to hit any moment). This has something to do with the level of perception-action coupling (self vs. non-self).
- *Core presence*: "the intuitive perception of successfully acting in the external world toward a present object." As the ball is landing on our side of the court on the screen, we are about to take a swing as we see the ball

jump up again. We take the swing in the living room (not against a physical but a virtual object), and we see the result of our action on the screen. The ball is crossing the net to the other side as if we have hit it in our physical environment. This relates to the level of vividness (self vs. present external world).

- *Extended presence*: "the intuitive perception of successfully acting in the external world toward a possible object" After we have hit the ball, it is landing on the other side, and it is a winner. Now we not only won the point, we won the virtual Wimbledon finals. This now becomes the most distinct layer of presence. The level of relevance for us is so high that we are feeling joy because of our triumph and we are looking forward to hearing our name in the same breath with former winners of the "real" Wimbledon tournament (self vs. possible external world).

When immersive experiences are designed to offer all three layers of presence, audiences are able to reach a feeling of extended presence. That is when the actions they perform as a mediated self actually matter to them as a physical and psychological sensation, while at the same time they constitute who they are in the mediation through action and the feeling of presence.[119] As Gabriella Giannachi and Nick Kaye point out, presence is not only an experience, it is a performative practice of the self "realized in performative encounters with images, objects, technologies, bodies, sites, acts and events."[120] As a result, presence does not only exclusively occur during an encounter with technology. For instance, while participating in an immersive theater play, we might be able to reach a feeling of extended presence if the actions we perform as a character in the play become relevant to us.

Finally, I propose the following definition for the feeling of presence in mediated experiences:

> *Presence* is the conscious state of immediacy during an encounter with a mediated experience. This form of immediacy occurs often but not exclusively with the help of technologies. Presence can only be experienced through the creation of a mediated self and the performative practices that are realized during the consumption, transmission, and reception of mediated content. A level of extended presence can be achieved if the actions performed by the mediated self are considered relevant during the experience.

I argue, however, that immersion is something different.

Immersion has been described in various metaphors and theoretical concepts over the years, which very often confuse the phenomenon with presence. Besides "illusion" and its literary meaning *in-lusio* (to get or be drawn into the game),[121] there are also "*l'effet de réel*,"[122] "absorption" or "getting lost in a book,"[123] "entrancement" or "fascination,"[124] "recentering,"[125] "make-believe,"[126] "enchantment,"[127] "involvement" and "psychological participation,"[128] "engagement,"[129] as well as "transportation."[130] In recent years, the prefix "deep" is used to describe the psychological sensation of uncertainty regarding whether a mediation is real or not, as in "deep media,"[131] "deepfake,"[132] and Facebook's facial recognition algorithm DeepFace. Marie-Laure Ryan lists even more theoretical concepts dealing with possible worlds, semantics of fictionality, flow, mental images, and mental simulations, which try to capture the aesthetics of gamification, self-reflexivity, and interactivity of audiences.[133]

In fact, theories of immersion are as diverse as the "the history of Western art [which] has seen the rise and fall of immersive ideals,"[134] especially "the strong anti-immersion stance of postmodern art and literature as well as poststructuralist theory in the 1960s and 1970s."[135] For many decades, mass media have been defined by the "picture-frame-stage," the return of the frames to control the viewer's gaze, e.g., the tradition of the window view in cinemas of the industrial age, and later the "mediated correspondent in the framed glass screens of televisions (and then computers)."[136] In her analysis of immersive strategies in literary texts, Ryan discusses the attempts to overcome the window view within post-realistic and abstract-art forms, and the resurrection of immersive ideals in the last decades. She recognizes certain similarities based on the consciousness of the reader: "Whereas inert objects, entirely contained in their material bodies, are bound to fixed location, consciousness can occupy multiple points and points of view."[137] Gordon Calleja adds: "The metaphors of immersion . . . are defined by their discontinuity from the real, physical world."[138]

There have been many attempts in relation to digital games to understand immersion as a multidimensional construct that goes beyond the feeling of presence.[139] For instance, in his book on engagement in digital games, Calleja proposes a framework that focuses on the involvement of players. His empirical model encompasses two constituent temporal phases: one representing "moment-to-moment involvement" during gameplay, and another representing so-called "offline-involvement." With the latter, Calleja acknowledges player involvement *beyond the experience of the game*.[140] Laura Ermi and Frans Mäyrä, on the other hand, describe "imaginative immersion" as the sensation of being mentally absorbed by a game's story, its world, or its characters—a sensation that often occurs as well outside of the game experience.[141]

Back to our triumph at Wimbledon. After we have finished playing the game and turned off the console and the TV, we should have realized that we merely won the Wimbledon finals in a video game. If we traveled to London and visited the Wimbledon Lawn Tennis Museum, there would be no trophy with our name on it. According to earlier theories of presence, this would be the time when the feeling of presence should have ended, as we fully returned to our physical environment and are no longer attached to either the technology or our mediated self in the game. However, there is something else going on. We might still experience a positive cognitive reaction after winning the virtual finals, a feeling of joy, success, and strength. We might be able to transfer that feeling to our physical reality, get an ego boost out of it, and it might even give us a feeling of success for the rest of the day. We have achieved something through our mediated self that contributes to our understanding of physical reality and selfhood.

This is *immersion*, the often imprecisely defined stage that lasts beyond the mediated experience. This occurs when our mediated self and the mediated experience matter to us even *after the mediation has ended*, when we still feel like a winner long after we have successfully finished a video game. Or when we still feel emotionally devastated after watching a dramatic movie. Or when we are not able to forget a particular song that was playing in the background when we had our first kiss, even decades after it happened. Or when we are not able to get that Tinder match out of our heads even if we have not met in physical reality yet. These mediations, among many others, result in the feeling of being immersed even outside an actual encounter with a mediated experience.

As Jan-Noël Thon explains, immersion is a shift of attention.[142] It might be the shift of attention to the unfolding of a story and the fate of fictional characters. It might be the shift of attention to your digital avatar on social media, or the memories of a rock concert you witnessed last summer. It might be the shift of attention to the parasocial relationship you have with a celebrity. These overwhelming experiences merge with what we perceive as physical reality and make it very often impossible for us to differentiate between the physical and the mediated.

Immersion distinguishes itself from presence by being long-term. Although in presence research those two terms are often used interchangeably, I argue from a humanities and social sciences perspective that they need to be separated and analyzed as different forms of engagement. This might be a provocative departure from the dominating discourse in the context of digital games. And even though I will build upon the extensive research on both presence and immersion, I hope to provide a more distinct interpretation of immersion in a vast range of mediated experiences. While extended presence means the mediated self is positioned within the immersive experience as a conscious state of immediacy, immersion is very often unconscious and not limited to the actual mediation. Instead it creates meaning beyond the frames of mediation. Immersion is the absorption of the self within an all-surrounding mediated experience even after it is disconnected from the devices. In the case of immersive technologies, this becomes the paradox of being neither completely here nor there, in a space between different modes of reality: "Theories of immersion commonly focus on strategies and techniques of resolving this paradox, but . . . it is precisely in viewers' experiences of immersive media that this paradox resurfaces as a challenge to the theories themselves."[143]

I propose the following definition for the feeling of immersion:

> *Immersion* is the (unconscious) state of absorption even after the mediation has ended. It is not limited to the actual consumption, transmission, and reception of mediated content. The audience's relations with both the mediated content and the mediated self stay intact beyond the mediation. In many cases, this leads to the creation of meaningful mediated realities for a long period of time, very often highly influential as physical and psychological sensations that make it difficult to distinguish between the physical and the mediated.

Ultimately, immersion results in a feeling that absorbs and transports our consciousness into a state of liquidity and uncertainty. The space in between different modes of reality is a *liquid space* in which the mediation is more than just an echo in our minds. Instead, the feeling of immersion is created in this space as a continuous, individual, inclusive, and fluid development of the mediated representation.

The definitions of presence and immersion are often confused, because we always think of the mediation itself, but rarely beyond it. It is a phenomenon that Roland Barthes describes when he writes that he felt ready "to fetishize not the image but precisely what exceeds it." In his short essay "Leaving the Movie Theater" (1975), he speaks of the experience of having two bodies while watching a movie: one that is inside the film, the other one sitting inside the cinema.

> . . . as if I had two bodies at the same time: a narcissistic body which gazes, lost, into the engulfing mirror, and a perverse body, ready to fetishize not the image but precisely what exceeds it: the texture of the sound, the hall, the darkness, the obscure mass of the other bodies, the rays of light, entering the theater, leaving the hall; in short, in order to distance, in order to "take off," I complicate a "relation" by a "situation." What I use to distance myself from the image—that, ultimately, is what fascinates me: I am hypnotized by a distance; and this distance is not critical (intellectual), it is, one might say, an amorous distance.[144]

## 2   Beyond the Mediation

**Liquid Spaces**

> Bedding Taylor Swift
> Every night inside the Oculus Rift
> After mister and the missus finish dinner and the dishes
> And now the future's definition is so much higher than it was last year
> It's like the images have all become real
> And someone's living my life for me out in the mirror
> —Father John Misty, "Total Entertainment Forever," *Pure Comedy*, Sub-Pop, 2017

Immersion is created in a *liquid space*.

Etymologically, *immersio* (Latin) means "diving in"/"diving under," and hence feeling both physically and mentally surrounded by a different environment. Immersion is both a tempting and unusual experience, very much like jumping into a swimming pool.[1]

Imagine how the water in the pool is rapidly covering the surface of your body. It wraps you up completely. All your senses are encapsulated and surrounded as you sink deeper and deeper into the pool. You can feel the water surrounding you, but you cannot grasp it. The water provides no stable surface. Instead, it slips through your fingers as soon as you try to hold onto it. It constantly moves around you, and you move with it. As humans, we have developed strategic and institutionalized mechanisms for moving securely within large bodies of water, while at the same time protecting our vulnerable bodies: swimming. If we haven't learned already, we learn how to swim during our time at school. However, a certain danger always remains, due the intense physical and mental adjustments we need to make while underwater. For instance, our bodies reorganize air supply and slow down the heartbeat to reduce the use of oxygen. This can be a terrifying experience for many people, especially for those who are not used to it or are not experienced swimmers. The first body-water contact is often preserved in our memories. Negative recollections of being exposed to water may even cause aquaphobia, the fear of large bodies of water, or feelings of anxiety while being too far away from land during swimming. In any case, we deal with the experience of being exposed to water long after we have left the pool and stand on firm ground again.

The body might appear to be a closed system, but it can be easily manipulated to become an open, vulnerable organism. We are not able to fully protect ourselves while immersed in water. If you opened your mouth and tried to breath underwater, you would suck the liquid up through your mouth and nose. Too much water inside your lungs would stretch their capacities and potentially lead to suffocation and drowning. You would literally *lose yourself* in the liquid mass. What started as a unique and alternative experience would end up producing a state of existential fear. When fluids are allowed to expand without limitations, they do not spare the materiality of any solid surface, not even of our own bodies. The liquid is controlling us, not the other way round. As Neil Postman said, we are becoming the "tools of our tools."[2] Immersion, just like water, is like a friend-made enemy.

The feeling of being surrounded by alternative realities in media, and being exposed to them like to water in a pool, is a key phenomenon through which to describe cultural and media practices in the twenty-first century. Immersion is a state of "all-overness," a totality of experience that pulls and absorbs audiences in it. Immersion eliminates any aesthetic distance. As with our first body-water contact, mass media are full of long-lasting memorable experiences of inception that overpower us for years to come. They might leave us defenseless, unprotected, and submissive without being able to ever fully grasp them completely. Boris Groys, for instance, describes the experience of sitting in a movie theater as a state of total paralysis and physical immobility. The images on the screen might move, but the audience is literally ripped out of their lives. The viewers lose control of their attention and their freedom of movement.[3] Immersive experiences have the ability to penetrate us, move within us just like water or air, and remain in our minds even long after the mediation.

We might know how to swim. However, we have not yet developed strategies to behave and protect ourselves within mediated realities. Very often we remain exposed, vulnerable, unshielded, and influenced by immersion long after the mediation has ended.

In many cases, immersion lasts forever.

The understanding of immersion as a liquid sensation fits into the concept of "liquid modernity." The term has been coined by sociologist Zygmunt Bauman to describe myriad phenomena of instability, complexity, uncertainty, and change in the age of digital media.[4] Bauman defines modernity as the age of fluids in which solids and stable conditions are falling apart and now appear unstable, unreliable, and uncanny. Since then, the idea of liquidity has been mobilized in discourses ranging from social theory to aesthetics, and from informatics to architecture. It describes a new relationship with the networked environments of life within global capital. Our world is now ever-fluctuating. Nothing is static. Whatever has once been solid is now unstable, unjust, unequal, and forever shifting. More specifically, within the studies of digital culture and digital technologies, we have seen an increasing turn toward affective relations, plasticity, resonances, and flows. For instance, images and sounds—no longer grounded in an analogical relation to the real—are seen variously as malleable, untethered, viral, or fluid.[5] Furthermore, Oliver Grau points out that mediations in digital media are forever shifting because "the concept of 'the original' is foreign to the computer. With regard to the data, there is no difference between an original and a copy, for the machine's system protects the structure from any intentional or unintentional modification."[6] However, as Burcu Dogramaci and Fabienne Liptay also argue, the term "immersion" should not be used exclusively in relation to digital technologies, but rather in relation "to any act or experience of plugging into something" and with a "wide variety of uses, especially in the English language, a baptismal font or a swimming pool, a chemical solution or a medicinal bath, the shadow of a planet or a foreign language can equally serve as immersive 'media.'"[7]

Liquid spaces are defined by *rabbit holes*. They are gate openers that allow us to cross borders into mediated realities, which lead from the solid (physical) to the liquid (mediated)—a metaphorical and liminal threshold of ambiguity and uncertainty. In fact, the ideas of dissolution, liquidity, and liquefaction refer not only to immersive media but to all kinds of mediated and cultural practices in postmodern culture. In all of them, we are crossing thresholds and borders between fictions, frictions, worlds, hybrids, narratives, performances, practices, and subcultures. There is a threshold between the moments we are feeling present while consuming mediated content, and the moment the fluidity of immersion kicks in.

In his *Arcades Project*, Walter Benjamin defines such a threshold (in German: *die Schwelle*) as "a zone. Transformation, passage, wave action are in the word *schwellen*, swell, and etymology ought not overlook these senses."[8] In theater and cinema, the threshold is the imaginary crossing of the fourth wall, even though we know we remain in our seats. In literature, it is the entrance to the diegetic world of fiction while we sink deeper into the page and disregard the fact that we are reading a symbolic representation on paper or a screen. In screen cultures, it is the penetration of the canvas in front of us, which we are looking at while disregarding it at the same time. Immersion is something we are in, after we have developed a feeling of extended presence within a mediated environment. It is the liquid space that inhabits us long after we are disconnected from the medium.

The container or pool, to remain in the metaphor of water, represents the medium. The media's shape and form frame the liquidity of its content. If the container cracks and breaks, the fluid spills over its frames and no longer knows any limitations. It no longer separates the physical reality from the mediation. Gundolf S. Freyermuth acknowledges the historical principle of a threefold separation across visual media established between the Renaissance era and postmodernism: firstly, the separation of one medium from the other as a pre-intermedia phase (in particular, the separation of the image space from the textual space); secondly, the separation from the audience (spatial distance and physical concealment with the help of frames, curtains, doors, stages, glass panes, etc.); and thirdly, the separation from the environment through hardware framing (such as VR).[9] In postmodernism, immersion means that the rules of separation, stability, and solids no longer apply. The content is no longer separated from the viewer. Instead, audiences are invited to step inside it, be a part of it, and to feel immersed by eliminating the barriers and "replac[ing] the separation . . . with sensory, full-body experiences."[10] Peter Sloterdijk argues that immersion is unframing images and vistas. It dissolves the boundaries between the physical and the mediated.[11] Immersion invalidates the security mechanisms on which audiences rely while consuming media. It suspends the frame of the painting, the stage of the theater, the screen in the cinema or on the mobile phone, and the headset of the VR experience.

In 1930, under the title "We are right in the middle!," film theorist and critic Béla Balázs described this fundamental new aesthetic experience: "But film has not just brought new material into view in the course of its development. It has achieved something else that is absolutely crucial. It has eliminated the spectator's position of fixed distance: a distance that hitherto has been an essential feature of the visual arts. The spectator no longer stands outside a hermetic world of art, which is framed within an image or by the stage. Here, the work of art is no insulated space, manifesting itself as a microcosm and metaphor and subsisting in a different space to which there is no access."[12] Expanded film spaces in CinemaScope, IMAX, 3D, but also the 360° cinematic worlds in VR push the frame of the image beyond the audience's field of vision to make it disappear. Mediation therefore becomes fluid; it flows and spills over the edge of containers and devices. The act of separation is replaced by experiences of immersing, melting, and plunging into the mediated space. Bauman argues that these fluid activities "dissolve some others and bore or soak their way through others still,"[13] while at the same time moistening and drenching the solids. Whatever we consider as solid elements within mediated representation, such as frames, devices, and shapes, as well as social, narrative, and technological conditions, are now repressed by the mediation. Media texts do not stay in their place anymore. They inhabit our physical reality and merge with it.

In 1902, Georg Simmel wrote that the frame "excludes all that surrounds us, and thus also the view as well, from the work of art, and thereby helps to place it at that distance from which alone it is aesthetically enjoyable."[14] This safety net is now about to disappear, and the mediation merges with the physical environment. As a result, often only the mediation remains as the dominant mode of reality in the perception of the audience—even more so in VR, as this kind of technology "is immersive, which means that it is a medium whose purpose is to disappear."[15] That is when the shift of attention starts, the merge with the physical world. This is when "the principle of separation"[16] initiates and conditions the modes of reception. Distance and isolation from the mediated experience are no longer available.

The medium mediates by making itself invisible. Jay David Bolter and Richard Grusin describe this as "immediacy," a state in which "the medium itself disappears and leaves us in the presence of the thing represented."[17] Oliver Grau calls this "mental absorption" and a reduction of critical distance. He refers to it as "regimes of perception."[18] This "technological iceberg" would be "inaccessible to the user who is unaware of it—that there will be an illusionary disappearance of boundaries to the data space."[19] The state of immersion might be a desirable feeling for many, but it ultimately results in a lack of control and exposure to the liquidity of the mediation.

The security mechanisms of mediated experiences have always been there to help us, defining mediation in the first place. Aleida and Jan Assmann famously argue that most human experiences depend on the use of media. Consequently, physical reality must be everything that is not media.[20] Frames have been implemented to function as "basic orientational aids that help us to navigate through our experiential universe, inform our cognitive activities and generally function as preconditions of interpretation."[21] As a result, "frames also control the framed."[22] Media have always helped us to define both the physical and the mediated by allowing distance as a prerequisite for any critical reflection. Theodor W. Adorno once said that "distance is the primary condition for getting close to the content of a work,"[23] and Arnold Gehlen added that "direct emotionality of experience is held to be alien to art, and rightly so."[24] Hartmut Böhme stresses the subject-constitutive quality of distance in relation to immersion: "All happiness is immersion in flesh and cancels the history of the subject. All consciousness is emancipation from the flesh to which nature subjects us."[25]

If all the frames, stages, and technologies dissolve now, we are about to confuse different concepts of realities, eliminate every distance to media texts, and lose our "cognitive guides of interpretation."[26] Those phases of liminality are ambiguous. They emphasize the dissolution of boundaries and control, of the "transgressive, boundary-breaking, all-eroding modernity."[27] Ultimately, they epitomize all the conventional boundaries between physical and virtual, analog and digital, natural and artificial, animate and inanimate, human and technology.

I argue there is a so-called *liquid space* in every mediation, when we are neither here nor there, when the physical reality collides with the mediated representation, and when we *dive in* and *dive under* as part of immediacy. This space is a risky notion as there is no emancipation anymore, no freedom, and no container. There are no frames. We are completely surrounded, exposed, and occupied by immersion. This is when mediated content materializes itself as an integral part of our identities. This is when the medium no longer merely activates emotional re-/actions from audiences, but (actively) constructs feelings that emerge already within mediated experiences. As a result these experiences replace, eliminate, and deconstruct whatever we have perceived as "real" before.

# Beyond the Mediation

> Liquid spaces are *moments of uncertainty, instability, and fluidity* in mediated experiences. They emerge as *thresholds between the physical and the mediated*. These spaces create the feeling of immersion even beyond the mediation by *eliminating critical distance and dissolving the frames of media*.

The recent hype around immersive technologies seems to fulfill the promise to make us feel immersed *all the time*. Everything is possible after putting on the headset: going to class, meeting friends, listening to music, watching a show, having sex. Mediation would never end, and "transparent immediacy," as it can be found in the excitement, liveliness, and realisms of today's digital images, would lead to the disappearance of the interface, "so that the user is no longer aware of confronting a medium, but instead stands in an immediate relationship to the contents of that medium."[28] Ray Kurzweil even once predicted that "[b]y 2030 you'll see full-immersion, shared, virtual-reality environments, or spaces, involving all the senses, where we can actually go inside our bodies and brains and tap into the flow of signals coming from our senses."[29]

Immersive technologies focus on the mechanisms through which humans are able to access their mediated selves more directly than through any other media technology before. For instance, when we are joining a virtual birthday party in social VR and are creating virtual selfies with other virtual avatars, this experience will lead to *actual memories through the state of immersion*. When a VR experience leads to "real" emotional reactions, they are first and foremost experienced within the mediated space before they get translated into physical reality. This ultimately changes the way in which we perceive our identities and the world around us. These multisensory experiences pull us again and again "from experiences of immersion into distanced positions of self-observation, reflection, or distraction."[30] This is the phenomenon that is often referred to as the ultimate *empathy machine*.[31]

Immersion is immediacy *beyond the mediation*.

Our orchestra hall example can help us to understand this issue better. While you are feeling cognitively *present* in a virtual orchestra hall as you listen to a live recording, *immersion* occurs in the liquid space, which eliminates separation and distance between the music and your living room. This feeling carries on after you have switched off the music. Immersion builds upon what presence research defines as the third layer: extended presence,

**Figure 2.1**
Facebook Spaces: a virtual birthday party creates real memories. Reprinted with the permission of Facebook. © 2017 Facebook.

the level of relevance of the mediated encounter. After you have turned off the record and put down the headphones, you are able to hold on to whatever emotions the experience has activated in your mind. Memories from the past, associated images, visualized sceneries. Whatever the music has emotionally released in you often remains for a longer time even if you might be already present in another (mediated) environment. The mediation means something to you. The need for immersion makes you return to the music again, even after years or decades. You are eager to reexperience the illusion of being present in the orchestra space again and to conjure the feelings you once had. So even while you are feeling present in the mediation with the help of technology, the feeling of presence ends as soon as you turn off the music. You are only able to feel presence as a conscious act as long as the mediation takes place. Everything beyond that is the feeling of immersion.

While we are binging a show on a streaming service, we are feeling present in the story and are emotionally invested in the fate of the characters. Their world matters to us. We are drawn in because "being transported into a story has strong cognitive and emotional consequences and leaves the viewer susceptible to change from the themes of a story they are experiencing."[32] However, even after we have stopped watching, we might still be concerned with the unfolding story arc, be worried about the outcome of the next episode, and read online reviews of the episodes we have just seen. Our imagination is activated, as we remain invested in the aesthetic illusion. There are no frames anymore, the show is everywhere, and we are within it. We might even take some concrete actions to buy posters, t-shirts, and mugs as rituals of our fandom. We might acquire the box set with all the behind-the-scene features, commentaries, and outtakes. We are part of the mediated experience long after the show has ended: "All of this enables us, and indeed renders us disposed to immerse ourselves in others' and our own (past) experiences, be they actual, represented or merely imagined."[33]

The liquidity of immersion involves the idea of alternative realities. Alice fell into the rabbit hole with nothing to hold onto. On the other side, she was introduced to the concept, depth, and complexity of an alternative world beyond the limitations of Victorian middle class. At the same time, we also find alternative realities in many media and cultural practices in our lives. They are always defined by a complex relationship between several interrelating influences that span across physical and digital environments. That relationship is certainly not limited to immersive technologies.

If we remember our example of *Harry Potter* and the idea of hyperrealities, we do not necessarily need to enter a studio complex or theme park to get engaged in the story universe. Fans also love to visit platform 9¾ at King's Cross train station in London,[34] famous for its magic brick wall, which brings Harry and his friends straight to the platform to catch the Hogwarts Express. In terms of Baudrillard's understanding, the platform is a representation. There is no physical platform with the name "9¾" at King's Cross station from which actual trains arrive and depart. It is nothing but a sign on a wall, but the semiotic codes, narrative acts, and performative practices associated with it place it as a form of immersive theatrical experience inside the physical space of the train station. Fans are required to queue for hours to dress up as a young wizard (including coat and scarf) and grab the handle of a big fake luggage trolley that is partly embedded into the wall. The performance act is always the same: fans are pretending to "jump" into the wall with the trolley, while a staff member is holding one end of the scarf in the air to make it look like an actual jump. Everything happens for the snapshot, which can then be bought in the merchandising shop right next to the experience. The photo/mediation claims and defines reality. Or to borrow a popular catchphrase from digital culture: *"Pics or it didn't happen."*[35]

There is a lot of *pretending* here. Surely visitors must know the act is not real. The whole experience is divided by metal gates from the surrounding environment in order to avoid interrupting the workflow of the physical train station. And while this is certainly a moment of presence and arguably a part of a transmedia story universe, it is also something more. There is a certain kind of physiological and psychological sense-making *beyond the mediation* of the books and films. This sense-making is established by semiotic codes (the famous platform name, costumes, trolley), narrative acts (the journey to Hogwarts), and performative practices (performing the role of a young wizard, the "jump" into the wall). At the same time, all of that is an experience within NBC Universal's ecosystem, the media conglomerate that owns the *Harry Potter* franchise. But most importantly, this experience is claiming reality. In that very moment fans forget about the performative acts. They believe in *being there* and are immersed in this alternative reality as a psychological and physical sensation beyond the mediation. In this, and many other cases, fiction supersedes physical reality to become an immersive experience.

Postmodern culture is full of these immersive experiences or "modes of immersivity."[36] If you have ever sighed, cried, or laughed while watching a film and showed signs of pity, stress, suspense, or even sexual arousal for the characters long after the film has ended, you are immersed in a

**Figure 2.2**
*Platform 9¾*: Visitors pretend to jump into the magic brick wall to catch the Hogwarts Express. Reprinted with the permission of National Rail and King's Cross. © 2012 John Sturrock.

mediated narrative. If you have ever, while attending a concert, taken out your smartphone to record the performance, you are hoping to immerse yourself again by watching the recording in the future.

Immersion is a defining term for all kinds of media and cultural practices in the twenty-first century, because the phenomenon needs to be extended *beyond the mediation*. We are merging with the liquid space that creates immersion after the mediation has ended, after we have closed the book, left the cinema, turned off the TV, switched off the computer or smartphone, and taken off the headsets.

If we use immersion as an interpretative lens to examine a range of mediated practices in our lives, we need to rethink our entire media culture. In fact, immersion then becomes the key to discussing mass media in the twenty-first century. Immersion should be understood as a sociocultural concept, which defines the sensation of all-encompassing engagement and involvement in all-surrounding mediated experiences.

## The 360° Gaze

Looking at it from a humanities and social sciences perspective, immersive experiences share similarities and specific repeating patterns. These recurring characteristics can be identified and analyzed with the help of a framework that allows us to discuss the liquidity of immersion with concrete terms and attributes. I call it the *360° gaze*. The framework of the 360° gaze captures the fluid paradigm of the immersive. It materializes the abstract concept of liquidity. When we are neither here nor there, oscillating simultaneously between physical and mediated realities, we need a compass. I propose this framework as a guiding principle to investigate immersive experiences in analog and digital media.

In addition, the 360° gaze framework offers new insights on postmodern theories and the history and logics of mass media culture. The framework acknowledges, transfers, and transforms the intersections, mutations, and transmigrations of the real and the mediated, immersive mass media, and digital media practices. It contributes new insights to the analysis of mass media in postmodernism by explaining how immersive experiences, from performing the self on digital platforms, fandom, and binge-watching to VR technology, have become essential characteristics of our times. This theoretical approach challenges the often misleading phenomenology and terminology of presence and immersion. By introducing a framework from a humanities and social sciences perspective, the methodological question of how to identify and analyze different kinds of immersive experiences can be tackled with the same theoretical approach. At the same time, the framework recognizes cultural-historical dimensions, functional and evaluative issues, and media-comparative point of views.

Within the framework of the 360° gaze, immersive experiences are generally characterized by (1) *semiotic codes and social activities*, a specific kind of sense-making through metaphors, rituals, routines, and narrative acts. In addition to that, there are also perceptual and sensual levels of (2) *performativity, processes, and practices*, while we are moving within the spatiality of these experiences. We also recognize the ways they manipulate the human mind and activate a certain desire for engagement, emotions, and involvement. Implications of (3) *psychology and reception* ultimately influence our beliefs, thoughts, and reasoning. Finally, we need to acknowledge immersive experiences as part of far-reaching (4) *cultural industries and digital ecosystems*, which uncover the power structures, logics, tactics, and dynamics of immersive media. This determines how immersion is used to target audiences and to engage them in mediated experiences that become essential tools to gratify their needs.

Epistemologically, by understanding immersion as an integral part of today's society, this framework reshapes the articulation of knowledge regarding immersion as a set of organizing signs, co-creations, gestures, and performances. It is transferable, adaptable, and flexible. Its components do not always appear in the same weighting or relevance. The framework builds upon a new understanding of spatial, cultural, and narrative practices beyond the politics of representation in which the symbolic is emphasized over and above the "responsive and rhetorical"[37] and practice is downplayed.

At the center of the 360° gaze is the *self*, a mediated representation of our selfhood, influenced by the interrelation of these four corresponding and constantly readapting characteristics. The self is a fetish in contemporary media culture. Over the past decade, the mediation of the self has affected people's informal interactions, as well as institutional processes and professional routines. The immediacy of immersive experiences has led to a variety of practices of self-management to possess, maintain, and curate identity like a consumer product. In digital culture, the conditions and rules of social interaction depend on how immersive the performance and management of the self is. We construct our mediated self depending on the experience. For

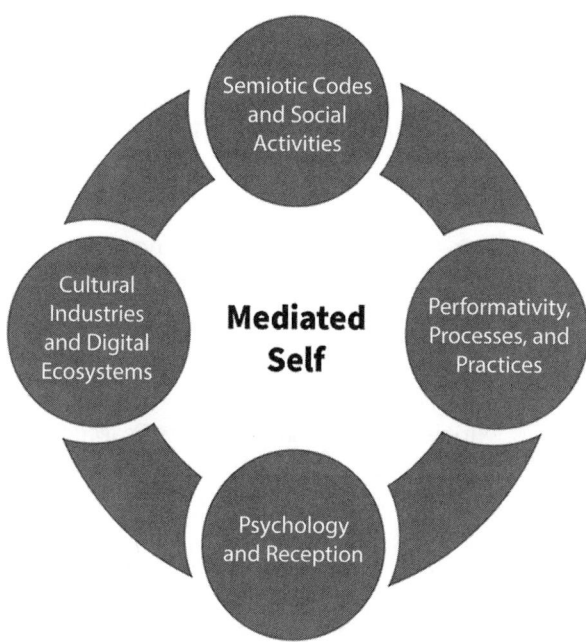

**Figure 2.3**
The 360° gaze. © 2020 by the author.

instance, we create digital avatars of our identities on social media to install the digital self as a substitutive reality. We might have different profiles on Facebook, LinkedIn, and Tinder. In fact, we are creating different *ideas of self*. Thomas Metzinger argues there is not a single self, but rather different "conscious self-models"[38] that we are not aware of and do not always recognize as we create them. I am following Metzinger's argument that no one ever has or had just a single self. Instead, we develop phenomenal selves, conscious ideas of how we think we are, and how others perceive us through mediated representation.[39]

The self is at the center of every immersive experience as it is ultimately an essential part of the mediation. In a digital game the mediated self might be a character to play with, in a novel it might be implemented through a first-person perspective. The mediated self is first and foremost affected, and immersion has an impact on the perception of our selfhood. However, the self is not a given, instead it is influenced by the characteristics of the 360° gaze framework. As Edward O. Wilson notes: "The self, an actor in a perpetually changing drama, lacks full command of its own actions. It does not make decisions solely by conscious, purely rational choice. Much of the computation in decision making is unconscious, strings dancing the puppet ego."[40] Even the idea of self becomes an immersive experience, as the first-person perspective is fluid and adaptable. We are able to adjust our profiles on social media, or enhance the skills of our character in *World of Warcraft*. We might even come to identify with Darth Vader instead of Luke Skywalker or Han Solo. We are able to develop a feeling of extended presence in the (mediated) representation of our selfhood. And probably more than with any other mediated representation, we are influenced by the mediated self beyond the mediation. Ultimately, the mediated self influences how we perceive our identities and bodies in the physical world.

We get to know who we are and what we want to be *while being immersed*.

Throughout the book, the idea of the self as both process and mediation will be the most essential principle to explain the human capacity *to exist* in immersive experiences. For instance, in playing a computer game or navigating through social media, the self is very often described as an avatar, the link between the physical world and the digital space. However, I argue that the self is a representation in any mediated environment, whether immersive theater, a streaming platform, or a theme park. The mediated self is not always a person or character, but in any case a perspective from a *first-person point of view*: "we perceive, or rather experience, the represented/created world (elements) 'from within,' as if we were eye and ear witnesses."[41] The mediated self might be a fully defined character in a computer game, or the cursor on your screen while you are navigating through websites. As soon as we enter the liquid space of immersion, the self becomes an incorporated and absorbed element of these environments. That is why metaphors such as "to get lost in a book" and "forget the world outside" are so accurate, as they include (without actually naming it) the self as a defining element of the experience.

> The *360° gaze* captures the characteristics of immersive experiences in analog and digital media. At the center of the framework is the *mediation of the self*. During the state of immersion, the mediated self is influenced by (1) *semiotic codes and social activities*, (2) *performativity, processes, and practices*, (3) implications of *psychology and reception*, and (4) *cultural industries and digital ecosystems*.

The framework is named "360° gaze" for several reasons. The term combines two different meanings related to immersive experiences. Firstly, "360°"

obviously refers directly, while not exclusively, to the current hype around immersive technologies, such as 360° videos and VR—all-surrounding audiovisual and digital content in which we can move, communicate, and interact. But it also underlines the state of absorption in any kind of mediated environment beyond the mediation. It refers to the "make-believe" of being involved, engaged, and attached as a psychological and physiological sensation activated by a mediated representation. The 360° gaze positions the self at the center as it is surrounded by the mediated experience. It refers to the act of stepping into the image (or any mediated representation) as a way of falling into the rabbit hole—not only to feel immersed, but actually to understand what else is around.

Secondly, since Michel Foucault's work *Discipline and Punish*, the definition of the "gaze" uncovers perspectives and dynamics of governmentality in disciplinary mechanisms within specific apparatus of power.[42] The gaze concept is part of what Foucault defines as the *dispositif*, a "heterogeneous ensemble consisting of discourses, institutions, architectural forms, regulatory decisions, laws, administrative measures, scientific statements, philosophical, moral and philanthropic propositions—in short, the said as much as the unsaid. Such are the elements of the apparatus. The apparatus itself is the system of relations that can be established between these elements."[43] Immersive experiences operate in a system of power, partly, though not exclusively, because they are often operated by powerful stakeholders that determine its conditions and surveil its users. But the other characteristics of the 360° gaze framework also exert power over audiences, as they are mutually dependent but constantly change their weighting and relevance in different immersive media. For Foucault, the *dispositif* is a net that spans across all elements, and therefore is an ideal template for the 360° gaze. It is a flexible, adaptable framework, integrating discursive and nondiscursive acts that influence our beliefs, hopes, fears, thoughts, reasoning, imagery, and emotions. The 360° gaze is a symbolic reflection of the relations within immersive experiences. It identifies how we modify our behavior within them and how we constitute the mediated self under the impression of surveillance.

Academic discourse on the gaze can be found in many disciplines, but the understanding originally used in film theory to describe the male[44] and female[45] gaze as modes of power is now broadly used in media studies as well. It refers both to the ways in which viewers look at images in visual media as well as to the gaze of those depicted in visual texts. In the case of mass media texts, the viewer can always look at those depicted in the text but cannot be seen by them. Several visual processes and immersive phenomena, which have been identified in relation to self and selfhood, have been related to the concept of the gaze: Jeremy Bentham's 360° panorama setting of the Panopticon;[46] Sigmund Freud's mixture of partial identification with desire;[47] or Jacques Lacan's look of an infant that is reflected back by a mirror ("mirror stage"/"ideal-I") as a coherent but illusory image of the self.[48] They all refer to immersive experiences: the all-surrounding spatiality of the panorama, the psychological implications of the subconscious, and the impossible objective perception of our own selfhood.[49]

While most of the gaze theories refer to visual media, the concept of the 360° gaze is less visual than perceptual and sensual. Perceptual, cognitive, and emotional levels are introduced in the framework (*psychology and reception*), because immersion is not only a visual but mostly a receptive and perceptual process. The gaze therefore constructs a system of semiotic codes and social activities, which situates the particular mediated experience in a complexly interwoven network of discursive and nondiscursive acts. It enables an epistemological shift away from the preserved knowledge of the archive of immersion to the embodied knowledge of the "repertoire": studying semiotic codes, performances, psychological and receptive implications, and tactics of industrialization as modes of power within emerging, interactive, immersive experiences.

The 360° gaze framework does not focus exclusively on one specific aspect, but rather integrates interrelating characteristics and corresponding questions. To make the theoretical and philosophical approaches more accessible, I want to discuss them with several different examples and a specific focus on immersive technologies and social VR.

## Semiotic Codes and Social Activities

> Which metaphors, concepts, rituals, routines, narrative acts, and artifactual meanings are associated with immersion? Which semantic units and communicative practices are used in immersive experiences and how are they becoming social activities? How do we co-create meaning in mediated realities through participation, interactivity, and performance? How are we constructing and communicating social knowledge on immersion?

## Beyond the Mediation

We are able to communicate in immersive experiences. What and how we communicate has much to do with the use of signs, symbols, and semiotic systems within a mediated reality.

Looking back at the history of immersive media, or media in general for that matter, we find a history of material-reproducible implementations of the human symbolic faculty and of ways of encoding and decoding symbolic functions that are the core of human cognition and communication. In fact, being human is intricately connected to the symbolic capacity and the use of signs. Cultural memory and collective cognition are interlinked with the semiotic systems of mediation. They are heavily influenced by media techniques such as language (oral, written, symbolic), images (pictorial, representational, abstract) in paintings, drawings, photography, film, and other audiovisual media, as well as recording and performative practices in theater and music.

Our attempts to construct, maintain, transform, and repair reality through semiotic systems are observable in the history of human communication: in art, science, religion, journalism, and mythology. In fact, Kant argued that knowledge cannot exist without sensation, and that the cognitive process is inherently linked to the understanding of reality.[50] In his cultural approach to communication, James W. Carey adds: "There is reality and then, after the fact, our accounts of it. We insist there is a distinction between reality and fantasy; we insist that our terms stand in relation to this world as shadow and substance. While language often distorts, obfuscates, and confuses our perception of this external world, we rarely dispute this matter-of-fact realism. We peel away semantic layers of terms and meanings to uncover this more substantial domain of existence."[51] Semiotics has its foundation in analyzing language, but the idea of "augmented linguistics"[52] is a shift toward the aesthetic experience and how meaning is constructed beyond representation. In the context of contemporary imagery, the art historian Hans Belting explained that semiotic image theory also erases the physical, medial corporeality of images.[53] Gottfried Boehm argued along the same lines. He wanted to understand how images create meaning apart from language through a "power of showing/appearing/presence."[54] The epistemological shifts to the "iconic" and "performative" turns in semiotics have been challenging the exclusive status of interpretation and representation, and instead focusing on the "non-hermeneutic," co-creation, and aspects of sensuality and intensity.[55] That is not to abandon the importance of "meaning," "signification," and "interpretation" of various connotations and denotations of signs, but to understand the aesthetics of semiotics as an interdisciplinary field in regard to organizing signs and symbols, co-creation, gestures, and performances. When Susan Sontag states, "We must learn to see more, to hear more, to feel more,"[56] one cannot help but think of the intensity of mediated experiences in moments of immersion.

The most common definition of communication—to transmit and send messages to others—entered everyday discourse in the nineteenth century. To communicate a message to someone means *to engage* the recipient in a particular mediated experience, and to make him or her *feel present* in a system of signs and symbols creating meaning. At the time, the term "communication" was used to describe both the movement of goods and people, as well as the transportation of information. Since then, communication has always been a process of controlling space and people through the power of systems, for instance, through language and mass media technologies. Whoever controls the mechanisms of communication within immersive experiences also controls the social activities that are affected by them: the rituals, routines, narrative acts, and artifactual meanings that constitute immersion. What, how, and with what effect we communicate in immersive experiences defines and ultimately diminishes the dichotomy between the physical and the mediated in our world.

People learn, adapt, and evolve media as a vehicle to produce knowledge. In effect, the roles within the process of mediation are reflexive, and people learn about the conditions of knowledge production through the acts of communication. Harold Lasswell's model of communication ("Who says what in which channel to whom with what effect?")[57] has drawn attention to what would later be discussed as "media effects." We are often reminded that it is the impact of communication we should discuss foremost, whether political, economic, or social. Audiences experience and contribute to a range of changes in cognition, attitudes, beliefs, and behavior after being exposed to media. Denis McQuail's typology of media effects recognized some long-term effects in light of digital communication, one of them being "reality defining."[58] This, and any long-term effect of communication, is naturally associated with immersion, because these effects still occur after and beyond the mediation.

But what makes signs, symbols, codes, and meanings immersive? Communication is a reflexive feedback process. It is about negotiating the foundations of mediated realities. Be it telling a fairytale to a child, or playing

*World of Warcraft* with millions of other players at the same time, it is all about creating a stable semiotic system consisting of *semantic units to frame actual or potential meaning*. Those units can be identified as distinct elements such as signs and symbols that *appear/are presented/are co-created* in the mediated reality of an immersive experience. They become immersive once they are associated with the production of knowledge and the understanding of the "real."

Vilém Flusser points out that any form of communication is always an attempt to create, (re-)produce, capture, maintain, and obtain the idea of the "real."[59] Flusser's *Kommunikologie* is an important milestone in understanding the semiotics of immersion, even though his text mainly focuses on the rise of technological images. Flusser describes the communicative use of signs and codes as a constant reaction of being alienated and distanced from the world. He argues that humanity's deepest desire had always been to preserve reality through mediation to ultimately feel immersed beyond the representation. For Flusser, media are a sort of time machine, a way to stop the clock, to move back in time, and to transport us into a new formation of reality.

In his historical analysis, Flusser outlines several forms of *Verfremdungen*[60] (alienations, or in our context: the dissolution of frames, containers, and devices) over the course of time in which humans were failing to develop a feeling of presence in the mediated representation of physical reality. Humanity's struggle started with image media to "overcome the gaping abyss"[61] between humans and the world. Images have a decisive meaning, in both prelinguistic and literary cultures. Flusser names the cave of Lascaux in southwest France and its drawings of particular hunting scenes as an example from prehistoric times,[62] while Oliver Grau starts his analysis of immersive image strategies in virtual art with wall paintings from the late Roman Republic, in particular the Great Frieze in the Villa dei Misteri at Pompeii.[63] Both examples have something in common: they emphasize the role of the semiosphere,[64] the symbolic-cognitive abstraction connected to human social behavior. One semiosphere is showing abstract hunting scenes in Lascaux, the other the mythical scene of the procession of Dionysius in Pompeii. Observers are transported into the image space not by interpreting the artifact but rather by experiencing a sensual connection with the mediation. They ought to feel present in the particular scenery as an "almost physical reality."[65] Both images have a spatial dimension, which wants "to break down barriers between the observer and what is happening in the images on the walls."[66] When Grau describes the "dialogic communication between the figures,"⁶⁷ their "hair still wet from the ecstatic dance," and the meaning of the words *Ek-stase* and *En-thusiasmós*, "the physical and psychological immersion of the individual in the god," he is not only analyzing the image with the help of semiotic codes; he is also contributing to the social knowledge on immersion by transferring the visual techniques into the modes of language.

The origins of transporting audiences both mentally and physically to another place can be traced back to religion. Religion is, in fact, a kind of sensual and intensive process of *world-building*, a way to transport its followers to both meanings and places through the use of semantic units. As Carey points out: "This movement in space was an attempt to establish and extend the kingdom of God, to create the conditions under which godly understanding might be realized."⁶⁸ Religious art, especially in form of visual works in stained glass, paintings, sculptures, mosaics, metalwork, and embroidery has been used to establish the church as a gathering place in which to feel immersed in the divine. Pieces on stained glass are semantic units that tell stories and illustrate, supplement, and portray religious mythology. In his major philosophical text *Process and Reality*, British philosopher Alfred North Whitehead calls this "presentational immediacy,"⁶⁹ "an occurrence or 'occasion of experience' without clear boundaries, implied by our immersion in the world."⁷⁰ Physical senses as well as the conscience are influenced by religion's pervasive semiotic system, and more importantly followers are feeling immersed in its messages even after they have left the church. Jonathan Koestlé-Cate describes how that phenomenon influences the perception and apprehension of reality: "Therefore, whatever appears to perception is always surrounded, coloured, and permeated by a blurred nimbus of sensation of which we are barely, if at all aware."⁷¹ This unconscious sensation is called immersion. The creation of an "alternative reality" is essential to the semiotics of religion "giving us other worlds pretending to be ours."⁷²

In the last decades, digital technologies have established a range of semantic units to communicate, which ultimately are "a filter to all culture, a form through which all kinds of cultural and artistic production is being mediated."⁷³ They extend the understanding of semiotics with ideas of cocreation, participation, interactivity, and "meaning through performance." Lev Manovich describes the interfaces of digital technologies with the example of the window of a web browser, which replaces the "cinema and television screen, a wall in art gallery (sic!), a library and a book."⁷⁴

# Beyond the Mediation

The browser window is nothing but a rabbit hole, and the mediated environment that opens up for the user is "all at once, the new situation itself: all culture, past and present, is being filtered through a computer."[75] Manovich perceives the language of new media technologies through the lens of semiotics, the codes used, and the ideologies embedded. By discussing the role of computers, screens, and interfaces, he argues that the semantic units of digital communication are made up of other, already established semiotic systems that show us how they fit into larger structures of meaning that particular societies and cultures have created and enforced. That ultimately means that any semiotic code and social activity in digital media—posting, sharing, liking, tweeting, swiping, sorting, counting, rating, marking, filtering, swapping, snapping—are the culmination of other semiotic codes and social activities in nondigital forms. Freyermuth uses the tablet as an example: "What appears to be a radically new technology is in fact quite traditional in its media form: the tablet as a portable storage medium for images and text can be found at the very beginning of media culture,"[76] while he further mentions the social activity associated with it: "The 'biological naturalness' of the user interface that touch tablets offer thus correlated to the cultural familiarity of their physical form."[77] That is certainly true with other digital activities as well. The social actions of producing and spreading messages (post and share), showing approval (like), exchanging (swap), moving on to something or someone else (swipe), and assigning value to something (rate) all have been known before the rise of digital media. Immersive experiences mainly draw meaning from an already existing repertoire of semiotic codes and social activities and are shared throughout history.

To study immersive experiences from a semiotic and symbolic point of view means to examine the actual social activities in which they were created, used, apprehended, and distributed in networks of social mediation. The list of examples is endless. For instance, how and in which context we read books has much to do with how they are written (and vice versa). How we listen to music has much to do with how we have adapted to the activities of consuming it (and vice versa). For instance, the production of music in the age of streaming has adapted to the use of streaming platforms as a majority of listeners search for "songs" instead of whole "albums."[78] How we watch TV shows has much to do with how streaming platforms have conditioned our viewing habits (binge-watching) and tastes (and vice versa).[79] How we communicate with digital technologies has much to do with how the "iconic turn" of images (such as emoji) conveys universal emotions to immerse others in the age of globalization (and vice versa).

Emerging technologies and experiences in immersive media also establish semiotic systems. On the day Facebook acquired Oculus, Mark Zuckerberg posted on his Facebook profile: "Imagine enjoying a court side seat at a game, studying in a classroom of students and teachers all over the world or consulting with a doctor face-to-face—just by putting on goggles in your home. This is really a new communication platform. By feeling truly present, you can share unbounded spaces and experiences with the people in your life. Imagine sharing not just moments with your friends online, but entire experiences and adventures."[80] What Zuckerberg calls a "new communication platform" is connected through several social activities, of which he lists a sports event, education, and medical consultation—all of them related to a specific set of signs and symbols.

In late 2020, Facebook's social VR app Horizon is still in a beta phase. However, some of its semiotic system is already outlined, as the platform is in line with Facebook's other social media apps like WhatsApp and Instagram. Those have been around for many years. It also helps to look back at the designs of Facebook's abandoned VR platforms Spaces and Rooms to understand how the company introduces new semiotic codes in VR, while still relying on a range of established ones. Traditionally, Facebook's VR apps use social activities such as sharing and posting in the three-dimensional space to enhance the communal experience. In Spaces and Rooms, avatars were connected with the inner and outside world through a set of new symbols: a blue control desk, a blue interface with different environments to choose from, and a blue wristwatch. Early reports claim that Horizon plans to use blue dashboards and shields on the avatar's left arm as well as different teleport pads.[81] Metallic blue is the color most used for visualizing interactive components in VR environments.[82] The symbolism of the color reveals a lot: blue is the color of both arousal and comfort.[83] On one hand, users should feel aroused and excited by the different environments they can choose from; on the other hand, chatting on the Moon should be a comfortable and pleasant experience in VR. Users are able to post messages, as well as use likes and emoji to express emotions. Five empathetic icons—Facebook calls them "reactions" in VR, and they are clearly similar to the ones used on its social media platform—can be attached to 360° videos and photos without interrupting the momentum of the experience: "[S]eeing your friends' reactions may make the experience a little more social."[84]

## Beyond the Mediation

More so, because VR avatars are three-dimensional, their facial expressions and gestures become essential tools for virtual communication. While the users' countenances cannot be translated into the digital space yet, body movements are recognized and transferred by the room-scaling devices. Michael Booth, Facebook's head of social VR describes the process like that: "Shake your fist and your VR avatar's face will turn angry. Put your hands on your face *Home Alone*-style to express shock. Triumphantly thrust your hands in the air and your virtual self's face will show joy."[85] Obviously, this has an impact on the perception of the self, but also on how users perceive the possibilities of digital communication. Booth argues that, in addition to "reactions," the avatar's face is a "VR emoji"[86] linked to specific semiotic meanings: "We're coming up with a language that triggers your avatar to make certain emotions."[87] We know that communication is not produced within a single system but within the intersection of several interrelating sensual systems. If you are in a conversation and someone does not understand what you are saying, you would recognize it from the confused expression on his or her face. That would give you a hint that you should explain it differently, slow down, or provide more background. This momentum of physical conversations could now be translated into the virtual environment: you could shrug with your palms up and your avatar would make an expression of confusion. That is particularly interesting since research has shown that people tend to respond in VR as they would outside of it.[88]

Facebook even aims to recreate VR avatars from the users' photos posted on their Facebook accounts.[89] This would be the ultimate illusion. When photorealistic avatars resemble the physical being, their use as communicational tools would not be determined by the human body but by the rules and conditions of the platform provider. Facebook would decide how and to what extent users are able to communicate through their virtual faces. In that regard, the face of the VR avatar would transform into a semantic unit. The photorealistic VR face would become an emoji, a globally used icon that, as a result, could be also used by others as a communicational tool.

The "grammar" of immersion is embedded in the ways we communicate about it, and how we create mediated experiences in literature, films, and music, among others. For instance, when Thomas Elsaesser describes the tactics used to navigate through digital spaces, he writes: "pictures on a computer screen are not something to look at, but to click at—in the expectation of some movement taking place, of being taken to another place or to another picture space. The idea of a digital photo as a window to a view (to contemplate or be a witness to) had been . . . replaced by the notion of an image as a passage or a portal, an interface or part of a sequential process—in short, as a cue for action."[90] Elsaesser uses the metaphors of "being taken to another place" and of a "passage or a portal" to intertwine the semiotics of immersion with their performative practices. At the same time, though, he explicitly borrows from the rhetoric of immersion. In the end, "passages" and "portals" are nothing else but synonyms for rabbit holes.

The rhetoric of immersion is materialized in the ways in which we construct social knowledge on immersive experiences with metaphors such as "transporting," "diving in/diving under," and "absorption." Terms like "liquidity," "teleporting," "make-believe," "recentering," "being there," "falling down," "getting lost," and "sense of presence" are uncovering the preexisting ideologies, and the social, cultural and political functions of immersion in our lives, and how we are changing and adapting them over the course of time.

Semiotic codes and social activities are intertwined through their performative dimension, and therefore they are linked to "Performance, Processes and Practices" in the framework of the 360° gaze. The ways in which semiotic codes and social activities *appear/are presented/are co-created* in immersive experiences say a lot about how immersion works. Even though semiotic systems draw from a large repertoire of already established modes of meaning, each medium establishes its own semiosphere. We also understand the mediation of the self through its modes of communication, and the usage of signs and symbols in a particular semiotic system. Once new semiotic codes and social activities are about to become introduced (such as the face of VR avatars), they ultimately become part of the knowledge through which we understand the world.

## Performativity, Processes, and Practices

> How do we perform the mediated self within immersive experiences? Which perceptual and sensual levels of performativity do we accept in the constitution of immersive environments? Which particular role do immersive spaces play in the understanding of our performative capabilities as human agencies? How do we use technology in hybrid forms to extend our perceptual and sensual abilities? In which way do bodies merge with technological agencies?

# Beyond the Mediation

We are not only communicating, we are also performing in immersive experiences. Performativity in relation to immersion is always a performance of the self. It involves multiple presentational and communicational dimensions of performance-related processes and practices. The mediated self needs to act, react, and respond to the experience's temporality, spatiality, and choreography. It needs to *perform* in liquid spaces.

The notion of performativity is often seen in opposition to representationalism. Performativity does not represent what already exists. It rather determines what is real through the power of mediation, and therefore questions and destabilizes predefined conditions and practices. In their book *Performing Presence*, Gabriella Giannachi and Nick Kaye note that the sense of "being there" is produced in the dynamic, mediated, and tensive encounter between the self and "the other."[91] The authors acknowledge that the idea of self is responsible for how audiences develop different levels of presence. In other words, *immersion is an active process*, not a result of passive media consumption. The "other" is defined as the media, and ultimately, as the environments and experiences in which audiences are required to perform. They emerge from "an ecology, a network of processual positionings in time and space in relation to pasts, futures, elsewheres, and others."[92]

To comprehend immersion through its performative practices basically means to apply an anthropological understanding of space. The significance of space is determined by its sensuality, politics, and poetics.[93] It is defined by how human agencies, including what we have identified as the "mediated self," are performing in it. Turning right, for instance, defines what is perceived as "right," and by the rules and knowledge of signification, it also defines what is "left" at the same time. By creating "worlds of signification"[94] we are making sense of the environment through actions, interactivity, and participation. Any mediated environment falls into this category: the image space (starting with landscape wall paintings and the nineteenth-century panorama—the word literally means "to see everything": Greek *pan* "all" and *orama* "view"),[95] the textual world, the theatrical space, the spatial dimension of games, social media platforms, and the sense of three-dimensionality in VR environments. Michel Foucault calls these places "utopias," as they are "in an unreal, virtual space that opens up behind the surface; I am over there, where I am not, a sort of shadow that gives my own visibility to myself, that enables me to see myself there where I am absent."[96] Foucault uses the example of the mirror because the image we see in it is not actually real, yet still defines us: "I come back toward myself; I begin again to direct my eyes toward myself and to reconstitute myself there where I am."[97] None of these mediated environments exist in physical reality, but by setting actions and performing within their spatial dimensions we are still defining them as "real." This is how performativity is linked with semiotic codes and social activities within the 360° gaze framework.

Immersive spaces are essential features of the world we live in and shape our culture deeply.[98] In her book on the aesthetics of immersive image spaces, Laura Bieger argues that spaces play a significant role in the formation of liquidity, as world and image blend into each other, and the self is invited to step in and move within the image space.[99] The reality of the mediated self (for Bieger: the materiality of the mediated body) is determined by the reality of the mediated space. She even speaks of a "choreography" of immersion to evoke feelings of being "in-between."[100] We are not merely creating a sense of the environment through performance, we are also shaping the mediated self through performance-related processes and practices.

In her book *Hamlet on the Holodeck*, Janet Murray proposes the fun house as an example of immersive experiences, especially in relation to how the self is positioned and performing in it. She uses the metaphor of the "visitor" in the same vein as Jean Baudrillard does in explaining the idea of hyperrealities with the example of Disneyland. Audiences are similar to visitors of foreign territories as they are entering the closed environments of theme parks, amusement parks, studio tours, or fun houses. These immersive environments are defined by their concrete dimensions of space and temporality, and the different practices one must perform while being in them. On the example of the fun house, Murray recognizes the structure of a narrative world: "The fun house has an entrance and an exit that mark the beginning and the end of the story. As the visitor progresses on a moving platform, the dramatic tension builds from small surprises and hints of danger; then there are thrills and a mounting sense of threat or terror, which culminates in a big finish such as a free fall or an attaching beast. . . . A fun house is a movie made into a machine that you travel through."[101] Tony Bennett describes the self being exposed to "a number of ritual assaults" within the fun house experience: "[Y]ou are buffeted by skittles, the floor shifts beneath your feet and you have to cross a series of revolving discs."[102] Robin Curtis adds: "When one enters a fun house, one enters an environment in which the standard rules regulating the configuration of space and self's relationship

## Beyond the Mediation

to that space are overturned; this one encounters not only an opportunity to transcend these rules of spatial negotiation but also an opportunity to enter into a slightly different world."[103] This often leads to feelings of uncanny awareness, fascination, and intense reflexivity.[104] Béla Balász even recognizes a "feeling of vertigo," when audiences are confronted with the immediacy of their presence in mediated environments: "The greatest catastrophe depicted in a pictorial space that is separated from our own space will never have an impact comparable to the image that places us on the very edge of an abyss that opens up *before our very eyes.*"[105]

The self is defined by the performative actions within a mediated space. This becomes obvious when we are thinking of game experiences such as *The Sims*. In the world of the game, the ultimate goal is to fulfill the character's lifetime wishes, for instance, to become an astronaut, CEO of a megacorporation, or professional author. At the same time, "success" is not only defined by a profitable career, but also by the achievements in private life, such as finding eternal love, marriage, and having children. In the context of the game, the self ultimately needs to complete the picture by owning several status symbols such as cars and houses. The spatial dimension of the game allows players to constantly monitor what they have achieved (for instance, by offering a bird's-eye view into their homes). They need to act and react on the stage of the game, switch between specific roles (CEO, father, husband, friend, etc.), and use the appropriate set of performative skills to achieve their goals. Ultimately, performance in the reality of the game will influence what audiences would normally define as a "successful life" in physical reality.

Looking at the example of immersive theater emphasizes this even further, as performative acts are already established as parts of the immersive experience. Immersive theater is defined by the resonance of the transformed space and the extension of the playing area. The setting involves the interaction with the performers, and audiences are required to inhabit an imaginary world. Audiences need to perform to feel present in the experience. They need to switch into the roles of co-performers, which blurs the lines between spectator and event, actors and audiences, physical and imaginative worlds. Theater ensembles such as Punchdrunk[106] and Les Enfants Terribles have made the genre popular. In their productions they create performative spaces outside the traditional theater environment to interact with their audiences and engage them in the performance to a new extreme. For instance, Punchdrunk's production *Kabeiroi* (2017) has been a six-hour-long performance for only two audience participants at a time. As part of the performance, which was inspired by Greek tragedies and the remaining fragments of one of Aeschylus's lost plays, participants were required to travel to multiple locations across London and had no idea what and when something would happen to them: "At times you are closely instructed about how to proceed; at others you are granted enough autonomy for doubt to creep in. At one point I gave (sedate and self-conscious) chase to a woman I was sure was shouting my name. She turned out to be extremely drunk and absolutely nothing to do with the show. [sic!] The encounters with actors tend to come out of nowhere, which adds to a feeling of paranoia."[107] The blurring of physical and mediated spaces leads not only to the diffusion of actors and audiences. It also requires audiences to develop a new understanding of the mediated self as a participant in a theatrical play. As they do not know what will happen to them, everything is ultimately part of the stage, and conversations around them (which are not part of the event) get integrated on a perceptual and sensory level. Participants need to draw from their knowledge from prior theatrical experiences and react on the spot: whatever is required from them needs to be done.

These examples open up two kinds of performative environments, which are essential for performing the self in immersive experiences: the "performance space" (which, in the example of *Kabeiroi*, is ultimately extended to the size of a whole city) and the "narrative space" (whatever is determined as mediated content that gets adapted, extended, and changed through involvement and engagement of the participants). This distinction comes from the way immersion is discussed in regard to computer game experiences. For instance, Jesper Juul draws a distinction between "world space" (the narrative, fictional world) and "game space" (the mediated space, in which users can interact with their avatars).[108] His argumentation follows the idea that during playing, users are presented with a "performance space" in which they are able to move and interact with others. At the same time, though, there is a "narrative space" building up, one that is not always visually represented in the game space, but is always relevant in order for users to feel present and immersed in the game.[109] As Grau points out, though, Western societies have developed "a movement that seeks to blur, negate, or abolish the differentiation [between reality and mediated worlds]."[110]

While the narrative space remains essential in immersive experiences, the performance space often merges with the physical environment. As we

# Beyond the Mediation 83

have already established, immersive experiences are determined by the dissolution of their frames, even beyond the mediation of content. It does not make a difference if we are reading a book, watching a film, or listening to music—in all of these mediations we establish an idea of self that is ultimately becoming part of the narrative space and the performance space at the same time. For instance, the act of reading is a performative practice as it is defined by the idea of developing a sense of presence through performing the reading process. Reading takes a significant amount of attention and focus and is a private, creative, introspective, and almost intimate act.[111] It becomes clear how this is linked to the semiotic codes of this act: we are creating images in our heads and designing whole worlds based on words and their meanings. For Richard J. Gerrig, we perform what we read in the theater of mind.[112] Ultimately, readers (represented through their mediated selves) adopt specific perspectives within the experience because of how the text is constructed and develop sensual and perceptual reactions to the text. The level of emotional investment is supported by how the mediated self is characterized as a link between the audience and the characters in the story.[113] Listening to music, on the other hand, might happen in social, performative environments (concerts) or in intimate, personal settings (recording), but in the end both require a performative reaction. Audiences at concerts often fall into an almost trancelike state. They close their eyes, sing the lyrics of songs, or even show a strong emotional reaction to the performance on stage (which often is not limited to the stage, as performers walk into the audience space, etc.). This is a way of feeling immersed through the performance of the concert, and if one decides to have a chat at the bar instead of listening to the show, he might not have the same feeling of immersion. In fact, mediated environments are designed for us to perform in them. Performers on stage often encourage audiences to sing with or instead of them, to dance, jump, or clap their hands to become part of the experience. Through these performative practices audiences are not isolated and passive, but actively involved in the delivery of the performative act and the "stage" dissolves even further. All of these performative acts make the experience more immersive.

Immersive technologies, and especially social VR, replace the experience of the physical environment with the mediated one in an all-surrounding experience. These technologies often introduce a new set of performative practices for audiences to learn and adapt to. For instance, after putting on the Google

Daydream View VR headset, the self is able to perform actions with the help of the Daydream controller. The device has only three different options (touchpad, app button, home button), but they are linked with specific performative practices: using the "home button" to return to the Daydream home screen and to recenter the view and cursor; using the "app button" to show menus or pause a specific app; and using the "touchpad" to swipe and scroll, for instance, by throwing a virtual basketball or a virtual stick for a virtual fox to catch. By repeating these practices, users get comfortable within the environment, incorporating them as ways to navigate, interact, and communicate.

Facebook's social VR apps also introduce performative practices, which intertwine the self with the environment. While Spaces and Rooms offered three-dimensional locations based on their level of attractiveness and exceptionality (Moon, Mars, ocean, etc.), performances were set in a comfortable tone. Avatars did not have to "swim" on the bottom of the sea. They could pleasantly stand, sit, and chat with each other, or play a set of cards without getting influenced by the underwater world. Avatars were able to produce selfies and post them on Facebook by using the blue watch on their wrist, which connected them to the environment outside VR, mainly Facebook's social media platform.

Facebook's new platform, Horizon, uses a performative practice already well known from VR games: teleportation. It allows users to transport their avatars to different locations and let them move within those environments while ignoring the restrictions of their physical space. While their physical body movements are mainly focused on head and hands, teleportation is a way to move to a specific point in the environment by pointing to it with the controller (visualized by a beam—again in blue)—a performative practice already well known from Google's Street View on Google Maps and teleportation in VR games on the HTC Vive. There is much debate as to whether teleportation diminishes the feeling of presence,[114] but ultimately it is a performative practice that users have to adapt to in order to move within virtual environments. Whatever is required from audiences to perform within immersive spaces also depends on how technologies enhance their performative capabilities as human agencies.

In the age of digital performative practices, we are not only merging with mediated experiences but also with technologies. Human-machine hybridization means that the medium penetrates the body and finds loopholes to nest inside it. It reorganizes the boundaries of the organic and inorganic and redefines the notions of selfhood, identity, and human capability. Ultimately, that also changes the way in which we need to perceive immersion through performance. For instance, we are able to extend our sensual and perceptual abilities through wearable technologies. We measure and optimize the self with apps to condition our sleep, fitness, and nutrition with the help of smartphones and smartwatches, glasses, and contact lenses for AR experiences, and the Internet of things. The way we merge with technologies has much to do with how the data that is materialized by measuring devices, and their mediated messages, affect our thoughts, behavior, and mindset as we "become one" with the medium through performative practices. This does not happen in an ad hoc fashion. Instead it is a subtle, hidden, and often quite enjoyable process.

Performativity is a relevant component in the 360° gaze framework as the self (as a mediated representation) is entering a mediated environment, and by doing so, becomes absorbed by it. The spatiality of the immersive experience allows the mediated self to set actions, processes, and practices. Through them the mediated self engages with both the narrative and the performative dimensions of the experience, be it participating in immersive theater or using technology to interact in mediated spaces. As soon as technologies move closer to the human body and merge with it, the set of performance-related processes and practices needs to be reevaluated. When certain aspects of performativity can only be executed through the merging of the organic and inorganic, it also changes how we perceive the performative dimension of immersion.

**Psychology and Reception**

> How does immersion influence the human mind? Which factors activate feelings of presence and immersion? How is psychological reasoning related to mediated immersion including engagement, involvement, and emotions? In which way is our desire for immersion defined in various immersive environments? How can one differentiate immersion from other cognitive responses to mediated experiences?

Immersion influences the human mind. One of modern psychology's central findings is that our mind is constantly shaped and reshaped by a host of external factors while being often completely unaware of it. Our environment, including technology and other humans, unconsciously influences our behavior and defines our identities. As human behavior is context sensitive, it adapts to fundamental changes in the environment. We cognitively adjust to any form of mediation and therefore also to the psychological and receptive implications of immersive experiences. Hans-Georg Gadamer describes such an aesthetic experience as a form of transformation, as it "changes the person who experiences it."[115] This reaction is (to an extent) also part of enculturation. Evolutionary psychologists like Brian Boyd suggest that "we are unable *not* to imagine and respond to the characters and events of a well-told story, even if . . . [we] . . . know that the story consists of mere words, . . . or of contrived images, projected into a flat screen in pixelated form, of artificially costumed, made-up, and illuminated actors . . . we cannot stop conjuring up and responding to the story's invented people and predicaments, and even, if occasion prompts, weeping tears at character's fates."[116]

Immersion dissolves the "cognitive frames," which usually enable us "to interpret both reality and artefacts and hence other concepts that can be applied in perception, experience and communication."[117] Instead, immersion opens up cognitive "gaps" that audiences fill with sense and meaning. The idea of "gaps" within artifacts, *blank spaces as manifestations of liquidity*, is inspired by the hermeneutics of reading, and the aesthetic response of readers while they engage with a literary text.[118] It basically means that as soon as we read a text, we are confronted with its textual structure, including everything that is written and unwritten. The text comes to life, because we use our imaginations to fill those unwritten blank spaces. That is when we *engage with the uncertainties* within the text and make it a personal, individual experience. These gaps or blank spaces are thresholds, moments of crossing borders into liquid spaces and immersion. These thresholds are not only typical for reading texts. Audiences fill gaps in any mediated representation and always project meaning, sense, and interpretation onto them.

In all of these responsive acts, the human mind reacts with a spectrum of cognitive effects activated by immersion, from mild effects (sensory arousal, attention-getting, or increased entertainment value) to extreme ones (collaboration with other human-embodied avatars, illusion-forming, mediated voyeurism, obsession, and addictiveness).[119] Edward Castronova refers to the latter as "toxic immersion"[120] in the form of a negative perception of being part of a mediated experience. That is especially noteworthy in relation to mediated experiences that create a powerful illusion for the human mind and are able to manipulate it quite profoundly. Crossing borders to liquid spaces means to lose touch with solid ground like sinking down in a pool of water. We are confronted with the uncertain, the uncanny, the unreal, and ultimately the psychological effects of long-term immersion: "Immersion can be an intellectually stimulating process; however, in the present as in the past, in most cases immersion is mentally absorbing and a process, a change, a passage from one mental state to another. It is characterized by diminishing critical distance to what is shown and increasing emotional involvement in what is happening."[121] In that way, immersion enables effects that the psychologist Mihály Csíkszentmihályi describes as "flow"—the sense of becoming absorbed in the emotional and cognitive experience of one's own actions.[122] However, Jay David Bolter and Richard Grusin argue that these experiences of mediation "[do] not necessarily commit the viewer to an utterly naive or magical conviction that the representation is the same thing as what it represents."[123] While filmgoers of the earliest film by the Lumière brothers could not be fooled into thinking that the train they saw on the big screen would suddenly materialize and crush them, they still needed to process "the discrepancy between what they knew and what their eyes told them."[124]

Immersion results in audiences committing to a concept of an alternative, mediated reality: the diegetic world of a film, book, TV show, game, and VR experience, and the performative practices of immersive theater and social media. When looking at immersive experiences within the 360° gaze framework, we have to ask what exactly audiences commit to while being immersed, and what are its psychological and receptive effects.

The long-term psychological effects of immersion are the ones that influence the 360° gaze framework the most. They are grounded in a form of make-believe contract between those who produce and those who consume the mediated experience. The lack of distance, as mentioned before, is part of theatrical performative strategies within immersive experiences. As the human mind can be manipulated to question the difference between reality and fiction, natural and artificial, real and virtual, we have the desire to enter the performativity of mediated environments, e.g., role-play in games and immersive theater, to experience substitutional realities.

Take Madame Tussauds as an example. Madame Tussauds is well known for its wax figures of famous and historic personalities, including fictional characters. A visit to one of their museums is a popular tourist stop in many destinations, from New York to London to Tokyo. Hardly anyone, though, visits them to admire the delicacy of the wax figures, the creative craft that went into manufacturing them, and the love of detail. Instead, visitors are willingly accepting a make-believe situation outlined in the mediation of the experience. As visitors wander through the detailed and enriched *mise-en-scène* of wax figures, they not only admire the three-dimensional representations of well-known film scenes and locations, they also imagine all of this to be real. A mere spectator of this scenery may think the same, but participating in the make-believe performance with others has direct effects on how visitors perceive themselves in the mediated environment. Visitors are able to approach the figures, touch their bodies and (real) hair, and take pictures with them. They pretend to take selfies with the *real* George Clooney (as they know him from coffee commercials and the *Ocean's Eleven* films). They pretend to sit on a sofa with all four members of the Beatles (as they know the band from images from the 1960s). They pretend to be in the *real* Führer's bunker (as they know it from photos and documentaries). This is further encouraged by the tagline "Who do you want to meet?" in the company's marketing campaign. Madame Tussauds also constantly upgrades its wax museums with famous personalities based on their significance in specific locations. Now we might argue, why should anyone take a photo of something *unreal* and share it online, pretending this is the "real" celebrity instead of a wax figure? The answer goes back to the psychology of immersive experiences.

As I have discussed earlier, the idea of the self is defined by its actions within mediated experiences. This is what consciousness research is calling the "unit of identification,"[125] the conscious idea that we currently perceive as our "self." Thomas Metzinger points out that the self can be characterized and determined by any representational and mediated content within a conscious model of reality.[126] This can be witnessed on social media, where users pose and engage through their avatars in social interactions, as well as in games in which players define themselves through their avatar's ability to act and by the challenges they have to accomplish. But it can also be seen in parasocial interactions with celebrities—and what else is a visit to Madame Tussauds? Even though visitors are merely "meeting" three-dimensional wax representations of famous personalities, whom they only know from media—this is the substitutional reality they were looking for. Madame Tussauds is, by definition, an immersive experience, as it is shifting the focus of the self to the feeling of immersion beyond the mediation. I am not talking about the mediation of the wax figures installation at Madame Tussauds, but rather the mediation of films, images, songs, and TV shows that audiences watched before and were influenced by before entering the installation. Visitors are not interested in meeting the real Johnny Depp—they may not even have enough imagination of him as a private person to construct a concrete image in their minds. Instead, they want to meet Captain Jack Sparrow, the pirate in the *Pirates of the Caribbean* films. And this is what they will get: an almost identical, three-dimensional *mediation of a mediation*.

Later in this book, I will discuss the phenomenon of the *immersive parasocial*, the belief in a close encounter with media personalities. But for the moment it is enough to understand that audiences *want* to believe the illusion presented to them as it serves their need for immersion. In addition, they are accepting the meditated reality, when others within that reality accept it as well.[127] Mediation is not an experience aimed at one single individual, it also involves and affects those around us. As there are other visitors taking selfies with Johnny Depp's representation of Jack Sparrow within the mediation of the Madame Tussauds installation, the ongoing social cognition establishes both the degree of realness of the experience as well as of the individual's perception of self.

But what are the reasons for accepting this sort of make-believe? What do we as audiences gain from entering these experiences and feeling emotionally invested in them?

In an increasingly technological society, which stretches the binary contrast between the real and the virtual, escapism is often used to explain the human's desire to enter and remain in immersive experiences. Escapism is usually perceived as something negative and an opportunity to avoid the "real," as if we desperately want to get away from it all: real friends, real work, real facts, and real problems. Apparently, we are so bored with our lives that we would rather engage with mediated characters in books, films, TV shows, and games. But such an answer is too simple and one-dimensional.

In fact, our identities—represented through various mediated ideas of self—are influenced through media. It is an essential part of being human to explore who we are and what we are capable of through simulated and mediated scenarios. Escapism is a way to feel immersed and be fully engaged by focal activities.[128] Yi-Fu Tuan links escapism to the defining elements of culture and humanity: "A human being is an animal who is congenitally indisposed to accept reality as it is. Humans not only submit and adapt, as all animals do; they transform in accordance with a preconceived plan. That is, before transforming, they do something extraordinary, namely "see" what is not there. Seeing what is not there lies at the foundation of all human culture."[129] The mediated, the virtual, and the unreal are necessary for the human condition. Andrew Kuo et al. even claim escapism in mediated experiences provides a form of stress relief through psychological avoidance.[130] However, we do not always submit to mediated experiences to escape our stress. In fact, we are often putting ourselves deliberately in mediated scenarios of discomfort and anxiety to pleasurably experience every shade of our existence. Horror films, for instance, are more than a cheap thrill. They fundamentally confront us with humanity's deepest fears and let us experience our emotions in a safe environment provided by the mediated narrative. Nothing happens to us, but the feeling of being present in the mediated environment evokes actual physical and mental reactions. Getting lost in the story of the film, while sitting in the darkness of the cinema, is both a thrilling and satisfying experience for many people. To feel immersed and take that feeling beyond the mediation is a way of coping and dealing with trauma, drama, catastrophe, and death. It is a safe way to gain experiences outside of physical reality. Apart from social learning and observing others, we learn the most through immersing ourselves in mediated scenarios and experiences.

The long-term effects of immersive experiences on the human mind are still unknown, though. Most research so far has only investigated brief periods of immersion (and therefore, was focusing on the experience of presence). But how the human psyche is affected by immersion decades after the mediation remains unclear. For instance, VR headsets are mostly worn for a few minutes up to a few hours only. However, already short experiences of simulating kineticism and kinetosis might lead to discomfort and motion sickness, often named "virtual reality sickness." It remains a blind spot "how long a human can use immersive technology and remain isolated in a fully-immersive system" and what impact that would have on "the human perception, behavior, cognition, and motor system during and after using such interfaces . . . for a longer period of time."[131] It may very well be that the duration of the immersive experience, its content, and the user's preexisting psychological profile might impact long-term effects and cause or bring back undesirable personality traits and mental problems. In addition, children and teenagers are the primary target audience for immersive technologies.

## Beyond the Mediation

However, most VR research has been conducted on adults. How immersive experiences affect younger users, who are still in the process of developing their emotional abilities, is a matter of further investigation. I admit that appreciating the psychological and receptive dimension in the 360° gaze is a descriptive understanding rather than a fully analytical one. We simply do not know enough yet about how immersion influences our minds, and how it changes the way we think and act on a long-term basis.

But we are not completely clueless, either. In fact, long-term effects in regard to media reception have been researched from many perspectives, notably in relation to addictive behavior (e.g., Internet addiction)[132] and the understanding of selfhood and sense of agency.[133] In some cases, the latter could have profound effects on depersonalization disorders such as chronic feelings of unreality of self or body.[134] Long-term effects of immersion could cause damage to the perception of reality and one's own body, and ultimately lead to the desire to shift one's senses exclusively to mediated environments. This all plays a significant role in sociocultural forms of immersion, as I will explain later, with the examples of binge-watching and the configuration of the self in digital culture.

We need to understand how the cognitive dimensions of immersion influence our idea of self in particular immersive experiences. Robin Curtis adds: "In the same fashion as the fun-house attraction, immersive experiences can also offer the viewer or user the opportunity not only to transport the intact self to another pre-constituted place but also to challenge and transform the parameters of the self."[135] Building up relations to characters in games, films, and literature cannot be limited to identification only ("How does this relate to my life?") but has to be extended to a broad range of experiences that emphasize our emotional alignment ("How am I positioned in a particular viewpoint to understand the character, and how does this influence the perception of myself and the world?"). This has much to do with an understanding of involvement and the "aesthetics of empathy" (named *Einfühlungsästhetik* in the discourse of the early twentieth century). That means that aesthetic objects in mediated experiences are not simply a realistic representation, but "facilitate a heightened corporeal engagement with the world, its objects, and its forms by reminding one of the various ways in which one experiences one's own vitality in embodied experience."[136] Curtis refers to the self as not just being inserted in a three-dimensional space, as a doll is positioned within a dollhouse. Instead, we should consider how the environment influences the self and how selfhood constitutes itself through it. Take the example of virtual pets for mobile devices, a pet only visible through the screen, a virtual *Tamagotchi*. Users build up a deep connection with virtual animals as if they were real, "[y]ou will have to feed it, play with it regularly, give it baths, and brush its teeth, just so you can keep the little critter happy and growing."[137] We learn about the world and us, while we are developing feelings of affection and empathy for mediated characters. Immersive technologies heighten this emotional engagement even more as the distance between characters and viewpoints becomes blurred with the illusion to not only take the viewpoint of the character but instead to *be* the character.

A practical way to understand the relation to the self is to find out how emotional experiences are defined in other mass media. Gary Bente and Ansgar Feist note that conditioned emotional experiences within mediated environments are based on four characteristics: personalization, authenticity, intimacy, and emotionality.[138] These characteristics have been discussed in relation to "affect TV," a genre of television that crosses borders of privacy and intimacy (e.g., reality TV). However, in relation to emotional alignment, they could also be discussed in regard to other mediated experiences.

For instance, take Facebook Horizon. The experience is personalized for us through the use of avatars and the established sense of agency. VR is often referred to as the "empathy machine," because empathy means to have the same feelings as the character portrayed. In this case, though, the character is the user, and whatever the user feels in the VR space he also feels outside of it. The level of personalization is extremely high, even though the avatars are animated and not photorealistic (yet), something that has much to do with the authenticity of the experience created. Nonhuman characters can activate the same emotional alignment and level of acceptance as human characters. Elena Kokkinara and Rachel McDonnell confirm that even though photorealistic imagery supports acceptance and engagement, authenticity depends much more "on the levels of perceived ownership and sense of control (agency) we feel towards this virtual character."[139] Because users are able to perform the self in the VR environment of Horizon, the experience establishes a sense of authenticity. In other words: celebrating virtual birthday parties in VR does not give a realistic representation of physical birthday parties but refers to them by incorporating the same social activities and performative practices (presenting a cake, singing "Happy Birthday," etc.). At the same time, the experience is extremely intimate and emotional. Imagine this birthday party is the only possible way to celebrate with your loved ones who live in other countries all around the globe. To share such an intimate experience with others in the VR space will generate memories. If you want to return to them later, every picture or video of it would refer to the VR space, and not to physical reality. In that way the emotional experience activated by a VR birthday party will not differ much from a physical party. However, the physical party has never taken place, which ultimately creates psychological sensations only triggered by a mediated experience with no reference point to physical reality.

Ultimately, emotional reactions to immersive experiences have long-term effects, especially if they constitute the self in a way that audiences experience as desirable. That is when the human mind develops addictive behaviors in relation to immersion. This has much to do with our desire to dissociate ourselves from the here and now and imagine being elsewhere. Victor Nell even argues that "consciousness change is eagerly sought after" and refers in this respect to immersion being similar to the effects of alcohol.[140] In relation to binge-watching, I will outline later how addictive behavior is transformed into an accepted cultural practice, especially when a majority of users perceive it as something positive.

The human mind is influenced by the psychology and reception of immersive experiences. Even though the long-term effects of immersion have not been fully investigated yet, we know about the make-believe situation and the loss of critical distance. Audiences willingly engage with mediated realities, and the human mind reacts with various cognitive effects, ranging from mild to extreme ones such as illusion-forming, the loss of sense of agency, and addiction.

## Cultural Industries and Digital Ecosystems

> How do cultural industries industrialize and exploit immersive experiences? In which way do digital ecosystems integrate immersion in their market strategies? How do digital ecosystems expand through immersive environments? What are the logics and tactics for immersing audiences in the productions of cultural and creative industries?

In the summer of 2017, AltspaceVR, one of the first companies offering social VR services such as organizing live events in VR—from karaoke nights to eSports tournaments—announced it would officially be shutting down its platform. In a blog post entitled "A Very Sad Goodbye," the venture stated financial troubles as the reason for the closure: "The company has run into unforeseen financial difficulty and we can't afford to keep the virtual lights on anymore. This is surprising, disappointing, and frustrating for every one of us who have put our passion and our hopes into AltspaceVR."[141] The end of the startup might have been the result of the general slowness of the VR market and the hesitation of potential investors. Several attempts to save the company have failed, including discussions with Oculus cofounder Palmer Luckey. However, only a few weeks later, Microsoft came to the rescue. The multinational corporation acquired AltspaceVR to introduce social VR services to its portfolio. A press release stated: "Microsoft is excited to incorporate communications technology into our mixed reality ecosystem. . . . Microsoft is most interested in preserving the current community that uses AltspaceVR to connect and interact with new and old friends."[142] While branding and platform design remained the same to the AltspaceVR community, Microsoft's engagement with social VR attempts to compete with Facebook's social VR apps Horizon, Parties, and Venues. While Facebook already has a powerful social ecosystem, Microsoft is about to build up a platform for immersive experiences called Microsoft Mixed Reality, a blended version of VR and AR applications, accessible through four Microsoft branded HMD (developed by hardware partners Acer, Dell, HP, and Lenovo) and a standard feature on its Windows 10 operating system.[143] Microsoft steps into the market of immersive technologies and the services provided by AltspaceVR are important assets to challenge Facebook's social VR strategy.

Immersive experiences—no matter how meaningful they are for audiences—are part of an emerging market system. A range of evolving industries and products are trying to satisfy the needs of their consumers: "Immersion belongs to the side of the imaginary—this much has been established. On the side of the maker, it is project, plan, idea, and intention; on the side of the audience, it is wish, dream concept, and projection."[144] In this quote, Karl Prümm defines both the maker and the audience, and through it the importance of supply and demand. He adds "[i]t seems logical to locate this agenda of immersion in the industrial modern era . . . [that] brought forth the techniques and instrument-based media necessary for a perfect creation of illusion . . . [as well as] a heightened and relieving expectation of immersion among the public as a counterpart to the ubiquitous, capitalist working world, which had command over the human subject."[145]

Already early panoramic exhibitions in the nineteenth century were designed to meet the criteria of the market. Panoramas were mainly exhibited in financially powerful cities such as London and Paris, and international market strategies were developed to attract tourists. To maximize profit, opening times were extended, and prices depended on the day of the week and the type of panorama.[146]

Not much has changed in the fragmented digital media market of the twenty-first century. Multinational corporations such as Facebook, Google, Disney, HTC, and Sony are developing market strategies to create powerful global brands. They are targeting a growing startup scene working on immersive technologies to ultimately acquire them and introduce their "project, plan, idea, and intention" to their portfolio. But why is this relevant for immersive experiences and the 360° gaze framework?

The feeling of immersion does not emerge without any context nor can it be isolated from the impact of cultural, social, and especially economic influences. Global media industries are powerful players, and their products are mediated experiences, which are industrialized, packaged, sold, distributed, and exploited as formats suitable to specific markets and target audiences. In fact, immersive experiences are part of an economic value chain, particularly in the cultural and creative industries. The economic value of literature, music, games, films, and television programs is often embodied in physical and tangible products such as books, records, DVDs, and Blu-rays (with a solid surface). And even though their status as economic goods is generally obscured by their intangible value for audiences (their liquidity), they are ultimately content packages distributed to different national and international markets. No matter how meaningful the story of a film might be for its viewers or how influential and life-changing a record might be for its listeners, both are the result of the work of very often profit-oriented businesses and corporations. In digital economies, in particular with streaming experiences that no longer offer physical products, intellectual assets in the form of patents, copyrights, and brands are exploited globally in a variety of activities. In that understanding, immersive experiences are influenced by industries, platforms, and ecosystems in which immersion is among the criteria for media products to be successful on the market. The more immersive the content is, the more likely audiences will buy it, and the more likely it can be exploited throughout the value chain.

Take *Star Wars* as an example, one of the most successful film franchises with a global fan base across all age groups. The intellectual property (IP) spans across several films, games, books, comics, animated series, VR experiences, and theme parks, and its semiotic codes can be found in a wide range of merchandising products. If you are into *Star Wars*, you will find plenty of ways to get immersed. When the Walt Disney Company announced a deal to buy Lucasfilm in 2012, it also acquired ownership of *Star Wars*, including several operating businesses of Lucasfilm, such as live-action film production, animation, and visual effects. The goal was to introduce the quite successful IP to a new audience of millennials who were too young to have experienced the release of the original series *Episodes IV–VI*, while at the same time immerse those audiences again who did grow up with it. However, having learned from mixed reviews of earlier attempts to relaunch the franchise (*Episodes I–III*), the first installment of the sequels *The Force Awakens* (*Episode VII*) had been approached differently. With a new creative team under the lead of director J. J. Abrams, who had already successfully reintroduced the *Star Trek* franchise to the millennials market, Disney produced a story that merged new and previous cast members. The return of beloved characters Luke Skywalker, Han Solo, Princess Leia, and Chewbacca (portrayed by the actors of the original series) had already been teased in a global marketing campaign that included exclusive first images on the cover of *Vanity Fair*, and a teaser trailer in late 2014 generating a record of over 58 million views on YouTube in its first week. The teaser was specifically aimed at the fans of the original series, an uncounted mass of ultra fans, waiting for that moment for over 30 years to see their heroes on the big screen again. In the last seconds of the trailer, an off-camera voice utters the three little words "Chewie, we're home" before Han Solo and his lifelong furry companion Chewbacca appear on screen together for the first time since *Return of the Jedi* in 1983. It was the rabbit hole for hardcore audiences to fall into, to evoke old memories of their favorite characters and get excited for the sequel.

In fact, the strategy paid off for Disney. On the day of the release (December 18, 2015), fans worldwide queued for midnight screenings. *Star Wars: The Force Awakens* went on to break numerous box office records in the US and worldwide, including the highest-grossing film in the US of all time. Ultimately it kicked off a string of new *Star Wars* episodes and spin-offs, including *Rogue One* (2016), *Episode VIII: The Last Jedi* (2017), *Solo* (2018), and *Episode IX: The Rise of Skywalker* (2019). Apparently, Disney has plans to extend the *Star Wars* universe with multiple more films and live-action series, such as *The Mandalorian*, for its streaming platform Disney+ until the year 2030.[147]

I have outlined how deeply connected immersive experiences are to our understanding of being human and the idea of selfhood. But in what ways are immersive artifacts exploited within the parameters of industrialization? The industrialization of culture framework introduced by Timothy Havens and Amanda Lotz[148] analyzes media industries and their specific economic, social, and cultural implications under different parameters: (1) mandate, (2) conditions and practices, and how they influence and are influenced by (3) media texts, (4) the public, and (5) social trends, tastes, and traditions. These parameters are interconnected, intertwined, and depend on each other, and tackle the industrialization of successful story worlds as global formats in the media industries. Multinational conglomerates like Disney, NBC-Universal, and Time Warner spread their content packages in multiple formats through different channels. Disney, for instance, not only owns the *Star Wars* IP, but also *The Muppets, Winnie the Pooh, Pirates of the Caribbean,* Marvel (*The Avengers, Captain America, Iron Man, Thor,* etc.) and the Pixar imperium (including *Toy Story, Monsters, Inc.,* and *Cars*). There is hardly a successful film within the last decade that is not part of a franchise: *Star Wars, Harry Potter, Lord of the Rings, Pirates of the Caribbean, Toy Story, Ice Age, Transformers, Mission Impossible,* just to name a few. Referring to Havens and Lotz, these successful story universes are immersive by definition, otherwise they would not have the ability to last for such a long period of time. Only when a content product offers audiences the possibility to engage with its semiotic codes and performative practices on an emotional level, can it be exploited within the value chain—from books and films to games and theme parks.

Books, for instance, are very often distributed in various editions by large publishing houses. To "get lost in a book" means to first understand the publishing market and the tactics it uses to identify authors and texts that are going to stand out in the marketplace and engage large audiences. To generate and spread excitement about authors such as J. K. Rowling and Stephen King, media industries offer opportunities to immerse audiences in both old and new material. For instance, King's novel *It* (1986), about the terror of the monstrous clown Pennywise, had already been a massive cultural phenomenon leading to a two-part film adaption for television (1990). But it was the two-part feature film adaption (2017 and 2019) that set numerous box office records, including the highest-grossing horror film of all time, and made King again relevant for younger audiences. King's brand as an author was able to reach new audiences and led to several film adaptions (e.g., *The Dark Tower, Gerald's Game, 1922, Doctor Sleep*), TV shows (*Mr. Mercedes, 11.22.63, The Mist, The Stand*), and a Hulu show by J. J. Abrams spanning its plot across several books of the King story universe (*Castle Rock*). The publishing market is constantly looking for material that could potentially be part of a larger ecosystem focused on immersive experiences (e.g., film and theater adaptions).

Speaking of the performing arts: the most successful musicals could outperform a Hollywood blockbuster anytime. No film has ever grossed $1 billion at the box office in the US, but three musicals did: *The Lion King, The Phantom of the Opera*, and *Wicked*. Broadway and the West End in London are huge markets for popular shows, often using celebrities or well-known story worlds to reach large audiences and create immersive experiences: *School of Rock, The Bodyguard, The Lion King, Evita, Thriller* (Michael Jackson), *Tina* (Tina Turner), *Mamma Mia* (ABBA), *We Will Rock You* (Queen), just to name a few. Well-tuned productions are able to deliver stable profits, while at the same time make the commercial aspect of the artistic production less obvious.

TV is not much different. For instance, *Breaking Bad* is arguably one of the highest-rated TV shows of all time in the US, and its story has universal appeal (terminally ill lead character; issues of morality, family, and crime), and therefore could be licensed to several international markets. Even though the show has already ended, merchandising (box sets, T-shirts, mugs, cookbooks, etc.), licensing deals (rights for streaming platforms, television, etc.), and a successful spin-off (*Better Call Saul*) and sequel film (*El Camino: A Breaking Bad Movie*) keep the IP alive. In addition, emotional attachment in the form of nostalgia for media products and the enduring love that romanticizes the past bring back reboots of successful TV shows to audiences that want to be swayed by a reanimated chunk of their childhoods (*The X-Files, Fuller House, Gilmore Girls, Roseanne, Will & Grace*, etc.). Disney, on the other hand, revives several of its successful animated films as live-action adaptions (*Alice in Wonderland, The Jungle Book, Beauty and the Beast, Dumbo, Aladdin, The Lion King, Mulan*, etc.).

Mark Zuckerberg and others are at the forefront of cultural and creative industries dealing with immersive technologies. David Hesmondhalgh notes:

"The development of the internet and the web, and the entry of IT firms into the cultural markets, has certainly brought considerable change in the everyday cultural experiences of billions of people."[149] Through the creation of immersive platforms, which are embedded in powerful digital ecosystems, we are consuming information, knowledge, and entertainment in immersive spaces. Whether posting images on Instagram, texting on WhatsApp, reading newsfeeds on Facebook, or visiting birthday parties in Horizon, we are navigating our mediated self through Facebook's ecosystem. Our perception of mediated knowledge has much to do with how the performance of the self can be distributed, accessed, and processed on and through digital platforms. Companies like Facebook, Apple, and Netflix are incorporating immersive strategies and practices to ensure users remain in their ecosystem. Datafication of audiences becomes the most relevant tool within digital ecosystems to create immersive experiences such as binge-watching: "The matching of factors such as age, sex, race, income, occupation, education and area of residence with viewing behavior variables (e.g., amount of viewing and choice of program) result in the statistical determination of relatively stable 'viewing habits'—a set of imputed behavioral routines that form a perfect merger of the objective and subjective."[150] To persuade viewers to watch multiple episodes of *Stranger Things* in one sitting requires more than a compelling narrative. In fact, platform providers install digital regimes with immersive tactics and strategies. Many of these allow, for instance, data mining and the surveillance of user activity. Platforms implement rating systems and viewing mechanisms, and use the data to create profiles of individuals and their tastes to predict future desires and needs.

Computer-mediated interaction, mobile communication, and social media have changed the media landscape with a set of new economic, technological, and sociocultural implications. Digital ecosystems from multinational corporations like Google, Facebook, and Amazon transform user behavior and business models. The tactics and strategies of these platforms are very similar to each other and essential for immersive experiences. All of these platforms integrate a form of human-centered design that allows users to personalize their accounts to access both content and advertising. For instance, users access these platforms through a so-called "first screen," often a mobile device like phone, tablet, or laptop. However, as audiences use several devices to access these platforms, the ecosystem is able to spread across all of them. For instance, as soon as you pause a show on Netflix on your tablet, you are able to later continue it on your phone. All of these experiences are personalized as the platform offers an infrastructure of individual accounts and profiles, as well as rating systems. José van Dijck and Thomas Poell have generalized these tactics of digital platforms with the term "social media logic," which describes existing mechanisms on digital ecosystems.[151] The authors focus on social traffic on social media and define four main characteristics of its strategies: programmability, popularity, connectivity, and datafication. While scheduling and programming for mass media are often visible, programmability through algorithms, protocols, interfaces, and platform organization on digital platforms is not: "While algorithms are nothing but sets of coded instructions, it is important to observe how social media platforms shape all kinds of relational activities, such as liking, favoriting, recommending, sharing and so on. . . . The power of algorithms . . . lies in their programmability: programmers steer user experiences, content, and user relations via platforms."[152] At the same time, digital ecosystems feed the audience's desire to get immersed in content defined by its popularity. Platforms implement "distinct mechanisms for boosting popularity of people, things, or ideas, which is measured mostly in quantitative terms."[153] Connectivity allows digital ecosystems to track individual connections and behavior, while at the same time connect users with advertisers. The final principle of social media logic is datafication, a term coined by Viktor Mayer-Schönberger and Kenneth Cukier to describe the ability of networked platforms to render into data which has never been quantified before, such as metadata from smartphones, GPS locations, preferences for entertainment content, but also biometric data like fingerprints, iris and facial recognition, and even human DNA.[154] Datafication is invisible to users as methods for personalization are often inaccessible to public or private scrutiny.

In a 2017 article, I applied and extended Van Dijck and Poell's understanding of social media logic to immersive technologies.[155] I have argued that user experience and platform sociality need to be considered as key characteristics for immersive experiences on digital platforms. But not exclusively, as personalization strategies (e.g., first-person perspectives) and interactive features can be found in immersive theater, film, and music as well. Participatory cultures establish the audience as part of the experience and exploit interactivity and immersion as new forms of labor: "[Consumers] are not so much participating, in the progressive sense of collective self-determination, as they are working by submitting to interactive monitoring. The advent of

## Beyond the Mediation

digital interactivity does not challenge the social relations associated with capitalist rationalization, it reinforces them and expands the scale on which they operate."[156] Mark Andrejevic points out that immersion through participation and engagement results in users not only taking an active role in the mediated representation, but actually contributing to the commercial success of media and cultural industries. If no users posted on Facebook, the platform would cease to exist. If no viewers rated titles on streaming platforms, datafication would not be effective anymore. If no audience member accepted the performativity of an immersive theater production, the show would not be able to continue. Effectively, we are not merely the "mediated self," "identities," or "entities." We are also target audiences and consumers. We accept immersive experiences as a form of reality in the same vein as we neglect that we have paid money for them, engage and identify with global brands, and support the success of multinational corporations. These aspects are essential in the understanding of the 360° gaze, as we are not only living in the age of immersion, but also in the age of industrializing immersive experiences.

The characteristics of the 360° gaze framework are constantly adapting. How they relate to each other depends on the specific immersive experience. Their overlaps are important. The four interrelating characteristics cannot exist in singularity. Semiotic codes influence performative acts, and they are embedded in immersive experiences created by cultural industries and digital ecosystems. All of them have psychological and receptive effects on the human mind. By relating them to each other, the framework can be used to understand how audiences perceive immersive experiences, co-create meaning and practices, and embody the attributes of immersion as a sociocultural phenomenon.

That way the history of mediated practices becomes a history of immersion.

# 3  Intermedia Immersions

No, can you believe how far we've come
In the New Age?
Freedom to have what we want
In the New Age we'll all be entertained
Rich or poor, the channels are all the same
—Father John Misty, "Total Entertainment Forever," *Pure Comedy*, Sub-Pop, 2017

**Texts as Worlds**

The history of immersive media and immersive experiences often begins with the rise of image-based art. Long before written texts, the image has been influencing human's perception through semiotic representation. Early imaged-based art created powerful mediated realities that influenced how people understood the world around them. Oliver Grau traces the history of immersion in the visual arts back to antique murals. Wall paintings such as the ones in the Roman Villa dei Misteri near Pompeii eliminated the boundaries between the spectator's space and the illusory space of the painting by creating images that seemed to step out of their frames.[1] Early attempts of parallel worlds were placed like rabbit holes in the physical reality of the viewer by either occupying the physical space partly or even entirely, such as the Sacred Mountain of Varallo in Piedmont and the *trompe-l'œil* churches of baroque Europe.

The public attraction of the panorama, in particular, influenced many other immersive experiences to follow. Patented by the English painter Robert Barker in 1787, these 360° circular paintings installed the viewer within the picture. The panoramic view filled the audience's entire field of vision by overwhelming it with surrounding images of battle scenes or other historic events. As Grau points out in regard to the panorama: "All distance disappears as the observer is involved physically and mentally in the depicted events."[2] The panorama allowed audiences to create a mediated self within the mediated image space. The perceptual illusion introduced several characteristics of immersive media: the dimensions of real time, the absoluteness of presence, and the abnegation of mediality.

Immersive spaces in visual art also had an influence on literary texts. The rising popularity of the written word has established the process of "world-building" that would later become essential in digital games. Marie-Laure Ryan maintains that "for a text to be immersive . . . it must create a space to which the reader, spectator, or player can relate and it must populate this space with individual objects. . . . For immersion to take place, the text must offer an expanse to be immersed within, and this expanse, in a blatantly mixed metaphor, is not an ocean but a textual world."[3] Ryan mentions everything that we have discussed before in relation to mediated realities. Her use of the "world" metaphor contradicts the widely established metaphor of the ocean in which audiences "dive in/dive under" to immerse themselves in mediated worlds and alternative realities. A "world" is not so much defined by its limitations but by the relations of the objects placed inside of it.

The turning point for world-building in literary texts was the late eighteenth century. The period, which went on till the first two decades of the nineteenth century, has retrospectively been named romanticism, and it influenced radical developments in art, philosophy, literature, and music.[4] It opened up the door to a new understanding of the sublime, the unconscious, and the meaning of the real. Everything we now associate with immersion—the dissolution of frames as constitutive parts of mediation, the construction of alternative realities, the idea of selfhood in liquid spaces—had been part of a paradigm shift that explored both the aesthetics as well as the social implications of a new philosophy. It is no surprise that one of the most popular panoramas, *The Battle of Waterloo* (1815) by Robert Barker, falls exactly into that period.[5]

The main characteristics of romanticism are captured quite beautifully in one of the visual masterpieces of that era, Caspar David Friedrich's painting *Wanderer above the Sea of Fog* (1818). In the foreground, we see a lone man from behind, standing on an outcropping of rocks and gazing, perfectly still, into a landscape full of ridges covered in thick, gray fog. He looks into the faraway distance, which reveals silhouettes of forests and mountains, while the horizon becomes indistinguishable from the cloud-filled sky. We as viewers are hidden spectators of the scenery: we can see the wanderer, but he cannot see us. At the same time, the sublimity of the natural environment, the power of what vast nature can accomplish, is influencing the wanderer on an emotional level. He contemplates the vastness before him, as he inextricably becomes part of the surrounding in an immersive experience. We as viewers feel with him as the window-like image space dissolves in front of our eyes. The whole image represents a romantic understanding of longing (in German: *Sehnsucht*) to turn the inner perspective to the outside and direct the emotions from the soul to the environment. It is a borderline experience between safety and danger, and between loneliness and the world outside. The painting is more than a traditional use of symbolism. It creates a connection between the viewer and the mediation. Many other of Friedrich's works capture a similar meaning (e.g., *Woman at the Window*, 1822), but *Wanderer* is quite special: the spectator has been placed inside the scenery as there is no clear separation of the image space from the environment.

The romantics challenged the establishment of their time by emphasizing the importance of the individual self and what it might be capable of with the power of imagination. They were not interested in following imposed conventions and rules of those in power. They believed in the limitations of reason and argued that deeper, subconscious appeals are influencing people's understanding of reality. They believed there is something in the human mind that we simply are not able to comprehend—a darkness that is both appealing and threatening. They claimed that this is where our dreams come from, our nightmares, and our wishes to be someone else—ultimately, everything that humanity tries to reflect on with the help of literature, drama, music, and other art forms. Philosophers and scholars like Johann Gottlieb Fichte and Gotthilf Heinrich von Schubert discussed the inexplicable dimensions and horrors of the human mind: the creation of selfhood and self-cognition, dreams, sleepwalking, divination, mental illness, and spiritual events. It was the time of gothic novels, fantasy, horror, the *Doppelgänger* motif, and the creation of human-like creatures. Long before the invention of 3D printers, artificial intelligence, and genetic reproduction technologies, authors captured the idea of the unreal, the uncanny, and the artificial in their texts. It was the time of Mary Shelley's *Frankenstein, or The Modern Prometheus* (1818), a novel about the creation

**Figure 3.1**
Caspar David Friedrich's *Wanderer above the Sea of Fog*: the disappearance of the window-like image space. © 2020 CC BY.

of a humanoid, a human-like monster consisting of leftover body parts. Only two years earlier, German author E. T. A. Hoffmann published his short story "The Sandman" (1816) about a godlike robot named Olimpia ("she who comes from Olympus"), whose uncanny appearance drives the main character Nathanael into madness. Hoffmann, a lawyer in his day job, was forced to write most of his texts in the middle of the night. He was fascinated by the obscure temptations of immersion, when no one knows what is real and what is not, especially not the reader (e.g. *The Devil's Elixirs*,

1815). The texts of this period influenced many literary classics to follow, from the uncanniness of human nature in *The Strange Case of Dr Jekyll and Mr Hyde* (Robert Louis Stevenson, 1886) to the existence of parallel realities in *Alice's Adventures in Wonderland* (1865).

Romanticism also began to question the limitations of mediated representation and the aesthetics of separation. The romantics looked for a supreme spirit, a kind of poetry and art that lives on beyond the painting, the literary text and written word, *beyond the mediation*. Their so-called "progressive universal poetry" (*Progressive Universalpoesie*) differentiated itself from other expressions as texts were considered to be in a constant state of perpetual becoming. In their opinion, a text could never be entirely finished. Instead, it would expand beyond its own limitations. One of the most popular literary genres during that time, therefore, was the fragment. Hoffmann's fragmentary book *The Life and Opinions of the Tomcat Murr Together with a Fragmentary Biography of Kapellmeister Johannes Kreisler on Random Sheets of Waste Paper* (1820/21) is famous for playing with the aesthetics of the literary frame by mixing up two biographies in one book: one written by a man, the other one by his cat. Very much in the spirit of Caspar David Friedrich's *Wanderer*, Hoffmann's text is breaking down the fourth wall of mediation. The fictional characters, even the cat, talk directly to the readers and ask them to bring the confusing mix of the two biographies into the right order again. This results in an aesthetic play with the literary medium that challenges the expectations of the readers as to what exactly constitutes a biography. By doing so, Hoffmann experiments with the design of the printed format and the materiality of books. The already established use of printing technology at the time allowed not only the mass production of texts, but also the transformation of specific narrative concepts into design and typography.

The most influential novel in the romantic era from the viewpoint of immersion is *Henry of Ofterdingen* (1800/1802) by Friedrich von Hardenberg, also known as Novalis. In the form of a fragment published after Novalis's death, the text introduces shifts between the diegesis of the narrative, parallel dreamlike worlds, and several poetic and philosophical prophecies. At first, *Henry of Ofterdingen* appears to be in the literary tradition of the *Bildungsroman,* or coming-of-age novel, a genre focusing on the psychological and moral growth of the protagonist from youth to adulthood through a metaphorical (and very often literal) journey. But it is so much more than that, because the locations in *Henry of Ofterdingen* are not necessarily "real," as far as referring to physical counterparts. Instead, they are parts of dreams and imagination. They are creating a mixed reality world altogether. The recurring shifts between various representations of reality, both in the world of the story and in dreamlike surroundings, reflect on one another. The mediations in the text (dreams, songs, poems, memories) are as real as anything else.

The main character, Heinrich, is loosely based on a medieval bard of the same name, and before starting his journey he foresees his travels in a dream that becomes more real than anything he experiences later. The main focus of his dreams is a blue flower, a recurrent image that does not leave his mind for the rest of his journey. It lives on beyond the mediation of the dream: "I yearn to get a glimpse of the blue flower. It is perpetually in my mind, and I can write and think of nothing else. I have never felt like this before; it seems as if I just had a dream or as if I had been transported into another world in sleep. For in the world where I otherwise lived, who would ever bother about flowers?"[6] The blue flower is a symbol. It is the most central allegory of the novel and the whole romantic era, for that matter. It represents indescribable desire, the infinite, and unreachable. It changes its meaning throughout the book and assigns to whatever meaning and purpose Heinrich needs. The blue flower opens up a door for Heinrich into a new world, a dreamlike replica of a future and a world yet to come, but one he wishes to be in.

The blue flower is one of literature's first and finest rabbit holes.

The romantic novel is the kickoff for a new understanding of literary texts and marks a significant period of immersion in literature. Romanticism is now regarded as more than just an aesthetic movement. At the time, it was a social statement against the reasoning of the Enlightenment and in favor of a constructivist viewpoint of the self. To feel immersed by those texts meant to believe in a world of the unknown and accept that this world influences audiences as much as anything else. While pre-Renaissance art was excluding audiences from mediated experiences such as texts or images through frames and storytelling techniques, instead assigning them the role of pure spectators, viewers, and readers, post-Renaissance literature turned the audience "into the direct witness of events, both mental and physical, that seemed to be telling themselves."[7]

From that period on, authors focused on the projection of a three-dimensional space onto a two-dimensional surface. They started to create a *text as world*, a diegetic story universe, one that positioned readers in the text through various narrative techniques, spanned beyond the coordinates of the paper, and required audiences to experience the text as a story world that was no longer limited to the written word. Readers found themselves in a text full of references that directed their mind outside of it and stimulated their imagination to co-create the story world. At the same time, silent reading gained immense popularity, making the reading act a private, introverted, and personal experience.

The act of reading "transports" audiences to fictional worlds. It occupies readers completely through a process of thorough mental absorption that impacts them long after the book has been finished. The fictional world is experienced beyond the mediation. For instance, Goethe's *The Sorrows of Young Werther*—though not strictly part of the romantic period—had been published around the time, in 1774, and led to a series of suicides among its readers.[8] To "get lost in a book" is not only referred to in Jasper Fforde's novel *Lost in a Good Book* (2002). It can be also found, for instance, in Michael Ende's *The Neverending Story* (1984), in which a young boy escapes the cruelty of his childhood (the death of his mother and being bullied in school) by reading a book, which connects him to the fictional magical land of Fantastica. The story world influences his physical being, even to the extent that the mediated characters in the end cross the border to his physical reality. This is how we imagine the act of reading to this day, and this is when we associate it with immersion.

In the 360° gaze framework, the reader has a representation in the world of the text: the mediated self. Literary theory, especially the tradition of reader-response theory and cognitive poetics, recognizes the reader as an active agent in constructing the text by combining stylistic and narratological methods with insights from cognitive linguistics and psychology. The Germans even have a name for that kind of reader: the *implied reader* (*implizierter Leser*). The implied reader is another term for the mediated self. It is a mediated construction. When an author composes a text, he or she imagines a reader, a communicative recipient on the other end—not a specific physical person, but a mediated version embedded in the textual structure to which the reader responds. This implied reader is able to understand the story exactly the way the author intended it. This makes the reader's point of view the center of the experience. Every contribution by the narrator, the plot, and the characters is going back to the "vantage point by which he [the reader] joins, and the meeting place they [the elements of the narrative] converge."⁹ Readers are invited to fill several "narrative gaps" or "blank spaces" while reading a text (another term from the Germans, the so-called *Leerstelle*). Marisa Bortolussi and Peter Dixon call them "reader constructions," as they are "subjective and variable mental processes"¹⁰ in response to the text, and different to the "textual features," which are "objective and identifiable characteristics of the text."¹¹

For instance, have a look at *Harry Potter* and how the main character is described in the first book of the series: "Harry had a thin face, knobbly knees, black hair, and bright green eyes. He wore round glasses held together with a lot Scotch tape because of all the times Dudley had punched him on the nose. The only thing Harry liked about his own appearance was a very thin scar on his forehead that was shaped like a bolt of lightning."¹² Obviously, the popular image of Harry Potter is very much inspired by visual representations in mass media, in particular by Daniel Radcliffe's performance in the films. However, if we stick to the text, the description in the form of semantic units gives us merely a few details about his face (thin, black hair, green eyes, glasses, scar) and the conditions of his knees, but really nothing more than that. The rest leaves many narrative gaps that want to be co-constructed by the imagination of the readers. This process becomes immersive because of *how* readers fill up those gaps.¹³ It is a process of intense personalization. We need to assume that J. K. Rowling left these narrative gaps in her text intentionally for the implied reader to fill them. Based on the description in the text, we might not all imagine Harry Potter the exact same way, but he is probably always a shy, reserved, but very smart and courageous young boy, with only one physical abnormality on his body (the scar on his forehead). And this is exactly how the author wants us to see him. Obviously (and thankfully) the implied reader is just a placeholder and physical readers may arrive at different textual interpretations. They consciously and unconsciously respond to immersive features in the text and consciously and unconsciously leave out others. To understand the story world of Harry Potter, readers are not only relying on the semiosphere established in the text, but also incorporating other modes of meaning outside of it. With the help of their cognitive imaginations, readers construct the text by using their knowledge about the world and their own experiences. What Julia Kristeva describes as intertextuality, a formation of references to decontextualize and recontextualize a text, is the beginning of the postmodern understanding of immersion in literary texts, the deconstruction of the "author" as the commanding authority (famously described in Roland Barthes's "The Death of the Author"),[14] and the start of a dialogical co-creation of meaning.

> Literary texts often introduce *story worlds*. Readers co-create the world in the text by filling up *narrative gaps*. They extend the *text's semiotic system* with other modes of meaning through cognitive imagination.

The semiotics of storytelling in literature, which incorporate everything from the creation of characters with their internal and external conflicts, the structure of the plot, the story arc, and the modification of time and space, have been theorized and established over centuries in several defining works, from Aristotle to Syd Field. In her book *Narrative as Virtual Reality*, Marie-Laure Ryan gives an extensive overview of how literature took on the playful strategies of visual arts, the spatiality of worlds, intertextual references, and the text's self-referentiality. She also emphasizes how "[t]his evolution split literature into an intellectual avant-garde committed to the new aesthetics and a popular branch that remained faithful to the immersive ideals and narrative techniques of the nineteenth century."[15] Readers "accept, and experience, the world-within-the-text . . . as if it were 'a' world-without-the-text."[16] When we are reading a book, we become aware of the collaboration with the text and the production of meaning within an illusionary experience. We as readers look through the text toward the referenced world and are not paying attention to the materiality and structure of the plot.[17] Ryan argues that "[i]f readers are caught up in a story, they turn the pages without paying too much attention to the letter of the text: what they want is to find out what happened next in the fictional world." Immersion results in "the blocking out of the physical world"[18] while readers fill up the narrative gaps within the textual structure: "And if readers experience genuine emotions for the characters, they do not relate to these characters as literary creations nor as 'semiotic constructs,' but as human beings."[19]

Literary and narrative theories have been concerned with the collision of different realities in the past and dealt with it in relation to several narrative phenomena. Just one example is the analysis of time. Time is culturally constructed, because we think of ourselves in relation to its flow. Remember that, when you think that time is either "running out too fast" or "too slow." Narratology analyzes the use of time in relation to the representation of time in the text and the duration of the reading process. "Story-time" is the time that passes in the story (the mediated), while "discourse-time," on the other hand, covers the length of time used up for reading the story (the physical).[20] In texts, the meaning of time structures the story, but to analyze it means not only to take into account its representation in the fictional world, but also the time within the physical reality of the reader. As it is typical for all immersive experiences, this refers to the moment of crossing borders between the physical and the mediated.

Many novels since the nineteenth century emphasize the feature of narrative world-building and have immense commercial success that way (think of *Lord of the Rings*, *Harry Potter*, *Twilight*, and *Game of Thrones*). Literary texts with universal story worlds are exploited in economic value chains through film adaptions, websites, games, merchandising, and theme parks. Electronic literature, experimental texts on Facebook and Twitter, and interactive approaches that ask the reader to participate as coauthors, are very often part of an extending multimedia ecosystem.

Marie-Laure Ryan argues that the "rise and fall of immersive ideals are tied to the fortunes of an aesthetics of illusion."[21] In that respect she identifies a performative turn in literature that "turn[s] language into a dramatic performance, into the expression of a bodily mode of being in the world."[22] With users taking on the roles of characters to establish a perspective from a first-person point of view, Ryan questions the role of the author, while she establishes the natural link to the performative practices of the medium. This kind of interactivity is used particularly in the postmodern aesthetics of electronic literature such as hypertexts, kinetic poetry, text machines, and interactive digital fiction that require the reader to make choices to participate in the expanding narrative branch of the story and engage in role-playing activities.[23] Companies such as Delight Games create "interactive gamification novels" for digital devices that introduce immersive strategies to the market of electronic literature. Interactivity works specifically well within the genres of fantasy and science fiction, which allow audiences the opportunity to co-create another world different to their physical reality. Henry Jenkins, Marie-Laure Ryan, Sam Ford, Roberto Simanowski, Alice Bell, Astrid Ensslin, among others, have discussed this phenomenon in relation to the idea of immersive story worlds,[24] transmedia storytelling, convergence culture, and digital fiction by analyzing intersecting storylines and long-term continuity across different media platforms.[25]

Several storylines, which merge into a single story universe with multiple rabbit holes as entry points across different media, as well as interactive elements and user-generated content, allow audiences to become attached to the narrative world.

> The *act of reading as a performative practice* activates *psychological sensations* that block out the physical world and materialize the mediation. The world of the text becomes an immersive experience that is further intensified by *interactive features* and various *economic activities*.

Postmodern literature in the twenty-first century breaks the immersion of the text by using metafiction, self-referentiality, irony, and the deconstruction of the illusionary experience, for instance, through nonnarrative elements such as lyric poetry.[26] However, by doing so, postmodern literature emphasizes the performative practices instead of the semiotic system in the text. The framework of the 360° gaze can be still applied, however immersion is not achieved through world-building but through deconstructing the limitations of the medium. Postmodern literature often challenges the readers' expectations of literary texts. For example, authors such as Jonathan Safran Foer (*Everything Is Illuminated*, 2002; *Extremely Loud and*

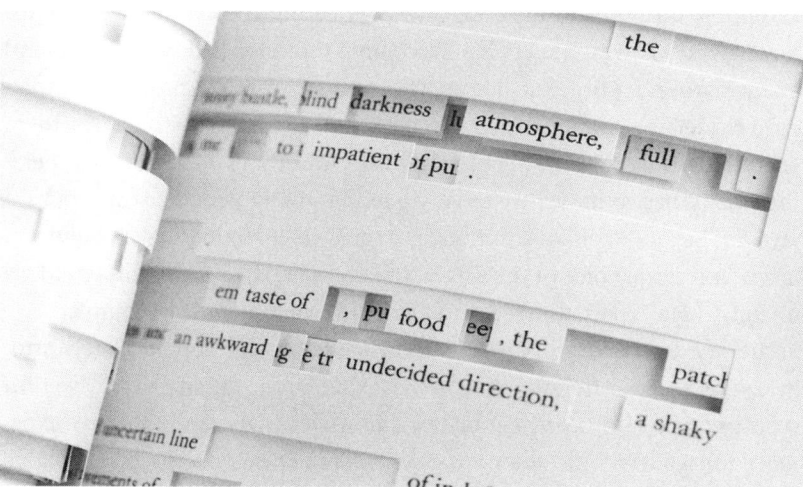

**Figure 3.2**
Jonathan Safran Foer's *Tree of Codes* experiments with die-cutting techniques. Reprinted with the permission of Visual Editions. © 2010 Visual Editions.

*Incredible Close*, 2005; *Tree of Codes*, 2010) and Mark Z. Danielewski (*House of Leaves*, 2000; *Only Revolutions*, 2006) refer back to the "progressive universal poetry" of the romantics and their unique approach to literary text. The format and structure of their books are unconventional, use multiple narrators, footnotes, columns, appendices, font and color changes, and even ask the reader to read it backward.

For *Tree of Codes* (2010), Foer took an ambitious approach. He experimented with die-cutting techniques and strategically cut out words from one of his favorite books by another author (Bruno Schulz's *Street of Crocodiles*, 1934). Due to the physical conditions of the printed layout that puts page over page, every cut out word left a blank space that revealed other words on the pages underneath. By applying that technique in a specific pattern, certain worlds would relate to each other and build a new narrative in an already established textual world.

*Only Revolutions* by Mark Z. Danielewski, on the other hand, is a Romeo-and-Juliet-like story told from the perspective of both lovers simultaneously. Readers are invited to turn the book 180° every eight pages to read the other perspective. In addition, it experiments with the design of the layout by constantly outlining several historic events printed in columns on the side of each page. This has quite a claustrophobic effect on the readers but immerses them in the story through the interactive experience of constantly turning the medium and connecting the narrative elements. Danielewski's newest project is a 27-volume project called *The Familiar* and revolves around a tiny, fragile cat. The author hopes the project will challenge readers with unexpected connections to global issues and let them redefine their understanding of literature: "And that's where I think literature finally has to move; we're very good at giving people a voice but we have not begun, strenuously enough, to give voice to that which will never have a voice: the voice of the waves, the animals, the plants, this world we inhabit."[27] He expects the series to take him over a decade to complete.

Literary texts with story worlds continue to be very popular. They introduce several rabbit holes for immersive experiences. The performative turn in postmodern and experimental texts, however, is also an immersive experience for readers. They are now much more focused on the performative dimensions of the medium and how the limitations and framings of the written text are deconstructed and overruled. The illusionary effects activated by story worlds no longer dominate. Instead, interactive experiences and co-creation reframe the mediated self and introduce different kinds of narrative gaps.

In any case, there are still enough blue flowers out there.

## Performing Arts and Sound

Charlie Kaufman's directorial film debut, *Synecdoche, New York* (2008), explores the human conditions of the self, and one's perception of reality through the notion of performativity. The main character, Caden Cotard (Philip Seymour Hoffman), is a depressed, miserable, and hypochondriac theater director. His existence falls apart on many levels: his parents, wife, and daughter leave him and he is terrified of death, while he perceives his body as a rotten wreck. His inability to accept both the reality and mortality of his own life makes him replicate his existence in a massive theater production. With the money from a prestigious fellowship, he recreates the settings of his entire life in a gigantic New York City warehouse—over a period of 40 years. He builds a duplicate of his house in it, and hires a cast of actors that are playing him and the ones he once lost. But the mediation does not end there. The actor who is "playing" Cotard in the production is following "real" Cotard all day long, eventually taking his place in physical reality by having an affair with Cotard's real girlfriend. And even more so, Cotard in the play wants to recreate his own life in the play by starting a gigantic theater production himself, and by hiring an actor who follows him. In the end, the scenery becomes a complex Matryoshka doll-like *theatrum mundi* in which different realities and roles intertwine, and the question of "what is real" becomes immanent. Not even Cotard knows in the end where the play ends and his physical reality begins. Ultimately, the massive production is never revealed to the public, and Cotard dies a lonesome death in the settings of his play.

*Synecdoche, New York* is a play within a play within a play (and so on), and an immersive theater production. Everyone involved in it devotes a specific period of time to the reality of the performance (here: over four decades!), and neglects the configurations of physical reality for the time being. Actors merge with their roles and neglect their physical existence. It is a sort of fictional game, as Jacques Ehrmann points out in his groundbreaking text "Homo Ludens Revisited": a game that defines both reality and culture for a given time. "To define play is *at the same time* and *in the same movement* to define reality and to define culture. As each term is a way to apprehend the two others, they are each elaborated, constructed through and on the basis of the two others, they are all simultaneously the subject and the object of the question which they put to us and we to them."[28]

This "play situation" is a make-believe contract between performers and audiences (Andreas Mahler calls it the "matrix of fictionalizing").[29] In a first instance, it applies to audiences sitting in the auditorium as passive witnesses of the events on stage through establishing a theatrical code of signification. Similar to literature, the ways in which theatrical performances establish "worlds on stage" have much to do with how these worlds are communicated as mediated realities. The early modern plays of English theater and Shakespearian drama, especially, were obsessed with exploring the possibilities of world construction through signs ("All the world's a stage" is the phrase that begins a monologue in Shakespeare's *As You Like It*).[30] Since the Renaissance and its devotion to anti-immersion and the framing of mediation, role-oriented acting has dominated the theater. The goal was to create a mimetic effect on stage to build an invisible "fourth wall" between the characters and the audience. Mahler points out that only the performing arts address this "essential doubleness of theatrical communication."[31] Only in the playful fictionality of the theatrical setting do we find semiotic codes for the external setting of the play (between actors and audiences), as well as for the internal setting (between the fictive instances, i.e., the characters on stage). This fundamental paradox addresses two levels at the same time with different semiotic codes—the ones characterizing the receptive process as a social activity (going to the theater), and the other ones establishing a world within a world, while neglecting any other world than the one represented on stage. Fernando de Toro and Carole Hubbard argue that theater uses signs "from nature (lightning, rain, age, youth, etc.), from society (gestures, linguistic terms, dress), and from other artistic practices (architecture, painting, mime, music)."[32] "All theatrical manifestations are then signs of signs or signs of things."[33] Auditory (words, tone, music, noise) and visual elements (mimics, gestures, movements, makeup, hairdo, costumes, props, set, lightning)[34] further support audiences feeling engaged in the mediation. This is what Roland Barthes means when he outlines his idea of "the effect of the real," for instance, when objects on stage present a certain time and space, and actors are indeed first and foremost characters.

In addition, Mahler stretches the consistency of the represented world and the frames of the play. The "heteroreferential function" of performance creates a three-dimensional environment "and does not distract from the believability of the fictive world-within-the-drama itself."[35] Traditionally, the picture-frame stage is separated from the auditorium in which the audience is seated. The setup takes away forthcoming configurations for audiovisual experiences such as cinema. The stage is well lighted, while the audience remains in the dark, to see but not to be seen. It is a form of *dispositif* (Foucault). The element of voyeurism emerges in a distinctive private and intimate setting and highlights that "in the special circumstances of the theater, in the privacy of the darkened auditorium, the individual may indulge in the 'gaze,' which is impossible in most social situations. Here no guilt is attached—indeed this what the spectator has, in a sense, come for!"[36] The creation of fictional worlds on stage can be repeated endless times as it is often the case with plays-within-the-play (such as in *Synecdoche, New York*). Mahler analyzes Shakespeare's *A Midsummer Night's Dream* as an example: "In other words, the (real) actors . . . play (fictive) characters . . . who are in turn (fictive) actors again . . . representing hypodiegetic characters . . . , who are watched by a fictive audience."[37] This complex interwoven setting "(explicitly) immerses and distances us at one and the same time."[38]

> Theatrical performances involve *two kinds of semiotic systems*: one for the external setting of the play between actors and audiences, and one for the internal setting of the play for the entities on stage. In immersive theater, these two semiotic systems overlap.

From the early days of Greek theatrical performances, immersion has been a constituent element of the theatrical space. Scenic paintings—named *skenographia*—were designed to subtly modulate the distance between reality and illusion and to give audiences a sense of depth of the theatrical stage. The gaze of the audience was directed toward the center of the performance on stage, and in return every line of sight was directed toward the auditorium. This offered audiences the illusion of being part of the scenery on stage. Greek theatrical performances were used to communicate ideologically and politically coded messages—a shared symbolic language. However, that has always been a paradox since audiences remained spatially separated from the performance on stage. Immersive practices such as illusionary scenic paintings were, therefore, necessary to minimize the distance in the aesthetic experience.

Audiences were strictly separated from the stage for several reasons. Besides aesthetic conventions of the picture-frame stage, the productions simply grew bigger over the course of time, and more people wanted to attend them. From an economic point of view, it would always have been possible to extend the auditorium with extra seating, but for several reasons (including safety) it was necessary to keep a certain distance between the auditorium and the stage. However, it remained essential to establish a sense of being "inside" the play for the audience. So it always has been one of theater's main ambitions to minimize the distance as much as possible.

In 1849, Richard Wagner envisioned a theater of the future in which the spectator would be "transplanted" on stage, while he "forgets the confines of the auditorium, and lives and breathes only in the artwork which seems to him as Life itself, on the stage which seems the wide expanse of the whole World."[39] In the early days of theater, different audience frameworks on multilevel platforms, ramps, balconies, and scaffolds surrounded the stage in a circus-like setup to give audiences a feeling of closeness to the aesthetic illusion (such as in the theater-design experiments of the Austrian architect Friedrich Kiesler). Projections, magic lanterns, sensory stimulation (such as smells), the play with light and darkness, and sound effects were used, in addition, to enhance the experience. For instance, in late eighteenth-century romanticism, the *phantasmagoria* had been a kind of horror theater that used projections of ghosts, skeletons, and demons to satisfy the audience's fascination with the obscure and bizarre. The images were projected right into the audience space to give the illusion that the stage had been extended. These aesthetics led the way to the Théâtre du Grand-Guignol, a French theater that specialized in shocking and terrifying its audiences in naturalistic horror shows. The technique of "breaking the fourth wall" was introduced for the same reasons. Characters turned to the auditorium, addressed the audience, and made it an accomplice of the events on stage.

In the first decades of the twentieth century, the idea of mimesis as the semiosphere in theater has been increasingly challenged, leaving room for what now is referred to as "postmodern theater" and "the problem of representation" in the second half of the twentieth century ("the crisis of drama"). This radical countertradition in theater (and literature) starts with Bertolt Brecht's anti-illusionist theater ("epic theater/*episches Theater*") in which naturalistic approaches to engender real human behavior are rejected. Brecht was not interested in absorbing audiences into the fictional world of the play, so they could use it as a form of escapism. Instead he believed in alienation and distancing effects (*Verfremdungseffekte*) to encourage audiences to willingly disbelieve the illusion on stage. Brecht's understanding of theater was a political one. He did not want audiences to lose themselves in the artificiality of the performance, but instead question historical and social implications through performative art: "Everything must be seen from the social standpoint. Among other effects, a new theatre will find the alienation effect necessary for the criticism of society and for historical reporting on changes already accomplished."[40] In his view, immersion would not lead to greater activism or political commitment, but interpellate audiences in the social, economic, and political regimes of modernity.

Postmodernism of the twenty-first century, however, embraces immersion again. The picture-frame theatrical process, which even Brecht felt obligated to refer back to, has been rejected in what is often called "immersive theater." In that sort of performative setting, the frames of the stage dissolve and the audience space merges with the performative space. Audience members are invited to actively participate in the play. In her book *Immersive Theatres*, Josephine Machon argues that the immersive turn in postmodern theater should be understood as a counter movement to the alienation, dislocation, and disintegration that audiences experience in digital spaces. Instead of getting lost in the anonymity of digital media, she sees immersive theater as a spiritual experience that reintroduces co-presence, collaboration, and a fuller sensual experience of theatrical embodiment: "In this way, immersive theatres can establish links across sensation, perception, emotion, and moral reasoning in form and content."[41]

The performative practices of immersive theater are concerned with a different understanding of the roles of audiences and the theatrical environment. For once, the stage is no longer defined as a location in a playhouse. There is no fourth wall. In fact, there are often no walls at all. Performative spaces can be extended to exterior surroundings, urban places, abandoned buildings, and locations that are not associated with theater spaces at all (like parking lots, large warehouses, and train stations). Physical environments outside the theater space get occupied by the rules of performativity, all atmospherically composed, as well as acoustically, olfactorily, gustatorily, and materially designed. Performance groups often use the entire space of a site and let audiences explore different rooms, halls, and corridors. These spaces are all-encompassing, intense physical experiences for those who visit them. Productions can last for several hours, often without intermission. They reflect on questions such as: What can be gained from considering the full sensorium that is brought to spectatorship? And what are the political stakes of blurring the lines between performers and spectators?

That leads us to the changing role of audiences. Within immersive performances, audiences are mobile and not statically positioned. Instead, spectators become nomads. They collaborate with professional actors and are often encouraged to interact with them. The invitation to constantly move and interact challenges the spectator's agency and his or her comfort as distanced theatergoers. Quite often audiences need to disguise themselves, wear masks, or appear in the fashion style of a certain historic period or genre. Their idea of selfhood is shaped by the theatrical performance as they are taking on a role to perform. This version of self is not identical with the knowledge of their physical selfhood (as audience member), but still not fully developed (as role in the play) to grant them individual sense-making. For instance, you might need to play the role of a wizard student at Hogwarts, and even get a fictional name, but the rest is up to you. Selfhood in immersive theater is defined by individual experiences on somatic, spatial, cognitive, and psychological levels in ephemeral landscapes, which are co-created by actors and spectators. In that sense, immersive theater shares kinship with popular tourist attractions such as Madame Tussauds and the Harry Potter Studio Tour, omni-theaters, and video gaming. It is inspiring spatial storytelling modes in VR environments in which audiences are positioned within an all-surrounding performative space and need to identify their selfhood through actions such as in social VR.

> In the picture-frame theatrical setting, different *performative practices and processes* give audiences the illusion of being "inside the play." In immersive theater, audiences *co-perform* with the performers in all-surrounding theatrical environments that merge the physical with the mediated. Both settings require the *psychological sensation of make-believe.*

Immersive theater reframes the theatrical experience to allow design and performance to maximize the sensory experience of audiences. By abolishing traditional realism it finds its origins in Antonin Artaud's Theater of Cruelty. Artaud's approach was a sensory one. Sensory disruptions such as shock and confrontation were used to both distance and immerse audiences. Building up on his ideas, theater moved outside in the late twentieth century. Immersive events such as the concepts of "happenings" and "environmental theater" were eliminating the distance between audiences and actors completely. To enhance the immediacy of the experience, Allan Kaprow and Richard Schechner were on the forefront of an aesthetic movement that constructed multiple-focus environments in which audiences were able to move around and participate. Spectators were allowed to come and go as they pleased, take part in the event, or move back to the surrounding crowd that watched the performance. This idea has been adapted by the performative practices of digital culture, such as the organization of flash mobs to perform an act for a limited time and then quickly disperse.

Immersive theater groups are a progression of this movement. Punchdrunk, Les Enfants Terribles, You Me Bum Bum Train, and the German Rimini Protokoll's Situation Rooms are opening up the aesthetic scope of the performance, while yielding up questions of interactivity, liveness, and participation. These groups are immensely popular, and because they accommodate only small audiences per show, they often sell out quickly. Their productions are carefully planned and exercised, but still open to theatrical improvisation. They are full of meticulously real tableaux, *mise-en-scènes* that activate a sense of presence as soon as audiences enter them.

Les Enfants Terribles, for instance, an immersive theater group based in London, produced a modern take on Lewis Carroll's story about parallel worlds in its immersive show *Alice's Adventures Underground*, played out in the seemingly endless tunnels of the Waterloo Vaults. As their mediated selves, audience members take on Alice's position as they enter the abandoned tunnels of London's famous tube stations. James Seager, artistic director of Les Enfants Terribles, explains the strategy of the play in our "moments of immersion" interview: "In our story, Alice is missing, she is being banished from Wonderland, and the audience takes her place. Creating that as a piece of theater draws parallels to video games, because at the beginning you get a tutorial to learn the rules of the game: 'This is your role in this immersive piece of theater, and this is what you have to do.'" At the beginning, audiences need to make an experience-altering decision: "Eat me" or "Drink Me." By the design of gamification, audience groups are then assigned to one of the four decks of cards: clubs, diamonds, hearts, and spades. Not only does that already introduce a minimal level of interactivity, it further allows each audience group to have an entirely different experience as they go on. As the experience continues, groups are whisked through sets of various kitchens, playrooms, studies, and courtrooms, populated by the famous characters of the story. At one point the Mad Hatter even invites them to a deranged tea party. Within the framework of the 360° gaze, the self is identified through actions in the performative setting, but is not fully developed, to allow individual experiences. Participants are put in Alice's shoes, but they are not playing the role of a young girl. The experience was designed to be comfortable for audiences, as Seager illustrates: "We do not pick on them, but we also do not rely on them. We try to introduce different levels of engagement. So if someone wants to really go forward and engage with the actors, they can do that. But if someone wants to hang back, they can do that as well and still feel involved." In regard to this sense of agency, Myrto Koumarianos and Cassandra Silver explain: "Our sense of almost unlimited agency resulted in higher stakes for our own participation and performance—our implication in the action begged our accountability to ourselves, to our fellow spectators, and to the performers. We remained autonomous, accountable, 'real' in some sense inside the space; yet we were simultaneously drawn into an identically faced

**Figure 3.3**
*Alice's Adventures Underground*: the White Rabbit performs in London's Waterloo station. Reprinted with the permission of Les Enfants Terribles. © 2017 Fiona Porritt.

and therefore faceless community, into a fiction over which we ultimately had no control."[42]

The "fiction without control" is the link to the ecosystems of cultural industries. These experiences, as Lauren Rabinovitz points out, need to be understood in the context of new industrial technologies and leisure consumption in a wide range of paratheatrical attractions in popular culture in which audiences are disciplined by three-dimensional, immersive environments and their surrounding ecosystem. The spectacle in immersive theater becomes part of consumer culture in postmodern societies by focusing on spatial experiences, hyperstimulation, and heightened self-awareness.[43]

This can be witnessed in various examples, such as the immersive cinematic experiences of Secret Cinema. The UK company is devoted to recreating cinematic worlds of popular movies in physical spaces, populating them with characters from the movie, reenacting several scenes together with the participating audiences, and—as the ultimate finale—screening the film for everyone involved. Former theatrical movie sites included reenactments of *Ghostbusters*, *Star Wars*, *The Shawshank Redemption*, *The Grand Budapest Hotel*, *28 Days Later*, *Moulin Rouge*, and *Blade Runner*. For *Back to the Future*, for instance, Secret Cinema recreated the fictional town square of 1955's Hill Valley, including several nostalgic shops and the famous clock tower. In preparation for the live event, audiences were assigned a fictional self in the story universe of the film—for instance, a student in Hill Valley's high school. They only got to know the exact address of the location on the day of the event, and they were advised to appear in clothes appropriate to the decade represented in the movie, the 1950s. In addition, mobile phones and recording devices were not allowed at the event, so as not to distract from the immersive experience and spoil others outside by posting images and videos on social media. During the film screening, actors replayed the on-screen scenes simultaneously on the town square, including the first appearance of the DeLorean time machine, and Doc Brown's nerve-wracking stunt from the clock tower. Presence within these spaces is developed through the liveness of the event, but immersion was already established before the event took place. People buy tickets to these events, because they are already fans of the movies and because they want to feel part of the story universe in a three-dimensional space. Universal Pictures, the film studio that produced the *Back to the Future* trilogy, had been aware of the cultural and economic significance of the films and supported the

Secret Cinema's events, while screenwriter Bob Gale made a guest appearance on site. In addition, the immersive events took place in 2014, one year before a musical was originally slated to make its world premiere,[44] and of course one year before Marty McFly travels to the future of 2015 in the first sequel. The film series is a huge immersive franchise, and the liveness and immediacy of immersive theatrical productions helped to reinstall corporate consumer culture and brands into the minds of its audiences.

> Performative experiences are exploited by *cultural industries* to create popular paratheatrical attractions. In these physical, three-dimensional environments, audiences are disciplined by *new industrial technologies* and the *spectacle of entertainment franchises*.

Interactivity and collaboration, such as in immersive theater productions, are characteristics of contemporary media. They play significant parts in performance art in the time of digital technologies, mediated spaces, and digital entities/identities. In postmodernism, digital technologies often play a key role in the aesthetics, content, techniques, and delivery forms of performative practices to enhance interactivity and collaboration in spatial experiences.[45]

Several immersive works cross these artistic boundaries and intermingle with the fields of painting, architecture, theater, film, photography, sound, and image art. The first wave of popularity of immersive technologies in the 1990s inspired interdisciplinary artists to create interactive image spaces designed through CAVE (computer-aided virtual environment) technology to allow users to penetrate the image sphere, like a virtual panorama, and interact with the virtual environment as a performing entity. Notable examples of developing interfaces, interaction models, innovative digital aesthetics, and performative environments are the works of Maurice Benayoun, Charlotte Davis, Monika Fleischmann and Wolfgang Strauss, Jeffrey Shaw, Simon Penny, Victoria Vesna, Dennis Del Favero, Blast Theory,[46] and Ed Atkins. Their pieces are "shaping very disparate areas, for example, telepresence art, biocybernetic art, robotics, Net art, space art, experiments in nanotechnology, artificial or A-life art, creating virtual agents and avatars, datamining, mixed realities and database-supported art."[47] The spatiality of digital environments is designed to challenge participants, create and question different versions of their mediated self, and reveal perceptional processes as performative practices. In recent years, the design of performative spaces has flourished with smartphones and newer headsets. AR, for instance, is used to reimagine and redefine physical spaces through the use of mobile screens. 3D digital art pieces are placed at various locations in the world and compel viewers to reflect on the relationship between art and location. Immersive technologies are also entering into the mainstream gallery system through highly critical art that questions religion, society, and art itself. For instance, Christian Lemmerz's *La Apparizione* (2017) lets visitors experience a melting golden Jesus figure looming above their heads by wearing a VR headset. Immersive technologies are also used to reimagine established artworks by transforming them into spatial experiences. *Dreams of Dali* (2016) allows users to move through and explore Dali's 1935 painting *Archaeological Reminiscence of Millet's Angelus,* while Fredrick Baker chose the Viennese secessionist artist Gustav Klimt for his imaginative VR remediation in *Klimt's Magic Garden* (2018).

Spatiality and performativity are also able to challenge the perception of other mediated experiences. Throughout this book, I mention several examples, but the most striking one is the perception of music. Sound has always been a spatial experience. Music has to make its way through environments first before its reaches our ears. As soon as sounds enter our bodies—like water penetrates us during swimming—we enter the fluidity of sonic landscapes. Concerts, in particular, are sonic experiences as the environment—whether a concert hall, jazz club, or rock festival—has significant impact on how we consume the music.

The Icelandic artist Ragnar Kjartansson experimented with the spatiality of performances by asking the US indie band The National to play their song "Sorrow" continuously for six hours in front of an audience at MoMA PS1 in New York City. Throughout *A Lot of Sorrow* (2013), the musicians subtly adapted their performance as the hours went by and fatigue slowly set in. Both band and audience members have been transported into a mantra-like transcendent state after the song has been played repeatedly for more than a hundred times. In the end, singer Matt Berninger lost his voice and decided to let the audience carry on instead. Ragnar Kjartansson takes the repetitive structure of pop music to an extreme and transforms it into an immersive performance. In our interview for this book, the artist elaborated on immersion as a three-dimensional experience: "When you repeat something, it becomes like a sculpture, a landscape—and stops being a narrative. Immersive experiences for me are like portraits, like a three-dimensional painting. They develop different levels, such as a landscape. At one time, for instance, I was performing songs by Schubert for ten hours in a row and that influenced the performance significantly. It becomes like a ritual. When you are witnessing that as an audience, it is almost like you are in trance. You forget about the world outside. You forget about that this is art."

The history of sound spatiality goes back to "the floating spaces of telephony and radio, enveloped by the neo-organic and neo-romantic discourse of the 'global village,' mapped by the stylus of phonography and tape recording, and now ejected into the semiotic ocean of digital sampling."[48] Karlheinz Stockhausen, Max Neuhaus, Maryann Amacher, Edgard Varèse, and others have investigated sound spatiality in several masterpieces of musical modernism and installations. Their influences are still relevant to this day. For instance, in 2018 the Institute of Sound and Music in Berlin opened a pop-up performance dome called Hexadome to demonstrate "the significance and power of spatialization for sound in complete harmony with immersive visual art and architecture."[49] Within the planetarium-like

**Figure 3.4**
*A Lot of Sorrow*: for Ragnar Kjartansson's immersive installation, the American band The National played their song "Sorrow" continuously for six hours at MoMA PS1. Reprinted with the permission of the artist, Luhring Augustine, New York, and i8 Gallery, Reykjavik. © 2013 Elisabet Davids.

panorama construction, participants listened to sounds coming from 54 speakers positioned around the room that, in addition, were surrounded by visual impressions on six digital screens. Artists like Brian Eno, Ben Frost, Holly Herndon, and Radiohead's Thom Yorke reimagined their music in 360° installations and performances with both artists and visitors positioned at the center. Bands such as U2 and Bon Iver experimented on their respective tours with 360° multilevel stage designs for immersive sound, and with the site-specific implementation of so-called Hyperreal Sound with the help of L-ISA technology from French manufacturer L-Acoustics that increases loudspeaker resolution across the stage to deliver spatialized audio experiences at large concerts.

Recent innovations in digital media technologies also redefine our understanding of spatiality in pop music. For instance, Dolby Atmos promises to be a new surround-sound system. Already used in cinemas around the world, it "spatializes" existing sounds by remastering them. The technology no longer constrains sounds to specific channels, but specifies where an individual sound originates. Sonic perception is transformed into a three-dimensional environment rather being a flat canvas. The acoustic *atmos*phere is immersive, as it "surrounds, includes, involves, envelops, and gives forth both the qualities of the environment and the experiencing human . . . in order to stand as atmosphere, a spatial arrangement needs to be experienced or imagined into being."[50] This leads to a cinematic experience of sound, delivered through speakers around the room, which places the listener in the middle of a three-dimensional sound world. R.E.M.'s 25th anniversary edition of their landmark album *Automatic for the People* (2017) was the first major pop record that has received this treatment. One reviewer wrote: "Rather than the sense of being at the heart of the action, it's a more painterly effect. Highlighting a little-noticed percussive section here, or long-buried string arrangement there, a vocal harmony can hide behind the listener, or an anthemic chorus punch from all around."[51]

To "step inside" spatial audio is also the ambition of Google's project *Inside Music*.[52] In collaboration with popular podcast company Song Exploder, audiences are able to walk into a VR environment lining out the different layers of song composition. Vocals, guitars, bass, drums, strings, and other layers are dissected within a 360° panorama. Once users move closer to a certain channel, they will hear it in more detail. Indie artists such as Phoenix and Perfume Genius already contributed to the project.

> *Performance art and music* in postmodern culture are defined by *spatial experiences* and the changing role of audiences through *interactivity, co-creation, and participation.*

In recent years, popular artists such as Beck, Gorillaz, Muse, Squarepusher, The Weeknd, and Jean-Michel Jarre have increasingly incorporated immersive technologies into their performance art and music. Icelandic artist Björk, in particular, has proven to be one of the most influential voices in using VR to create a three-dimensional canvas for sounds. As a singer, Björk has a long history of embracing visual arts and digital technologies through music videos by the likes of Michel Gondry and Chris Cunningham, and educational apps in relation to her album *Biophilia* (2011). But her transition to VR art seems to be her most ambitious transformation. Within the last several years, and in combination with her album *Vulnicura* (2015), Björk released several VR experiences, testing the grounds between 360° video productions and fully interactive VR for HTC Vive. Björk's music inspired real-time, morphing environments and surrealistic galaxies that deliver her soundscapes in a towering, ominous atmosphere.

One of her most striking productions is *Mouth Mantra* (2016). The experience positions audiences inside an intricate model of Björk's mouth, while she is performing the song. The first-person perspective is a haunting view as the audience penetrates Björk's mouth and invades the sacrality of the performer. The sensory overload is designed to be a horrific spectacle, which at the same time creates an intimate experience between the artist and her audience. As the singer's mouth, teeth, and tongue are moving, they create an unsettling but moving atmosphere in which to listen to the song. Björk's immersive work in VR has been presented in the exhibition *Björk Digital* in several museums around the globe.

Another of the first major projects of mixed reality startup Magic Leap had to do with the spatial and augmented perception of music. In collaboration with Icelandic art-rock band Sigur Rós, the company produced an experience called *Tónandi*. The word is a made-up language, as often used in the band's lyrics, but literally could be translated as "sound spirit." By wearing the Magic Leap One headset, the physical environment of the listener gets enhanced by a multitude of digital layers such as lifelike organisms that move and react to the music. While audiences are listening to the

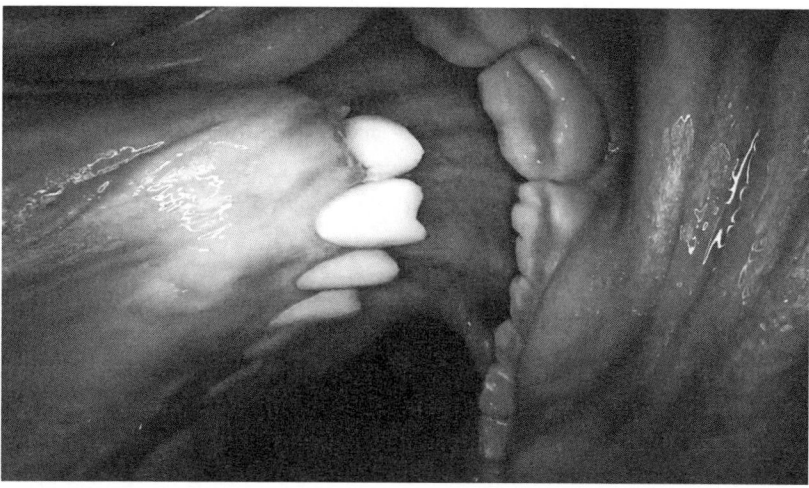

**Figure 3.5**
Björk's *Mouth Mantra*: inside the singer's mouth. Reprinted with the permission of Prettybird and One Little Independent Records. © 2016 Prettybird and One Little Independent Records.

compositions, they see dreamlike creatures and waves floating inside their actual living room. Unlike in AR experiences, MR allows users to interact with the digital elements. They are able to explore them with their hands, and reach out to change the shape of the waveforms. This causes not only a visual reaction, as the digital elements alter their positions and forms, but also a sonic transformation. Once audiences explore the visuals, the sounds react to them, and vice versa. One reviewer described the individual and realistic experience: "At one point, waving my hands near the table allows me to summon up the roar of lead singer Jónsi Birgisson's signature bowed guitar; in another corner, banging on virtual white puffy wisps releases different tones of what is unmistakably his voice. Whatever I'm seeing, it isn't just something pasted over my surroundings, but something that acknowledges those surroundings, and therefore seems more real."[53]

Quite a similar project has been attempted with the prototype for the "360 Cinema Concert" in Rotterdam. While a real orchestra was playing in the concert hall, the physical location had been augmented with digital images that could be seen through the Microsoft HoloLens. While watching the orchestra play Gustav Holst's composition *The Planets* (1914–1916), audiences saw a virtual night sky projected in front of their eyes. They could alter it by reaching out and changing the positions of planets and stars. That way they were able to create individual sonic and visual experiences.

Another augmented experience has been realized with *Fields*, a free AR iOS app that lets users create so-called "scenes"—different AR soundscapes and installations—in which users can create physical sounds in digital environments, record them, and then let them loop in different corners of their physical spaces.[54] In all of these experiences, audiences are represented in the form of a mediated self that is able to individualize the sounds and personalize their overall experience.

Immersive experiences such as *Tónandi* and *Fields* are descendants of Björk's and Radiohead's album-related apps (*Biophilia* and *PolyFauna*) and change the signification and perception of music. As audiences are invited to take part in the artistic process and the spatiality of the mediation, the lines between performers and spectators get blurred. Ultimately, the immersive experience challenges the audience's understanding of selfhood and creativity.

### *Moments of Immersion: James Seager*

James Seager is the artistic director of Les Enfants Terribles, one of the most innovative immersive theater companies in the UK. He has produced several award-winning productions, including the company's biggest show, the Olivier Award-nominated immersive piece *Alice's Adventures Underground*, which he co-directed and produced.

**How would you describe the experience of *Alice's Adventures Underground*?**

I suppose the one thing we are looking for is a kind of awe for people to lose themselves in the world. The whole point of immersive theater is to not only immerse audiences in another environment, but also in a story. When people go in, there is kind of a sensation towards the incredible set design, but then they realize they are part of the story. They get carried away and this is how we want it to be.

With immersive theater you always need to ask yourself: What is the audience's role, and is it clear? Because if that is not clear, you will not get engagement or total immersion. With *Alice*, we were very clear that the audience is essentially put in Alice's shows. When you are reading *Alice's Adventures in Wonderland*, you are experiencing Wonderland through the eyes of Alice. This is like a vessel for the reader to meet the mad hatter, the cat, and others. So at the beginning we have made the controversial decision to cut Alice from the show—but it was an important one and the right thing to do. In our story, Alice is missing, she is being banished from Wonderland, and the audience takes her place. Creating that as a piece of theater draws parallels to video games, because at the beginning you get a tutorial to learn the rules of the game: "This is your role in this immersive piece of theater, and this is what you have to do."

**What is the "tutorial" for the audience when they take Alice's place?**

Obviously that process of learning is very subtle. We do not give audiences an actual tutorial to read in advance. In *Alice*, the audience is divided into four groups—hearts, diamonds, clubs, and spades—and in each group you are experiencing a different storyline of Alice's world. In addition, you will receive tasks to complete. There is a level of thought process in it that involves all the actors, designers, and scriptwriters, but we always think about the audience and its role. The complexity behind it is immense, and if the audience understands some of it, it even creates another layer of engagement. What you should not forget is that audience members might not like to "perform." They might shy away from that. I have started out as an actor and even took part in immersive pieces as part of the audience. I hated it when I was asked to perform.

So we wanted to make sure that audiences are feeling comfortable. We do not pick on them, but we also do not rely on them. We try to introduce different levels of engagement. So if someone wants to really go forward and engage with the actors, they can do that. But if someone wants to hang back, they can do that as well and still feel involved.

**From your perspective, are there any long-term effects caused by immersive experiences in theater?**

People get really carried away, sometimes even too much. However, in our shows, I have the feeling that they always know that the characters they are talking to are actors. We aim for full immersion and believability, but I do not think people go for it 100 percent all the time. There is always that level of full acceptance missing that the performative world is real. I think audiences know that it is theater, but that does not detract from their involvement. There is a level of detachment. We had feedback that people wanted to go again and again.

However, we also have a clear beginning and an end. Therefore we are disrupting the feeling of immersion at the end of the show. VR is another thing, though. I can imagine we could totally get lost in another space like *Ready Player One* or *The Matrix*, even to a degree that we want to live in this world.

**As we see immersive experiences in VR and other digital media, will this be the way we experience mediation in the next years?**

I think it is the populist version of media. That is for sure. The currency of experience is taking over real currency. That comes from social media where people are showing off, and this kind of currency goes into what people want to experience. The whole thing of uniqueness and individuality is something we see across different media, not just theater. Think about binge-watching television and streaming music. The audience is a part of all of these mediated realities. That currency is quite big, and it will be the future of media.

## Moments of Immersion: Ragnar Kjartansson

Ragnar Kjartansson is a multidisciplinary Icelandic artist, celebrated for his playful, endurance-based performances and video installations. All of his works have a social immersive character, drawing audiences into live performances full of humor and sorrow. His works include *The Visitors* (2012), *A Lot of Sorrow* (2013), and *Me, My Mother, My Father, and I* (2014). He had solo exhibitions in the Barbican Centre London, the Palais de Tokyo in Paris, and the New Museum in New York, and he represented Iceland at the Venice Biennale.

**You have created multiple installations linked to performances. Audiences can walk through them, sit down, listen, and be surrounded by them. Why are you using immersion in your art?**

Immersion is like a tool for me, and as an artist you can use it. I have created many pieces that distanced audiences, but very often I go for the "come in and find out" approach. I got interested in immersion after reading about Karlheinz Stockhausen, his approach to spatiality, and how he created pieces that engage audiences. So I did this one piece called *The Visitors*. It is an installation with several video walls and you are listening to a song played by a band on that screens. As the audience, you are walking through it, sit down, lie on the floor, and listen to the emotional landscape that runs over you. The images do not even matter that much, because the music is everywhere. It is surrounding you. It is pure sentimental *schmaltz*, but damn, it can be immersive. The piece is not perfect, it is full of little mistakes—but that makes it so real for audiences. My good friend Björk once said that the best art pieces start out as a joke. And once you take the joke away, you have something really beautiful.

**Very often you are using repetition as a motif in your work. Once you asked the US band The National to play the song "Sorrow" for six straight hours live in the Museum of Modern Art in New York. Would you say repetition is an element of immersion?**

Absolutely. Immersion is a religious act. It is even inspired by religion, and if we are honest, religion is purely based on repetition and immersion. I look at it like an artistic tool as well. When you repeat something, it becomes like a sculpture, a landscape—and stops being a narrative. Immersive experiences for me are like portraits, like a three-dimensional painting. They develop different levels, such as a landscape. At one time, for instance, I was performing songs by Schubert for ten hours in a row, and that influenced the performance significantly. It becomes like a ritual. When you are witnessing that as an audience, it is almost like you are in trance. You forget about the world outside. You forget about that this is art.

**What do you think about Björk's VR work and the use of immersive technologies in art?**

I saw Björk's VR work, and I thought it was cool. These pieces work for a lot of people. But the thing is, when I am into virtual reality, I am so aware of reality. The only piece in that exhibition that really got me was called *Stonemilker*, the one in which she is on the beach in Iceland. I was totally blown away by it, because I know that beach in real life. That was mind-blowing, because that was the only piece in which she has used physical reality as part of virtual reality. The other ones were gorgeous, but they were always fantasy worlds.

I am obsessed with reality. I hate to play video games in which I am killing monsters. I love video games in which I am killing people. I am joking of course—but reality is an image as well. I remember vividly, before I was going to New York City the first time, I saw a lot of movies—the glamorous parts of it, as well as the drug and crime scene. Once I finally got there, it was exactly like in those movies. I had the feeling I knew this place so well. There were no surprises anymore. I knew it from made-up realities.

**What would you do with immersive technologies, if you had the chance to use them in one of your next projects?**

I would probably recreate reality to the point when the mediation would become totally ordinary and boring. You put on the headset and all you see is pure boredom. It is surely not a place you would like to go to, and no fun to be in. Like a parking lot. I think that would challenge a lot of people to think about these technologies and immersion.

It is like that Oscar Wilde quote: "In art, good intentions are not of the smallest value. All bad art is the resource of good intentions." I think you can say the same about immersion as well.

## The Extension of the Screen

In March 2018, Samsung announced a new generation of television flat screens: The Wall. Working with modular display technologies QLED and Micro-LED, Samsung's TV lineup is a mere 146 inches diagonal, ten feet wide, and six feet tall. Unlike traditional LED-based screens, QLED and Micro-LED use an array of millions of individual light-emitting diodes (LED). They make it possible to divide the screen in smaller portions to adjust its size according to the needs of customers. The screen can be as gigantic or small as costumers wish. But its size is not the only reason it is named The Wall.

What Samsung calls the "ambient mode" is a feature that lets the flat screen mimic the wall surface behind it to create a more seamless look. The

Wall is not only fixed to the physical wall of the room, but it reduplicates its actual color and texture through a virtual app. In addition, it is able to project images or display time, weather, news, and traffic information whenever the TV set is not in use.

The screen is not only attached to the wall. The screen *mediates* the wall.

Samsung aims for customers that are "mindful of the aesthetics of their space."[55] At the same time, the company changes the idea of audiovisual sensation that we have become accustomed to in the last decades. Wall-mounted TV sets are nothing new, but they are a symptom of a world that is focused on screens. Screens are rabbit holes: they are opening windows into mediated spaces. They are thresholds into mediated experiences, and they position audiences inside the *black mirrors* of their screens.

Look around, and you will find screens everywhere. We are spending hours per day staring at our phones, laptops, tablets, TVs, and computers. The architecture of big cities is merging with media art and visual advertisements through the use of gigantic screens, as can be seen in New York's Times Square or London's Piccadilly Circus. We are recreating screen-based versions of books and newspapers on digital e-readers and adding visual components to audio sensations such as music. Streaming music on Spotify and YouTube is always intertwined with a visual component that actually would not be necessary for consuming the music. Developments on the mobile-phone market such as Samsung's *dual edge display* for its latest Galaxy phones not only reposition the screen, they redefine the perception, usage, and knowledge of mobile mediation. In such a setting, the screen is no longer clearly framed and impenetrable, which would make it slightly positioned above the phone's frame. Rather, it curves over its boundaries, spills over the limitations of its container, and expands and redefines the phone's shape. It offers new aesthetics, spaces, and functionalities while eliminating some others. Dual-edge phones introduce a newly defined edge panel to access the most frequently used apps and to display text messages, news updates, and the weather forecast. In 2019, Chinese electronics company Xiaomi even unveiled the prototype for a fully immersive wraparound display that would surround the whole body of its Mi MIX Alpha phone. In addition, flexible-screen technology is used in the development of edgeless and foldable phones and allows superthin flexible displays to be used as wearable extensions on cars' steering wheels and dashboards, smart speakers, and even t-shirts, hats, and handbags.[56]

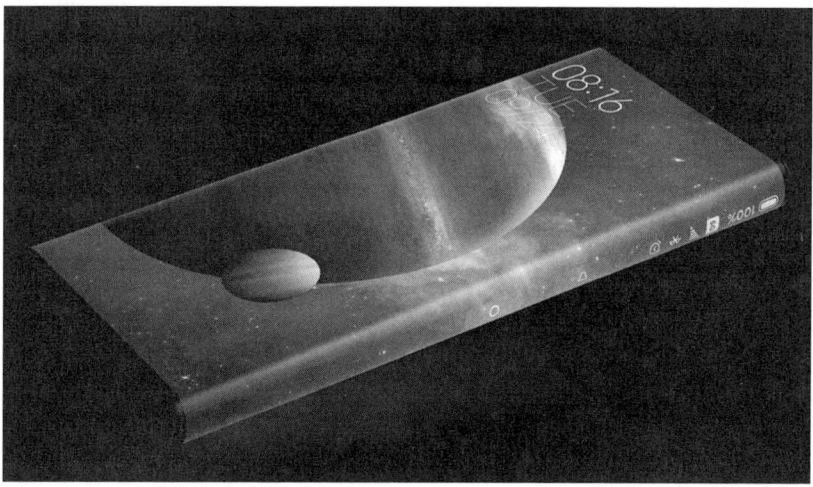

**Figure 3.6**
Xiaomi's futuristic Mi MIX Alpha phone: the screen curves over its boundaries and wraps around the whole phone. Reprinted with the permission of Xiaomi. © 2019 Xiaomi.

Screens are becoming more dominant in our lives as they merge with the physical surrounding. Very often they even blend with our own bodies. Screens on smart watches, phones, and VR headsets move closer to our eyes. As a result, our whole sensual and perceptual apparatus no longer recognizes the screens in front of us. Think about the reasons why we put our smartphones in cardboard glasses to access 360° video content. We are blocking out our physical environment, but ultimately we are blocking out the screens as well. They are positioned so close to our eyes that we cannot see their frames, textures, and limitations. That way, they filter and discipline our gaze and create a sensation of make-believe. Casey J. McCormick even speaks of "screen intimacy," a personal close encounter with technology: "The screen is likely closer to the viewer, perhaps even in her lap or bed, and this screen is the same one used for various forms of personal communication."[57] This is where Samsung's The Wall and other emerging screen technologies fit in. A display that merges with the physical surroundings does not make the screen disappear, but rather underlines that it is already inherent, always present, and always in use.

Beyond the screen, there is still immersion.

What I call in this chapter the "extension of the screen" refers back to a long history of visual sensations in screen- and image-based media. That means not exclusively the physical extension of screen sizes (although in some areas, such as telecommunication or entertainment, they do indeed get bigger), but it also refers to how rectangular screens and their content use up one's senses with sights and sounds, and how they define modes of perception and implicit social bonds. Béla Balázs recognized that in one of his early texts in 1938, when he articulated his thoughts in relation to the new emerging medium of cinema: "We no longer know how we have managed in these few years to learn how to see: how we have learned to make optical associations, draw optical inferences, to think optically, or to become so utterly familiar with optical abbreviations, optical metaphors, optical symbols and optical concepts."[58] Balázs described a change in the way visual images got produced and perceived. Legendary are the reports of how audiences supposedly reacted at the first screenings of *L'arrivée d'un train en gare de La Ciotat* (1896) by the Lumière brothers, fleeing the theater in fear of a train coming right at them. Tom Gunning and others have rightly claimed that audiences at the turn of the twentieth century could not have been fooled by that. They were astonished, not fooled. And what astonished them was that moving images could be so "real." That audiences were able to learn how to feel present inside the narrative of a film has a lot to do with configuring and disciplining their gaze and a specific cognitive understanding of moving images.[59]

The illusion of space and world creation have been influenced by the success of the awe-inspiring panorama, the developments in photography, and several prestages of the moving image, such as the stereograph by Eadweard Muybridge at the end of the nineteenth century. The disappearing separation between viewer and mediation can be found in many technological inventions that added depth to images, such as stereoscopic photography (1838), a setup later used as well in mobile VR solutions for cardboard kits. Other exemplary milestones in that field are the stereoscope (Charles Wheatstone, 1838), the lenticular stereoscope (David Brewster, 1849), and the View-Master (William Gruber, 1939). All of these devices placed the image closer to the spectator to activate feelings of involvement and presence. In 1877, an article in *Scientific American* envisioned a future with new technologies such as the stereoscope and the phonograph that would introduce an entirely new form of mediation: "It is already possible by ingenious optical contrivances to throw stereoscopic photographs of people on screens in full view of an audience. Add the talking phonograph to counterfeit their voices and it would be difficult to carry the illusion of real presence much further."[60]

But all of them were just preparation for cinematic images in film that were about to redefine audiences as the objects of control and immersion: "Today, however, there is no need to spell it out; we are all in the picture."[61] The immediate experience of watching a film creates a distinct version of the self within the mediated environment of the film space.[62] That self is different to any other phenomenological self that is created in immersive experiences. Time and space are shaped by the cinematic experience, and audiences are mentally pulled into the story as the screen acts like a kaleidoscopic space. Whatever abstraction might be represented, the screen's size and its role as light source are hypnotizing and blur the lines between the reality of the cinematic experience and the world on screen. For André Bazin this ultimately results in the design of a "total and complete representation of reality." Not as the extreme mimetic representation of physical reality, but rather as a fully perceptional and sensual cinema, the "reconstruction of a perfect illusion of the outside world in sound, color, and relief."[63]

There is a significant scene in John McTiernan's *Last Action Hero* (1993) in which Danny, a teenage boy, gets transported into the fictional world of his favorite action film series and is able to meet his hero, Jack Slater (Arnold Schwarzenegger). That scene materializes the perceptual sensation as Danny transforms into an immortal character in the film in the same way that Jack Slater becomes mortal when he later crosses the borders into Danny's physical reality. It is that liminal phase of thresholds, of stepping into the picture, which defines the mediated self in that particular moment. Scott MacDonald describes that blurred border between different spaces with the metaphor of the fog after watching the movie *Fog Line* (1970): "For a few moments at the beginning of this film, viewers cannot be sure that the image they're looking at is a motion picture. Indeed, it is only once the fog has thinned enough for an identification of the image to be possible that we can recognize that something other than the movie projector—the fog—is moving."[64]

The audiences of first cinematic public screenings were able to witness the new type of spatial experience that the cinema facilitated. The pure kinetic enjoyment of the filmic space provided a rhythmic intensity that was both fundamentally novel and astonishing at the time.[65] We might very well argue that immersive technologies such as VR and 360° video currently rely on the same kind of novelty as an emerging medium. But cinema is way beyond pure sensationalism and has developed a specific language for creating narratives, signifying meaning through setups, montage, editing,[66] and most importantly the camera, that "takes my eye . . . into the very heart of the image. I see the world from within the filmic space. I am surrounded by the figures within the film and involved in the action, which I see from all sides."[67] Balázs argues there is no viewpoint of the audience, but that the gaze identifies with the camera and therefore with the characters of the film. While this is often true, the camera mainly impersonates the gaze of the viewers, and gives them an unspecified point of view to make them feel part of the film space, even though they remain physically seated in the movie theater. Camera systems, dollies, and cranes, among other things, depersonalize the camera view and make it a neutral witness of events, with only one goal: to enhance the feeling of presence in the diegetic world. Karl Prümm adds in regard to the Steadicam system: "On the one hand, this results in a disembodiment and a deindividualization of the camera's gaze; on the other hand, the camera is more or less glued to the body of the operator, whose maneuverability and agility are exploited."[68] Different dynamics that emphasize the spectator's sense of motion, for instance, by disembodying the camera or moving the objects filmed, are what we could call first attempts of *aesthetics of immersion* in film. Besides different point-of-view shots, the unchain camera system has been established to draw the spectator into the scene. Unstable hand-held camera footage, that directly transfers body movements to physical inscriptions on the film material, is used to create intimacy and immediacy for audiences, as in horror films such as *The Blair Witch Project* (1999), *[Rec]* (2007), and *Cloverfield* (2008). In those films, characters are talking directly to the camera and carry a camcorder with them all the time—even in moments full of terror and shock—to establish a setting of factual reality. This is a formation of reality that makes the spectator both a witness to and a part of the filmic events. The camera represents the spectator. If the characters flee from the Blair Witch, we run with them, stumble in the pitch-black dark of the woods, and can hardly see our hands in front of our eyes.

There have been several attempts in film history to limit the view of the audience to a specific protagonist. The film noir adaption of Raymond Chandler's murder mystery *Lady in the Lake* (1947) is shot entirely from the viewpoint of its main protagonist, Marlowe, giving the audience an uncanny perception of events. The film begins with Marlowe telling the story in retrospect, but then the camera quickly dissolves into the action and ultimately takes the detective's viewpoint, a novelty that was not much more than an experiment. Seeing Marlowe's face through windows and mirrors made audiences realize who they are in the story, but it did not create a greater level of intimacy with the character. In fact, the rhythm of the narrative changed significantly through the subjective viewpoint, giving audiences no option for an information surplus. They simply could not know more than Marlowe, which worked fine for a whodunit detective story, but was not transferable to other films without losing essential story characteristics.

Only decades later, and with the success of first-person porn videos and video games, the protagonist viewpoint returned in films such the remake of the slasher film *Maniac* (2012), and *Hardcore Henry* (2015), an action film shot entirely with head-mounted GoPro cameras and innovative magnet-based stabilization systems. The film borrows from the aesthetics of first-person shooters and game design tropes as audiences take on the perspective of a prototype cyborg. The cyborg in the film acts like an avatar—no face, voice, no memories, and no need for complicated flashbacks that would destroy the first-person illusion. The structure of the film has been defined as "levels" and refers to the spatial design of video games involving the use of different challenges to keep the player's interest high.

While *Hardcore Henry* was an unusual experiment, postmodern film aesthetics—much like literature and the performing arts—began to deconstruct the rules of world-building. Contemporary film genres such as *mind-game films* are experimenting with the audience's expectations of time and space, withholding information, or presenting it in ways that break the illusion of immediacy in the story. Examples are Michael Haneke's *Funny Games* (1997), M. Night Shyamalan's *The Sixth Sense* (1999), Christopher Nolan's *Memento* (2000), David Lynch's *Lost Highway* (1997) and *Mulholland Drive* (2001), and Alejandro Amenábar's *The Others* (2001).[69]

---

Film has developed a *specific language* to create narratives and signify meaning through setups, montage, editing, and camera perspectives that create *the illusion of immediacy and immersion*. Postmodern film deconstructs the rules of world-building to challenge the audience's expectations.

**Figure 3.7**
*Hardcore Henry*: the visceral thrill of taking on a first-person perspective. Reprinted with the permission of Huayi Brothers Pictures, Bazelevs Productions, and Versus Pictures. © 2015 Huayi Brothers Pictures, Bazelevs Productions, and Versus Pictures.

Over the course of time, technological developments were mainly inspired by the physical dominance of the cinema screen and the spatiality of the experience—both often associated with the spectacle of cinema under the umbrella term expanded cinema.

An early influence of multisensory experiences has been Morton Heilig's Sensorama (1962), an arcade-style theater cabinet (or as he called it "Experience Theater") that stimulated all the senses, including sight, hearing, and smell. Heilig created five short films, all screened on a stereoscopic 3D display. While the participant was sitting on a vibrating chair, the simulation recreated the experience of being on a motorcycle. The road was projected on a screen in front, while the street noise was coming from different speakers placed around the head. The participant even smelled gasoline through the valves in the machine. Heilig wanted to deliver a full sensory experience for audiences by diminishing sensory input from the outside and activating human senses predominantly with the medium.

Spatiality in film, often discussed in relation to 3D technology, had been introduced in the early 1950s. The marketing poster of the 1952 action film *Bwana Devil* had already famously claimed: "The flat is gone! You—not the camera—but you are there! A lion in your lap! A lover in your arms!" The combination of spectacle, participation, involvement, and the (illusory) dissolution of the apparatus promised an immediacy that is still wished for today (e.g., the poster of *Spy Kids 3D* (2003): "You have to cover your eyes—to see it in 3D!"). Despite the commercial success of 3D films such as *Gravity* (2013) and *Avatar* (2009), the technology almost never added a substantial aesthetic quality to the narrative world. The illusion of letting the narrative space enter the physical space of the movie theater had often been used merely as a sensory trick: rollercoaster rides, crashing avalanches, gigantic waves, and snowflakes falling in the auditorium (*The Great Gatsby 3D*, 2013). For many, 3D remains a gimmick that punishes artistry in the name of the medium's requirements, and that molds stories to fit the technology instead of the other way round. Further developments to add other physical sensations such as seat movement (4D), special effects like wind, rain, and snow (5D), and interactive elements such as shooting with plastic guns (7D) do not necessarily enhance the sense of immediacy and remain visceral thrills and obscurities in shopping malls, arcades, entertainment centers, and theme parks. Interactive VR as a cinematic sensation (also sometimes referred to as 9D or "hyperrealism") could suffer the same fate as dedicated VR cinemas, which are often hidden in arcades and cafés and placed as side attractions in movie theaters.

Widescreen technologies, on the other hand, had more success in establishing themselves as mass-market technologies. The 1900 Paris Exposition introduced the Cinerama, a widescreen process that projected images simultaneously from several projectors onto a huge, curved screen. The technology had obviously been inspired by the spectacle of the panorama and the sensation of immersive spectatorship. It paved the way for commercially successful widescreen technologies such as IMAX (Images Maximum), with its giant, curved, semicircular screen. IMAX and predecessors of widescreen projections like CinemaScope offer audiences a godlike mode of vision, and it is no wonder IMAX films are primarily associated with documentaries, a genre that highlights the "real" and "actual": "Through IMAX's marketing hype about the immersive and mimetic qualities of the experience, we are invited to vicariously take up the challenge of performing superathletic feats such as extreme surfing or snowboarding; the implication here is that the experience of viewing the IMAX film will be so realistic that we will feel as though we are actually moving."[70] In her book, *Shivers Down Your Spine*, Alison Griffiths draws a comparison between IMAX technology and the history of the panorama. She argues that the visual spectacle in IMAX theaters is a form of immersive experience, as it is referring to "[t]he desire to *be* elsewhere without actually *going* elsewhere."[71]

The era of broadcasting domesticated the screen and created a form of mediated intimacy with distant and remote world events.[72] Television quickly turned out to be a window to the world. Audiences were able to gaze into the lives of others while remaining mobile in their own living rooms. The intimate relationship with television transferred a rabbit hole straight into their homes. Living room units were traditionally constructed with a place for the TV set; TV schedules fragmented the lives of viewers; and families gathered in front of their screens during prime-time slots to immerse themselves in alternative realities. With the rise of reality TV and casting shows such as *Survivor* and *America's Got Talent*, the world of the viewers collided with mediated representation. The world beyond the screen began to portray ordinary people as television personalities, made them part of the fabric of consumption, and reinterpreted the understanding of the public gaze. It was the beginning of mediated selfhood being commodified and accessible as content and data.

With the rise of digital technologies and mobile communication tools, and PCs and laptops in every work environment and home, screens transformed into access points to other realities that portray users, while at the same time act as the control desk for managing the digital self. Yasmin Ibrahim remarks on the relation of the self with one's personal screen experiences: "The screen no longer belonged to the powerful or the celebrity, it became a space in which one could view oneself and offer the self to others as an object and subject. . . . The inserting of the self into the screens, our ability to consume and narrate ourselves and equally to make ourselves content for consumption of friends and strangers extended the social importance of the screen in the digital age. The screen belonged to us and we started to belong to the screen."[73] Mobile screens establish an intimate relation with the mediated self and the digital publishing sphere to "re-cast with world events through the ability to partake in them through discussions, signposting, re-tweeting, blogs, image curation, etc."[74] Computers, smartphones, and the Internet have become part of screen cultures that are linked with film and television and modify the individual.[75] Screen technologies are now essential tools for several mediated practices in our lives. They are no longer merely rabbit holes leading to other worlds. We now need them to access our digital selves *beyond the screen.*

> *Media and communication industries* discipline *the performance of spectatorship* through widescreen technologies and the spatiality of the experience. Convergent and mobile media transform screens into access points for several mediated practices and *the curation of the digital self*.

In digital culture, screens influence the feeling of immersion. Their size is often designed to fill up the viewer's gaze, whether it is the screen in movie theaters or the smartphone display. At the same time, screens are a threshold between the physical and the virtual. For instance, Nintendo developed a cardboard kit called *Nintendo Labo*[76] to extend the possibilities of its game cartridge *Switch*. The device can be inserted into several DIY cardboard kits: a little house with cardboard buttons and a crank to slot in, a fishing rod, a motorbike, a piano, and a cardboard VR headset.[77] They all become fully functional once the device has been attached to them. For instance, the real string extending from the fishing rod transforms into a virtual string on the Switch screen to catch virtual fish. When the rod is tilted and twisted, the virtual string reacts. That way the screen does not only become the entry point to a mediated environment, but the user can participate in the mediation through DIY cardboard accessories to create a physical gaming experience. On its programming platform Toy-Con Garage, Nintendo asks its users to invent their own cardboard models in which the screen can be inserted, from cardboard guitars to cardboard guns.[78]

Another popular example is the success of the mobile AR game *Pokémon Go*. The uniqueness of the experience is the combination of two different realities through the use of AR layers and location-based gaming. The game asks players to be physically active in real-world settings in order to locate, capture, and battle in-game virtual monsters that are blended into the physical environment and are only visible through a mobile screen. Players need to visit physical sites called PokéStops scattered all around the world and explore different areas to display the locations of Pikachu and his friends. To catch a Pokémon, audiences have to throw a virtual ball at it.

Pokémon had been a market brand for decades, and that was certainly part of the AR game's surprising success. The appeal became immediately visible after its launch. The game gained over 20 million active daily users and more than 550 million installs in its first 80 days in the summer of 2016.[79] It defined AR for a broad public. Even though interest in the game faded again, it captured the attention of its users for a time. It shifted their

**Figure 3.8**
*Pokémon Go*: the AR game positions virtual characters in real-world settings. © 2016 Dalton White, CC BY.

attention to the game with such intensity that players had the impression that the screen-based mediation felt more immediate to them than their physical surroundings. Playing the game on busy streets has led to various accidents, injuries, and even deadly incidents because of players being distracted by the game.[80] Others felt so engaged by the experience that they were looking for Pokémon at different landmark sites, including even the United States Holocaust Memorial Museum in Washington, DC: "there were plenty of people inside the museum who seemed to be distracted from its haunting exhibits as they tried to 'catch 'em all,' as the Pokémon slogan goes."[81]

In screen cultures, digital technologies produce spatial, interactive experiences that create, curate, and modify the mediated self. Ultimately, they are becoming significant parameters in evaluating where we begin and where the mediation ends.

**World(s) of Warcraft**

Imagine sitting at a dinner table. You are staring at the plate in front of you. The food on it has been barely touched. Somewhere close you recognize voices. Someone is arguing. You are looking to your right and you discover two people sitting at the table with you, debating and shouting at each other. She is about to cry, he is hammering his hand on the table. A plate has been smashed against the wall. You notice the traces of spaghetti sauce on it. Her face is swollen up and blood is dripping down her nose. He must have hit her. She is taking a glimpse at you, and is faking a comforting look to ensure you that there is nothing to worry about. "Everything will be alright, sweetie, Daddy is just angry after work."

It might take a moment to realize that you are positioned in the body of a child, and that the two people are your parents having an abusive episode right next to you. You know nothing more about who you are yet, and what your "self" in the narrative is supposed to be. No name, no age, no gender. But the intimate, personal, and affective scene in front of you defines how you are put on the spot within the story. This is what will make you care about the characters, because you are one of them.

Under the influence of screen media and immersive technologies, a new kind of language for audiovisual content is emerging. Frank Rose calls it "deep media" in his book *The Art of Immersion*: "[S]tories that are not just entertaining, but immersive, taking you deeper than a one-hour TV drama or a two-hour movie or a 30-second spot will permit."[82] While Rose does not further elaborate on what he means by "taking you deeper" or how this is activated, he recognizes "the outlines of a new art form," while at the same time observing that "its grammar is as tenuous and elusive as the grammar of cinema a century ago. We know this much: people want to be immersed. They want to get involved in a story, to carve out a role for themselves, to make it their own."[83] Rose's book is about digital storytelling. Interestingly, it does not touch on new developments of immersive narratives in technologies such as VR. But the author uses the concept of immersion to outline the idea of liquid spaces in relation to storytelling. Rose recognizes that audiences approach stories differently through immersive media. Immersive technologies give them the opportunity to enter mediated realities, take on a role in them, and interact to some degree with the world and its (virtual) participants. This has implications for how we will think about representation, storytelling and spatiality, interactivity, and the role of audiences.

The foray of immersive technologies into the mainstream marketplace has made questions of presence, immersion, and agency even more important for contemporary storytellers who are grappling with the difficulties of narrative construction that the new medium presents. As with any new medium, it takes time to define its language, terminology, and functionality for storytelling and human experience. There is a lot of experimentation. The output celebrated at distinct "immersive" segments at film and media festivals such as SXSW, Sundance, Raindance, Sheffield Doc Film Festival, Marché du film Festival de Cannes (Cannes XR), Venice Film Festival, Vancouver International Film Festival, International Film Festival Rotterdam, Berlinale, and Tribeca Film Festival spans across genres such as cinematic VR (360° videos), live-action narratives, interactive installations, guided tours, empathy pieces, animated features, documentaries, and mixed reality pieces that incorporate elements of immersive theater.[84] Recent immersive productions are diverse, and they need to cope with rapidly changing technologies and high expectations from audiences. Many of them push boundaries of volumetric storytelling using photogrammetry and stereoscopic video capture techniques to create the perfect illusion of mediated realities.

When we look at immersive storytelling, we need to (re-)define the role of audiences, and the degree of interactivity in relation to the spatiality of the narrative presented, two topics that are familiar from the discourse on presence and immersion in digital games. Games are often viewed as shining proponents of cutting-edge virtuality that embody the alluring assumption that some other world exists on the other side of the screen. At the same time, games are associated with the opposite of seriousness and work and offer a rabbit hole into another space apart from the boring chores of ordinary, everyday life. In fact, games offer everything we expect from immersive experiences as outlined in the 360° gaze framework: a distinctive mediated self, specific semiotic codes and social activities, performative practices exploited by digital ecosystems, and psychological sensations that separate the physical space from the mediated world. Various successful games, such as the multiplayer battle royale game *Fortnite*, offer interactive storytelling and world-building experiences for millions of players every day.[85] Their narrative structure involves more than just repetitive mechanisms. Instead, they combine gameplay with engaging, interactive worlds, which are full of historical and mythological detail and populated with an emotionally realized cast of characters. Digital games often introduce story worlds full of rules, stakes, and architecture and let the stories emerge organically within the community.[86] MMORPGs (massively multiplayer online role-playing games), in particular, such as *World of Warcraft*, *EVE Online*, and *Final Fantasy 11*, provide cognitive, emotional, and kinesthetic feedback within the games process. At the same time they create hyperrealistic environments for players in which they can completely live their digital existences like in *Ready Player One*'s OASIS: "The average MMORPG player spends 22 hours a week playing the game. And these are not only teenagers playing. The average MMORPG gamer is in fact 26 years old. About half of these players have a full time job. Every day, many of them go to work and perform an assortment of clerical tasks, logistical planning and management in their offices, then they come home and do those very same things in MMORPGs."[87] Decision-making, problem-solving, and pattern-seeking are aspects of engagement within the game context that emphasize the similarity between games and a series of actions leading toward the resolutions of different objectives. That users find meaning in the mediated experience has positive effects. Feeling present in simulations such as *World of Warcraft* can influence cognitive skills and behavior beyond the mediation. For instance, job applicants who took on leadership roles in the game experienced positive effects from it when they performed their duties in senior management.[88]

The environment plays a significant role in the sense-making process within *World of Warcraft*. The creators have designed a landscape that appeals to a mass of users as they can roam endlessly without feeling the pressure to move forward. Players could actually spend most of their time in these environments, collaborate with others, and develop a sense of meaning beyond the game narrative, one that even leads to virtual friendships, virtual love, and virtual marriages inside the context of the game, and outside in physical reality: "Who knew a World of Warcraft subscription could deliver more romance than Match.com?"[89] But effects on cognitive behavior might not be always positive and could have lasting influences on the psychological profile of users, for instance, by fostering roles to manipulate others or by promoting the realness of sex and violence (especially virtual pedophilia and virtual rape). We still need more empirical evidence to find out to what degree certain thoughts and behaviors are encouraged by immersive experiences in VR horror and survival games such as *Paranormal Activity VR*, *Edge of Nowhere*, and the VR version of *EVE Online*, *EVE: Valkyrie*. Very often they are also part of an already existing transmedia franchise such as the VR game *The Walking Dead: Saints and Sinners*.

Most VR games like *Orbus VR* are taking on the idea of world creation and "living stories," as it is known in immersive theater. Users are inside an animated fantasy world, can communicate with others through the microphone in their VR headset, and need to collaborate in teams to complete certain tasks. At the same time, they can also simply *exist* in these worlds, have conversations, start a workout session, or go fishing.

Oculus, HTC, Sony, and Google use (multiplayer) games as entry points for audiences into the VR segment, but they also design experiences without narratives. Sony's *Perfect* (2016), for instance, is a VR experience for PlayStation VR that lets users escape into three relaxing locations: Northern Lights, Tropical Beach, and Mountain Wilderness: "Whether you need a quick five-minute experience to show the family on Christmas Day or want to immerse yourself for a whole evening, Perfect lets you leave behind the daily grind and slip away."[90] Sony even recommends users put some relaxing music on Spotify or enjoy the in-game radio while drifting off into a supposedly relaxing location like a tropical beach. By developing a feeling of presence on the VR beach, users are practicing their relaxation techniques with the help of technology. Similar to listening to music during meditative practices, gazing at a virtual setting with digitally created beaches, sunshine, and soothing ocean sounds tricks the mind into achieving calm and focus. *Perfect* wants users to feel relaxed even after they have taken off the headset and ended their meditation session. The illusion of the virtual environment aims to support physical and mental relaxation and is immersion beyond the mediation.

Another example of a nonnarrative in the form of an interactive VR installation is *Treehugger: Wawona* (2017).[91] In addition to using the HTC Vive, users need to strap on a SubPac vibrating backpack and hold two handheld controllers that are mounted on fingerless gloves. In the experience, users approach a giant cavity of a foam sequoia tree and are able to enter it, while the installation is blowing a woody fragrance at their faces to stimulate other senses. *Treehugger* is a marriage of technology and nature. It is a digital representation that, in this form, allows audiences to touch, feel, smell, and perhaps even understand the life force of a giant sequoia tree. It is about the feeling of embracing a symbol of life, and just like physically throwing your arms around the trunk of a tree, the virtual experience is supposed to calm you down, lower your heart rate, and rebalance your stress level.

As soon as one wants to establish a narrative in immersive media, though, two scripts have to be developed: one for the actors, and one for the audience. The concept of spatial storytelling requires the spectator to be inside a three-dimensional diegetic space, and it involves the hybridization of film with other storytelling formats like digital games and immersive theater. With the launch of 360° cameras at a reasonable price, and the massive success of game development programs such as Unity and Unreal UDK, multinational corporations like Disney, Facebook, and Google, as well as storytelling enthusiasts around the world experiment with audience engagement and the spectator inside the story. They investigate various formats, from panoramic images in 360° videos and volumetric photogrammetry to high-end image generation in real time through games engines. With Facebook and YouTube adding 360° video channels to their platforms, interest in the spatial format has grown rapidly. But the new storytelling world is still in its infancy and the audience's expectations are likely too high at this point. On one hand the new technology poses fundamental issues for the production. Not every 360° camera is able to capture surrounding images, which ultimately makes it difficult to develop a feeling of presence when half of the image is blacked out. In addition, the amount of data captured for a 20- to 30-minute video is immense and needs to be stitched in postproduction—often a long and painful process.

On the other hand, questions about composition, narrative structure, and visual grammar in immersive storytelling are often related to what is called the "interactivity paradox": "the integration of the unpredictable, bottom-up input of the user into a sequence of events that fulfills the conditions of narrativity—conditions that presuppose a top-down design."[92] The challenge remains how to balance a preauthored story with elements of interactivity. VR storytellers learn a lot from the intersection of film and video games,[93] and the convergence of gaming and cinematic storytelling has become an intense focal point.[94] Yelena Rachitsky, executive producer at Oculus, calls this the yin and yang of immersive storytelling: "Are you centered in your embodied experience and emotional engagement of a story (yin)? Or are you centered in your head of thinking about the strategy of your next action in achieving a goal in a game (yang)?"[95] She thinks of immersion as a state beyond the mediation, hoping it is "cultivating an experience that you have, and it's about the story that you tell yourself after you take the headset off."[96]

In 360° videos, the user's representation in the immersive experience is most commonly passively incorporated in the story: as a person, who is present, but not recognized by other characters of the narrative ("0th person"); or as a person, who is present, and is recognized by other characters of the narrative, but is not able to interact with them ("3rd person"); or as a person, who is present and also defined as a character, who is recognized by other characters, but not able to interact with them ("2nd person"). The most interactive approach is an emergent position inspired by games design ("1st person"), meaning the user is a character in the story and able to direct the narrative to a certain degree, as well as interact with other characters or users. This high-end VR experience is typically nonlinear, sensory, environmental, cooperative, connective, and community-driven. Paul Moody explains: "This is where the distinction between VR and 360° film becomes most apparent. While the producers of VR as a computer-generated form are right to seek to break down barriers between the user and the narrative, the viewers of 360° film expect some degree of pre-existing authorship, and appreciate that they cannot interact completely with the characters."[97]

Recent 360° cinematic experiences are often part of bigger story universe to reintroduce audiences to already well-established characters of film and TV shows. There is hardly a successful franchise that has not been further exploited through 360° cinematic VR, from *Star Wars* to *Blade Runner*. For the popular TV show *Mr. Robot*'s second season, USA Network managed to reactivate the show's audience with a transmedia marketing campaign using Facebook Live and various other social media platforms. *Mr. Robot VR*[98] is a 13-minute VR narrative that has been introduced as part of the campaign and was premiered at San Diego Comic-Con 2016. The audience (in a "0th person" perspective) experiences a flashback journey with lead character Elliot as he remembers an early encounter with his dealer and love interest Shayla. At the first global VR simulcast, fans were invited to a baseball stadium to watch the experience together with the cast at a specific time. Right after the simulcast the content disappeared from the Within app, the VR startup that distributed the experience created by VR production company Here Be Dragons. After that, fans were only able to access the production on specific VR platforms such as Google Daydream VR.[99]

> The *semiotics of immersive storytelling* is still a field of experimentation. It currently blends storytelling techniques from film and digital games by introducing a first-person perspective with emerging storylines. Audiences *perform a role in the story* and, depending on the technology, can interact with the environment and other characters.

Standalone 360° productions also gain more and more interest in both fiction and factual genres. *Dinner Party* (2018),[100] for instance, is based on the real case of Betty and Barney Hill, which led to the first reported UFO abduction in 1961. In 15 minutes, the viewer is a silent witness ("0th person") of a couple struggling with evidence of extraterrestrial life. Usually 360° video would require a static perspective, but innovative moving-camera work gives audiences the feeling that they are gliding and floating through the 360° scenes to reflect the movements of an UFO.

360° photorealistic images work well within the documentary genre and journalism. The *New York Times*, *National Geographic*, *ARTE*, and the *Guardian* have created multiple pieces and apps to access immersive documentaries. Productions such as *Clouds over Sidra* (2015), about the Syrian refugee crisis, and *Limbo* (2017), about waiting for asylum, are unframed experiences. Unlike a camera that frames the scene, in these instances the direction of the viewer's gaze is implied by creating sites of attention with the filmed subjects and sounds (also called "forced perspective").

Lynette Wallworth's *Collisions* (2016)[101] deals with the translation of a first-person perspective and the capacity to explore memories through immersive technologies. The Emmy-award-winning piece by the Australian director is a thought-provoking 18-minute VR production, which transfers the viewer into the perspective of the Aboriginal elder Nyarri Nyarri Morgan. For many years, Nyarri lived in the remote Pilbara desert in Western Australia with no knowledge of or contact with Western culture, until Great Britain started nuclear tests in the 1950s, quite close to where Nyarri and his tribe lived. It was his first contact—but foremost a shocking one—with Western civilization, science, and destructive technology. He kept his personal recollections and memories of these events to himself for over half a century until Wallworth translated his perspective into a 360° video production, which places the audience in his viewpoint. The result is a personal and intimate experience for viewers. Any distance to the mediation disappears, as Wallworth explains in our "moments of immersion" interview: "These new synaptic connections and indelible memories leave a longer trace. The trace of that experience, because we felt present, hold with us as a memory. Not as something we have viewed, but something we were in, we were with. Where that sits with us cognitively is completely different. It impacts us in a different way. We are no longer separated from it."

Another example is *Notes on Blindness: Into Darkness* (2016),[102] a VR experience that complements the British documentary *Notes on Blindness* (2016) about John Hull, a theologian who becomes blind and records the experience of losing his sight in audio diaries. His original tapes can be heard through headphones while the VR piece reimagines the experience of becoming blind. The acoustic space consists merely of Hull's voice and sound effects such as wind, rain, and footsteps. It acts as a guiding principle, while most of the visuals are evocative of visual impairment and only reveal silhouettes of a shadow world of blue and white shapes.

As soon as the element of interactivity is introduced to the narrative, the crossover with games is established and the self is an active participant within the story that is unfolding. This is where the experimentation comes in. Recent strategies follow a transmedia approach to integrating VR strands in an already existing story universe.[103] One of the first major productions in that field was the futuristic murder mystery *Halcyon* (2016),[104] designed by Canadian studio Secret Location for the US TV channel Syfy (part of NBCUniversal). The series included ten episodes regularly shown on TV, but interspliced with five interactive VR episodes accessible on Oculus Rift and HTC Vive. In the VR episodes, audiences took on the role of one of the detectives in the TV show to gather more clues and information about the murder in a gamelike setting in VR. They could explore the crime scenes in detail and obtain backstories to the characters in the show. Similar productions in the field simply recreated visual elements known from films in 3D spaces, such as in the VR production *Ghost in the Shell* (2017) for Oculus Rift.[105] The experience does not give anything away from the movie, but lets audiences jump off a skyscraper wearing a hip-hugging thermoptic camouflage suit, part of the movie's famous opening scene.

High-end VR is currently most effective, though, when it is targeting aspects of human empathy. VR is already used in treatments of disease, damage, and health-related issues, especially mental health, for example acrophobia, paranoia, fear of flying, PTSD, and even depression. It is also effective when attending to individual and collective trauma.[106] In Jordan Tannahill's *Draw Me Close: A Memoir* (2017),[107] the director is coping with the terminal cancer diagnosis of his mother by letting audiences experience the relationship to her in a VR installation. After putting on the VR headset, users see animated versions of Tannahill's childhood room and his mother. At the same time, an actress is playing "Mother" in an immersive theater-like approach in the physical environment. She is touching the user, holding him, drawing with him on the floor, and tucking him into bed. All of that is mixed with animated images of the mother through the headset. This creates an intimate setting framed by technology and visual aesthetics. It allows users

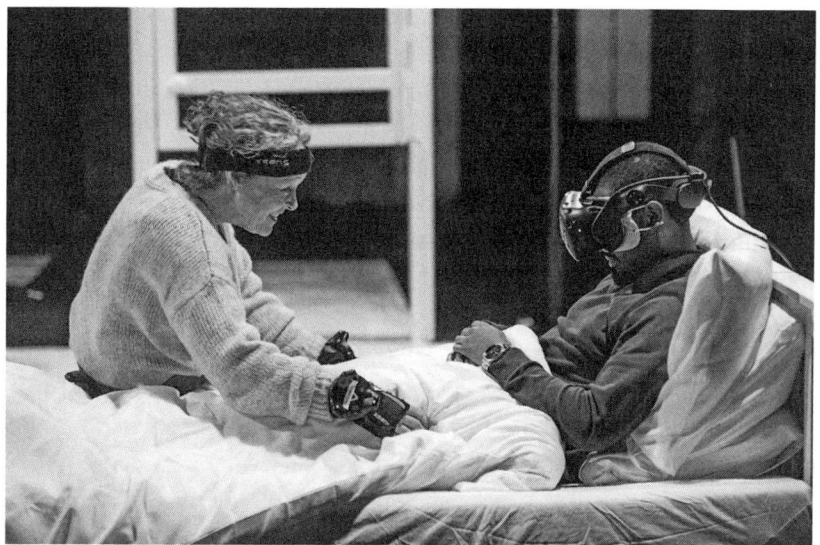

**Figure 3.9**
*Draw Me Close*: getting tucked in by "Mother" in VR. Reprinted with the permission of the National Film Board of Canada and the Royal National Theatre. © 2019 Ellie Kurttz for the National Film Board Canada and the Royal National Theatre.

to project images of their own mother onto the animation to cope with fears of illness and mortality. In our interview for this book, Tannahill is aware of the high ethical standards artists should apply when working with VR: "It is an immense responsibility for artists to consider ethical obligations to audiences. There is an added imperative with VR to take into account that a lot of audience members have never experienced these technologies before. VR is still this incredible novel form and a lot of audiences are still learning to see, walk, and negotiate through this virtual space."

> Immersive VR productions are able to trigger *strong physical and mental reactions*. They often position the audience in the middle of *empathy-related experiences* in a variety of fictional and factual genres produced by an *emerging industry of content creators*.

Another striking example is *The Last Goodbye* (2017),[108] a VR visit with Pinchas Gutter, one of the last Holocaust survivors, in the Majdanek concentration camp in Poland. The 20-minute-long experience combines 360°

video capture of Gutter within a photorealistic room-scale experience, so that audiences are able to explore the concentration camp on their own while listening to the emotionally harrowing memories of one of its last survivors. One of the reviewers describes the experience as an "emotionally painful journey": "It'd be one thing just to see these rooms in VR, but hearing Gutter's recollection from childhood makes it all the more tragic and meaningful. When he reveals that he can no longer remember anything about his twin sister, aside from her golden braid, it's hard not to tear up alongside him."[109]

*The Last Goodbye* has been called a "VR Holocaust Memorial," but unlike other memorials, it lets users experience, rather than merely visit, a version of that reality. The ethical implications of such an experience and the first-person testimonial of an actual Holocaust survivor are immense. What will happen in a few decades though, when all of the survivors have passed away, and VR experiences are all that remain to let us experience the horrors of the past: "You're not just seeing something displayed on a screen, or imagining it from a book; in some sense you're sharing an experience."[110] Who will be in control of deciding what version of individual and collective trauma we will experience in the future?

The future of immersive storytelling will further explore the mediation of the self in relation to new emerging and convergent technologies. In a few years time, audiences will be able to step into a holodeck experience full of virtual live-action role plays and will be surrounded by AI characters that personalize their experiences with in-depth psychological exploration within a merge of AR and VR technologies. Storytelling will once more become a transformational experience. One major concern, though, is that this could potentially influence how immersive media would contribute to our cultural memory, and that it could possibly damage the mechanisms of being immediate to our self and the world around us.

*Ready Player One*, all over again.

### *Moments of Immersion: Jordan Tannahill*

Jordan Tannahill is a Canadian theater director, author, and filmmaker. His VR experience *Draw Me Close: A Memoir* (2017) has been widely praised as a mix of live performance, VR, and animation. The 15-minute piece explores Tannahill's experiences in the wake of his mother's terminal cancer diagnosis.

*Draw Me Close* **is a mixture of immersive theater and VR production. In what way does VR enhance the experience of an immersive theater production?**

I was approached by the National Theatre (UK) and the National Film Board Canada that invited theater artists to a lab to explore VR within a performance context, trying to figure out whether or not some of our unique vocabularies and mythologies as artists would unlock something in VR and vice versa. So I went through this two-week crash course. In a way I was very skeptical at first, because the content for this technology felt like it was still in its proto-phase, almost proto-cinematic in a way that the early Lumière brothers films were or early documentaries that were just grounded in the idea of "being there," the pure spectacle without uncovering the layers of meaning that come with it. So these VR productions did not have any specific aesthetics, storytelling, or in-depth character studies. I was really underwhelmed. Theater on the other hand is such an ancient and timeless medium. It can do so many things much better than a lot of different technologies.

I had three non-negotiables to make a theater/VR piece. Those were a sense of liveness and real-time experience, a sacred relationship between the piece and the audience, and also a sense of corporal reality—for me as a theater maker, performances and voices are essential, and so is the audience. The physical presence as an audience member needed to be acknowledged even if the audience wears a VR headset. So I have created this very personal work of mine about the relationship with my mother in the wake of her terminal cancer diagnosis, which happened not too long before.

**It has a lot to do with how you translate your memories of this relationship to a mediated form. How does cognitive memory change when you pack it into a mediated form and create a different sort of perception of it?**

It becomes a new product, a new memory, even, in a way. It is refashioning the truth. It probably uncovers an even deeper truth. *Draw Me Close* incorporates many different layers: my memories of the actual event, and then my communication of these memories to the artists and VR technicians involved. This process rendered and nuanced my experience to make it accessible to an audience. It becomes this standalone piece that is divorced from my own experience. It is still very specific for me, but also universal enough for the audience who does not know my mother or does not know me.

**By mixing elements of immersive theater and VR you have a physical actress playing your mother in the piece. How do you create intimacy through technology that way?**

Going to a VR theater at the moment is like a shitty first date, because it is not a collective experience. You may not get laid after it. The fundamental attraction of theater and cinema is this communal reckoning, the ritual of negotiating the fundamental issues of our existence. It was not so much of an issue with *Draw Me Close* as it is a one-to-one piece. It was inspired by the late UK performance artist Adrian Howells, who used personal touch to create intimacy within his performances. So in *Draw Me Close* it was interesting to see how various actresses performed the role of my mother in various settings at different festivals. Their background is theater, and they were saying that VR reminded them of mask work, so that the audience member wore that physical mask with the Oculus Rift or HTC Vive headset.

On the other hand, the face that the audience member is seeing of the actress is a digital mask through animation. Her face does not change much through the whole experience, as communication is mainly delivered by the tone of her voice and touch. Masks are such an ancient and ambiguous sort of technology themselves, if you will, so the audience members were able to project their own experiences. In a way the VR headset created the same sort of intimacy as it portrayed an image of my mother, which allowed the members of the audience to project images of their own mothers onto it.

**How did audiences react after seeing your piece?**

It was pretty intense. A number of people would come out quite moved and were crying at the end of the piece. Actually, we have realized that it was irresponsible for us to inject that piece in that kind of trade fair, arcade-style environment of these VR festivals. In Venice, for instance, we have created this kind of chamber, where you could sit for a while and gather yourself after the exploration. It is that kind of understanding that mediated experiences do not end. They are transported into the physical reality. It is an immense responsibility for artists to consider ethical obligations to audiences. There is an added imperative with VR to take into account that a lot of audience members have never experienced these technologies before. VR is still this incredible novel form, and a lot of audiences are still learning to see, walk, and negotiate through this virtual space.

**If corporations such as Facebook create platforms within a three-dimensional digital space, does that mean that memories do not necessarily need to come out of physical reality?**

There are always different layers of reality that will further intersect and coincide with one another in the future, and they each offer us different possibilities for interaction and personal reckoning. I am sort of cautiously enthusiastic about the potential this holds, even though there are many things we need to safeguard. I still believe, though, that no virtual space can incorporate the values that I hold dear: democracy, love, art. They all can exist in the virtual space as long as the best minds are involved in its development and stay in conversation with one another. We should not allow any one shareholder and tech innovator in Silicon Valley to purely dictate the terms on which this space is created.

## Moments of Immersion: Lynette Wallworth

Lynette Wallworth is an Australian filmmaker and artist well known for her innovative use of emerging technologies to tell powerful nonfiction stories. Her works include the interactive video installation *Evolution of Fearlessness* (2006), the full dome feature *Coral: Rekindling Venus* (2012), and the VR narratives *Collisions* (2016) and *Awavena* (2018). In 2017, she has received a News & Documentary Emmy Award for Outstanding New Approaches to Documentary for her work on *Collisions*.

**What are you particularly interested in when it comes to immersive art, especially immersive technologies?**

I have been making immersive works since 2001 with different forms of immersive technologies. I would always use the most appropriate technology for the project. So I did work with AR before, interactive installations, as well as VR now. I had heard the fundamental story, which *Collisions* is about, probably three years before I did the work. At that time I was also looking for the technology that would be able to deliver that sensation.

So in that sense, it is about that pre-contact Aboriginal man named Nyarri Nyarri Morgan, who has seen something of Western culture that, in his world, he has no reference for—Britain's nuclear tests in the 1950s in the Australian desert. He waited almost his entire life to share that story. The challenge was to capture that abstraction in that moment and what that would feel like. Initially, I thought about doing an interactive installation first, but it remained a parallel frame that was separating me from Nyarri's direct experience. VR offered the capacity to see what he saw from his perspective as it places the audience in his viewpoint.

**How did you choose the appropriate storytelling language in VR to tell that particular story?**

It was quite intriguing that there has not been a preexisting language in VR. When there is no preformulated path in front of you, the capacity to move into that space is even intensified. I was not able to look at things that have been done before to reference them. In fact, all I had in front of me was this open space in which I had to navigate and find my way. That particular challenge suits my process. When you put that work in front of people for the first time, you are creating a synaptic connection, because the brain is experiencing something new and the memory is more indelible.

With Nyarri, it was not really as difficult as you might imagine. I had previous experiences with his tribe and acted as a translator, if you like, to transfer their worldview. Interestingly enough, everything in their world is linked to place and to country. Being in a country for them is an intensively, continuously 360° way of accessing knowledge—topographically from above, and internally and intimate at the same time. That is the way they see their world. VR was the technology to go for as it was exactly made to deliver the way they see the world. When I put Jaunt's state of the art 3D 360 camera in front of Nyarri, he said, "Oh it has 16 eyes!" He and others in his community could absolutely conceptually understand what that camera was seeing. That translation made more sense for him, probably, than the interactive installation I had in mind before.

**Would you say that VR storytelling always needs to consider the spatial dimension?**

Indeed, absolutely. Because of the previous work I did with the community there, they practically put me on a three-year training course to make me understand how they see their land. They took me on a hunting trip. They made me witness how they create paintings layer after the layer. It all made me understand what to translate to VR technology. And more importantly, that makes it personal and intimate for the audience. *Hold: Vessel 1*, a previous work of mine, contained a glass bowl with projected images inside for the audience to hold in their hands. That work had shown me how immersive experiences work, because it was personal. Someone walks into a room, and they are asked to hold a glass bowl under a beam of light. There is a response between the work and the participant, because the viewer is inside the work. The work is only activated by the viewer.

The complexity between the viewer and the work makes it personal. You are inside the work, it is responding to you in some ways, but you feel yourself to be present. There are multiple forms of immersion, for instance, when cinema started, people had never experienced that before and different synaptic connections in the brain were triggered. It was this parallel screen and the brain needed to process things that it had never processed before until it cognitively understood what it was. It had been a parallel world that was separate from everything else. But unlike in immersive media, as a viewer I am not in it. The personal is what makes it intimate and immersive.

**The concept of selfhood seems to be quite essential and the idea that we are penetrated by very personal and intimate mediated experiences.**

Absolutely, and that is exactly how I think about immersive technologies. These new synaptic connections and indelible memories leave a longer trace. The trace of that experience, because we felt present, stay with us as a memory. Not as something we have viewed, but something we were in, we were with. Where that sits with us cognitively is completely different. It impacts us in a different way. We are no longer separated from it. My work is driven by the desire to create a cultural point of discussion around significant issues like the rights of refugees or climate change. Those issues seem to be very often outside our personal experience. And how can we shift that? By making them personal.

**That sounds like an immense ethical responsibility for VR producers, especially if they are representing big corporations.**

When I have heard Mark Zuckerberg say that VR allows us to go everywhere and do anything, for me that was a horrifying statement, because there are no ethics, formalities, or protocols attached to that. There is none of the language of cultural meeting that develops over generation upon generation. None of that is embedded in his statement. It was just about access and availability, which is very much in line with resource extraction of multinational corporations. There is responsibility when you let audiences put on a VR headset. It acts like a portal. It carries your mind and body to another place. Colors and sounds are intensified. What you are talking about, and when I think about social VR and Facebook, indicates how powerful the form is.

I want to offer the possibility to offer these forms to other individual voices, who are able to activate other imaginative capacities within human beings, and not restrict and formalize them by synthesizing everything under a multinational umbrella. If these multinational corporations are now going to decide that the resource they want is our imagination, and what they extract is our capacity to dream, then I am going to stand in a different position.

# 4  Sociocultural Immersions

> When the historians find us we'll be in our homes
> Plugged into our hubs
> Skin and bones
> A frozen smile on every face
> As the stories replay
> This must have been a wonderful place
> —Father John Misty, "Total Entertainment Forever," *Pure Comedy*, Sub-Pop, 2017

## Un/masking the Self

"Why do you look like Justin Timberlake?" wonders a slightly confused Priscilla Chan after making a video call to her husband in Facebook Spaces. And indeed, Mark Zuckerberg's three-dimensional self in the demo of Facebook's social VR platform uncannily resembles the famous pop singer. Or rather what we would imagine Timberlake's VR avatar to look like: curly hair, slim torso, and the baby face. If not Timberlake, it could as well be Coldplay's frontman, Chris Martin. Zuckerberg's VR avatar has been designed to match his physical appearance, but at the same time it remains ambiguous, vague, diffuse, and suggestive. The other VR avatars around him also appear quite generic and synthetic. Their mimics and gestures are way over the top. We are only able to see their torsos, heads, and arms. Below the waist, there is nothing. No legs, no feet, no genitals. The upper half of their bodies is floating in midair, like anthromorphic marionettes. The virtual experience does not look like our physical world. Rather, it resembles a 3D cartoon, a Pixar world. Putting on a VR headset is like becoming a character in an animated movie. But an animation, like any mediation, is its own formation of reality.

How VR avatars are currently designed depends a lot on the users' psychological reactions while interacting with them. Research over the years has pointed out that "realness" in relation to digital avatars is not necessarily linked to their photorealistic appearance. The term photorealism has been used for decades to designate visual experience as it is captured by the photographic and film camera. Each advance in projection, lighting, and shading tried to bring computer-generated imagery closer to physical appearances. However, close similarity makes computer-generated creations look uncanny. This ambivalence is often described as the "uncanny valley," a term coined by Japanese roboticist Masahiro Mori in 1970 as he was observing people's negative reactions to android designs.[1] Mori created a valley-shaped graph that visualized his findings that humans are less accepting of robots as they become more human-like. He noticed a dissonance between the human-like appearance and their behavioral fidelity. For instance, skin tone and fingers of prosthetic hands may seem similar to human hands, but their mechanical movements might lead to feelings of unease. Similarly, if VR avatars resemble humans but not perfectly, they appear more threatening than if they looked like straightforward animations.

The "uncanny valley" phenomenon goes back to Sigmund Freud's seminal study "Das Unheimliche" (The Uncanny).[2] In his text, Freud refers to literary examples such as E. T. A. Hoffmann's short story "The Sandman"—a narrative full of robots and *Doppelgänger*—to describe the cognitive dissonance in deciding what is real, unreal, alive, or dead. The German word *un-heimlich* captures this eerie feeling as the prefix *un-* lets the meaning of *heimlich* (familiar/home) merge with its formal antonym *unheimlich* (spooky/uncanny). Freud assumed that the uncanny occurs as a revelation of the repressed: *heimlich* is also used in German for "secretly." The change between *heimlich* and *unheimlich* points to the tension between what belongs to *heim* (home) and what does not. We are more likely to feel eerie, strange or uneasy when the environment and its synthetic agents cause a disruption to our predefined worldview. In recent studies this phenomenon has been used in relation to VR avatars, because while they appear to be human-like characters, they are limited in terms of social interaction and affective signals. We might even feel a lack of engagement with them once our minds are forced to accept that they are human-like. In social VR, identities are perceived as controllable objects, and bodies are nothing but specimens that can be examined like corpses. To avoid the "uncanny valley," VR avatars are currently mainly vessels. The British artist Ed Atkins even calls them "surrogates" or "crash test dummies" in our "moments of immersion" interview: "A body that can go through something in my stead, so that I do not have to."

The other issue with VR avatars concerns the visualization of virtual bodies. Part of that is going back to the idea of the wonderland in which bodies are fluid and adaptable constructs. Alice's body in the story is part of a body swap illusion. It shrinks and grows several times as part of her experience. So when users are represented in virtual environments, they need a body and would like to shape it the way they want. They want to define it by their choice of clothes, haircut, tattoos, etc. However, avatar stylization remains a relatively new field. Research is concerned with the relation of physical and virtual bodies, what it means to have a body in digital environments, and how body image distortion affects our attitude toward our selfhood. If I could design my VR avatar to be slim, would that have an effect on my attitude toward my physical body? If others perceive my VR avatar as attractive, would the instant gratification have an effect on my attitude toward my physical appearance and self-confidence? The answers to these questions determine if and to what extent we feel engaged with virtual bodies. Therefore, Facebook tries to avoid any disruption and dissonance in relation to VR avatar stylization. It wants to keep users on the platform and avoid losing them by feeling disconnected from their digital appearance. That is the reason why VR avatars often miss the lower half of their bodies. The conventions of social interaction are mainly linked to faces, gestures, and hands, not feet. But Facebook also wishes to avoid any sexual connotations. Without the lower body parts, VR avatars appear nonsexual and refrain from being too concrete about gender construction and perception of sex. Gender identities and sexualities such as transgender, queer, intersex, agender, genderfluid, bigender/trigender, and nonbinary are currently not part of the VR avatar stylization sheet. In fact, in the promotional video for their newest social VR app Horizon, Facebook tries to overcome that issue by hinting at the fact that avatars would be allowed to have nontraditional gender features. At one point, the female host admires the moustache of a male avatar, and decides to put on a moustache herself, because "Horizon is not about 'rules,' 'limits' or 'pants.'"[3] It is a subtle hint with a punch line. Facebook hopes to empower all of its users and make the upper parts of their VR avatars appear more believable and suitable for social interactions.

**Figure 4.1**
Facebook Horizon: the appearance of digital avatars in VR. Reprinted with the permission of Facebook. © 2020 Oculus.

In April 2018, Facebook launched a redesign of its avatar system, which featured new head shapes, hairstyles, and body types: "We've worked on making avatars feel more present in the VR space with richer materials, better lighting and shadows. We've also fine-tuned the tech under the hood to make avatar body movements look more fluid and natural."[4] Social VR world-building platforms like Horizon or the 5G-enabled service XRSPACE,[5] but also competitive social VR meeting spaces such as Microsoft's AltspaceVR, HTC's Vive Sync, Mozilla Hubs, VRChat, Rec Room, vTime, High Fidelity VR, and Spatial, want their users to feel comfortable with their digital selves in VR. The more comfortable they feel, the more likely they are to stay on the platform. The social VR platform vTime, for instance, incorporates sociality elements in both VR and AR, and its users can design their avatars with a tool called New vYou. The tool allows users to create their VR avatars in more detail, including eyebrows, facial hair, nose, lips, and cheeks.[6] Katie Kelly, head of engagement at AltspaceVR, elaborates on the platform's avatar design in our "moments of immersion" interview: "We wanted our avatars to be as simple as possible, so as a user you know these are not human beings in front of you, and your brain does not have to do the magic trick to say 'Yeah, I am still ok with it.' We have four different robots and a humanoid. It is not a big variety yet. But we are creating an avatar system right now that people have a choice to choose how they want to appear in VR." She adds later: "Interestingly, I think the Internet has taken away ownership to be your real self. People do not have ownership of their own picture anymore. There is a long way to go, but we give people the chance to be their authentic self. I think we need to be aware of how we perceive reality and emotions in VR, but this is true with any kind of technology. Immersive technologies can bring us back to the roots of how to communicate as a community." At one point, platforms might even try to create VR avatars based on the images and videos users upload on social media—a scenario that would challenge the "uncanny valley" but seems plausible in regard to a wide-reaching ecosystem like Facebook.

The digital self is one of many *sociocultural immersions*—immersive experiences that are not limited to one specific medium and affect social and cultural conditions. The digital self influences sociocultural aspects of our being even beyond the limitations of digital spaces, for instance, for which jobs we will get an interview and which people want to go on a date with us. I have previously argued that the mediation of selfhood is anything but simple.[7] It is already difficult enough to understand *who we are* in physical reality. Even if digital platforms provide sufficient tools to curate the digital self, users first need to be able to fully comprehend their own selfhood and how others perceive it through mediation.

In his self-discrepancy theory, E. Tory Higgins defines three kinds of representation of the self: "(a) the actual self, which is your representation of the attributes that someone (yourself or another) believes you actually possess; (b) the ideal self, which is your representation of the attributes that someone (yourself or another) would like you, ideally, to possess (i.e., a representation of someone's hopes, aspirations, or wishes for you); and (c) the ought self, which is your representation of the attributes that someone (yourself or another) believes you should or ought to possess (i.e., a representation of someone's sense of your duty, obligations, or responsibilities)."[8] Higgins argues that we are motivated to reach a level of self-understanding in which our self-concept matches our personal self-guides. So whatever we think we are, we would also like to see that represented in mediated spaces. Our social media profiles, however, are the self-curated projections of the attributes we (or someone else) would like us to ideally possess. They do not actually represent who we are, but instead construct *our ideal self*. Higgins refers to the psychology of peer pressure: the self is not an autonomous, independent subject, but rather a socially influenced object that directs, if necessary, its human needs of belonging to other relevant realities.

This process of understanding one's idea of selfhood is determined by a complicated, reciprocal relationship of three interrelating fields—the "identity complex," the "identity-image complex," and the "semantic networks" that surround us. The *identity complex* is the result of cohesive and competing strategies of self-perception. For instance, identity can be a result, a process, a form of interaction, or an affiliation with cultural and social constructs like family, peers, nation, gender, social class, organizations, society, media, and culture. We are constantly trying to figure out how our perception of self fits into the identity complex, especially when we are young. Mediation plays a significant role in that process, e.g., the influence of music on our identities. These multiple and intersecting parts frame our identity in a broader social-historical context. The *identity-image complex* is a negotiation between how we see ourselves (intrinsic image), how others see us (extrinsic image), and how we should be seen in the context of others (meta-image). For instance, our social media profiles are designed as the result of the mediation between intrinsic, extrinsic, and meta-images. Any photo or video we choose to curate our digital self is influenced by how we see ourselves, how we want others to see us, and how others perceive us in the context of social peers. Finally, *semantic networks* mean all the self-organized networks around us, which influence our perception of identities, including social institutions such as families and friends but also organizations and media. For instance, we might be fans of *Harry Potter*, so we associate with groups of other *Potter* fans, which ultimately will have an effect on how we perceive ourselves as *Harry Potter* fans. In this complex, multidimensional field of forces, which is inextricably linked to matters of power, values, and ideologies, we constitute the perception of our selfhood and ultimately how we wish to curate the self through immersive practices.

Digital culture is defined by the mediation of everyday life and the practices of impression management. The trend is to display and curate the self. *To confess* on digital platforms is a requirement *to exist* in digital spaces. The digital self is the result of subject-centered techniques and immersive social practices such as selfies, lenses, face filters, face swaps, and snaps, which at the same time uncover the power relations in digital ecosystems like Facebook, Twitter, Snapchat, Instagram, and Tinder, and how much influence they have on their users' online appearances. The practices of visualizing one's existence are embedded in the routines of everyday life and digital communication. Posting, sharing, liking, tweeting, swiping, sorting, counting, rating, marking, filtering, and swapping are standardized forms of staging the individual. They are performed through the interface designs of various apps. These social practices and activities are immersive as they force users to focus on one particular mediated representation—*the self*.

In May 2013, *Time* magazine famously called the generation of millennials (those born between 1981 and 2000) the "Me Me Me Generation," stating that "the incidence of narcissistic personality disorder is nearly three times as high for people in their 20s as for the generation that's 65 or older."[9] The article claims that this "overconfident and self-involved"[10] group of people mainly uses self-centered techniques and immersive practices to stylize and idolize themselves. And indeed, we might think we are surrounded by billions of narcissists, who are shamelessly immersed in their own digital image, are promoting themselves on social media, and have nothing better to do than celebrate their breakfast cereal on Instagram. On the forefront of this movement are celebrities like Kim Kardashian, famous for her reality TV show *Keeping up with the Kardashians* and her book *Selfish*, a curated collection of her own selfies. There are uncountable snapshots online of faces and people in all sorts of contexts, a vast number of them also inappropriate and thoughtless, such those made at Holocaust and 9/11 memorials, and in dangerous situations, such as great heights.

In one of my previous works on media phenomena in digital culture, I have argued that the overload of self-portrayals, face obsessions, and impression management in digital spaces has nothing to do with the psychological disorder of a whole generation. In fact, this phenomenon is defined by a process called *automediality*. European scholars of life writing such as Jörg Dünne and Christian Moser have coined the term to describe how subjectivity is constructed through mediated practices like writing and photography.[11] Automediality refers to the creation of subject-centered representations (such as the self) through mediated practices. The self is framed by digital practices that mainly exist to possess, maintain, and curate our identities in digital spaces. David L. Jacobs had argued in the early 1980s: "We use snapshots to communicate to ourselves, and those around us, and those who will succeed us, that we in fact exist. With snapshots we become our own historians, and through them we proclaim and affirm our existence."[12] Over three decades later, Mehita Iqani adds, in relation to the selfie phenomenon: "The selfie is a very interesting type of snapshot. In it, individuals often represent themselves as at the peak of their own attractiveness, and then use this image either as a profile image, or put it out into the public realm through, for example, their Instagram feed. The selfie is a way of saying, "look at me," out loud, in a public domain, it is about getting attention but also about crafting the self as an object in a very particular way."[13] The face, in particular, is used to visualize the individual, a direct reference to the identification of selfhood. Strategies to portray it, mask it (filters, lenses), deconstruct and distort it ("sellotape selfies") are part of the semiotic system in digital spaces that establish the self as part of a mediated reality. Automediality moves the shift of attention from the physical self to the mediation of the self. Both are now equally relevant to constituting selfhood. And more importantly, we rely to a great extent on our digital self. It influences with whom we will connect in personal and professional relationships. It creates meaning *beyond the mediation*.

> The digital self is the result of *subject-centered techniques and social practices*. The process of *automediality*, in particular, as part of impression management constitutes selfhood in digital spaces.

The mediation of the digital self is a theatrical and performative process. In fact, to stage and perform the self is an immersive experience. In 1991, interface and games designer Brenda Laurel noticed a paradigm shift in the practices of digital culture. She understood human-computer interaction as a collaborative mediated experience that extends the geometry of dramatic interaction to interactive designs in digital spaces.[14] Laurel used the metaphor of the "vanishing interface" to describe the dissolution of the frames and the immersive processes to merge with the world of cyberspace. In fact, she applied the metaphor of the "stage" to computer-mediated experiences, building upon the dynamic, constantly changing, and collaborative tradition of theatrical performances from Ancient Greek to Shakespearean plays. If the whole world is a stage, then digital stages (or platforms) exist to enact our roles, scenes, and activities. Laurel's argumentation is inspired by Erving Goffman's theory of self-performance. In his book *The Presentation of Self in Everyday Life*, Goffman states that human beings have various sociodiscursive needs, which they reflect and enact through different roles and personas addressed to different recipients in various contexts.[15] We enact on several stages at the same time (from Facebook, Instagram, LinkedIn, to Tinder), and use various tools, roles, costumes, styles, languages, and speeches. Similar to theatrical plays that are divided into acts, our digital performances are divided in sections, sometimes separated by only short periods of time, sometimes without any intermission at all.

Enactment (for instance in theater) is an active process. It is the decision to play a role within a mediated setup (the play), and even to pretend no one is watching (the audience). Following Laurel's definition, immersion would mean that the performers would *stay* in their roles after the play has ended. Furthermore, they would not even recognize that there are things such as "roles," but instead perceive the role as their actual life. For instance, imagine watching *Romeo and Juliet* on stage. Romeo is played by an actor named Peter, and Juliet by an actress named Mary. Peter has two children and is going through a divorce. Mary is studying languages part-time at university, and is single. But while they are playing their roles of Romeo and Juliet, they merge with the mediated experience. They are actively feeling present while they play their roles as a loving couple. They might even forget what is happening in their lives outside of the theater.[16] Peter and Mary would go off stage, leave the theater, and still *be* Romeo and Juliet. Their lives would evolve around the mediated representation of their stage personas.

This is what happens in the immersive practices of digital culture. The signifying and performative practices of curating digital autobiographies ultimately mean that the mediated self becomes the definition of the real, the foundation of our existence. We become who we are online. Anything from a personal website to a variety of social media profiles and a Wikipedia entry can influence how others see us. Social media commercializes our desire to shift our attention to the online world, without foreseeing that this leads to the creation of a parallel world in which we do not exist while our avatars have taken our place. The cognitive connection with the world around us begins to deteriorate.[17] However, this also reflects on us in the physical world, and we all know that. To have a well-maintained and popular Instagram profile with aesthetically and socially appealing visuals might affect our relations in the physical world positively. On the other hand, posting inappropriate or offensive content such as sexually explicit images and harmful comments might lead to social penalties—negative feedback, blocking, and isolation both online and in the physical world.

The way we create our profiles depends a lot on for whom and for what purpose we curate the mediated self. For instance, you might not use the same photos on LinkedIn as you would on Tinder. LinkedIn profiles exclude the idea of a private self but have an immense impact on the perception of our professional selves. Headhunters and potential recruiters search for specific CVs and profiles, which shape "an idealized portrait of one's professional identity by showing off skills to peers and anonymous evaluators."[18] Any headhunter who is inviting you to a job interview would naturally expect the same person to show up as the one seen on the LinkedIn profile. That does not imply you are not telling the truth on the profile (although people do indeed facelift their CVs); rather, it means you will do your best to match the performance you have given online, hence you want to match the mediated representation. On the other hand, to decide if you might want to date someone based on the photos on a dating app profile influences how you will perceive that person at your first physical encounter. You will create an ideal image of that person long before you have actually met. Ultimately, your date might even try to match the person you have imagined once you meet, and you might do the same.

The performative process to mediate the self involves several theatrical practices. One of them is the use of masks. Masks reveal and hide something at the same time. They change appearances, portray what was not there before, and cloud what should not be seen. To *mask* the physical and *unmask* the mediated self is part of the aesthetics of immersive practices in postmodern media culture. Hans-Thies Lehmann argues that the disguise not only influences how others see us, but also how we see the world around us: "The pleasure in dissimulating oneself under the mask is paired with another, no less uncanny pleasure: how the world changes under one's gaze looking out of the mask, how it suddenly becomes strange when seen from 'elsewhere.' Whoever looks through the eyes of a mask changes his gaze into that of an animal, a camera, a being unknown to itself and the world."[19]

The aesthetics of digital masks often distort and change our physical appearance in mediated representations. Dogface filters, flower crowns, face swaps, frown faces, and voice filters are the mirror cabinet and freak shows of digital culture. Freak shows in the nineteenth century exhibited abnormal appearances as part of performances on fairgrounds and in taverns. The display of heavily tattooed, pierced, and extraordinarily hairy characters was more than just entertainment. Visitors wanted to confirm the normality and conformity of their own bodies in relation to those exhibited. Their bodies were perceived as normal, because they were not "monstrous" (Latin *monstrare*, showing, displaying). We find the same logic behind the practices of masking and unmasking the self in digital culture. Distorting the face on Snapchat with the nose and ears of a dog masks the appearance of the human face. But at the same time, it unmasks the limitations of physical appearances. For instance, Instagram has been pushing a set of facial effects since 2017, all of them created to let users look like they had various forms of plastic surgery, "including cheek and lip fillers, a nose job, brow lifts, and skin-smoothing botox."[20] These filters, promoted by celebrities and influencers, were setting idolized standards of body perception while at the same time questioning physical appearances. In another example, Snap Inc., the company behind the Snapchat app and its augmented overlays called "lenses," launched Lens Studio, a desktop app to create, publish, and share your own lenses. The portfolio of masks became part of the world of user-generated content and extended the meaning of its signifying practices.[21] Modified images are a defining element of mediated realities, no matter if we call them "masks" or "lenses." Lev Manovich points out that "[t]he paradox of digital visual culture is that although all imaging is becoming computer-based, the dominance of photographic and cinematic imagery is becoming even stronger. But rather than being a direct, 'natural' result of photo and film technology, these images are constructed on computers. 3D virtual worlds are subjected to depth of field and motion blur algorithms; digital video is run through special filters that simulate film grain; and so on."[22] Ultimately, with the help of computer-generated images we are crafting several digital avatars that become the gateways to our identities and selfhood.

> The curation of social media profiles consists of several *performative practices*. Users are provided with a variety of tools *to aestheticize the digital self* and influence how others perceive it.

The creation of *avatars*[23] is the initial step to becoming immersed in the conditions of social media. In 1993, Howard Rheingold stated in relation to digital media: "We do everything people do when people get together, but we do it with words on computer screens, leaving our bodies behind."[24] We are leaving our physical selves behind and instead are using mediated practices and tools to craft digital entities that represent us. Therefore, users spend a lot of time initiating and curating their digital existences, and bringing them to life to express a certain kind of individualism: "The ability to select an avatar's characteristics appears to facilitate expressions of self, social status, and intimacy."[25]

The term "avatar," mainly used in relation to computer games and social media, refers back to *avatāra* in Sanskrit: the descent and human incarnation of God during times of distress on Earth. In Hinduism, Vishnu, the God of Protection and one of the principal deities, created ten different avatars to empower the good and fight evil on Earth. Deities had the desire to walk among humans. Therefore, they switched into physical bodies to visit the world they had created. The parallels to digital avatars are obvious. Our avatars are digital vessels outside physical reality. They are like immortal, God-like *Übermensch* creatures that we have created in a Promethean act of life-bringing selfhood. We feel immersed in idolized versions of ourselves that allow us to go beyond the limits of physical reality.

Sabina Misoch argues that avatar creation is either an "open" or "fixed" identification process;[26] Daniel Kromand distinguishes between "open" and "closed."[27] A fixed/closed identification process means that the avatar is a predefined figure that users have to accept as it is without being able to adapt it. These avatar types "combine clear motoric control with a predetermined set of abilities"[28] such as Sonic the Hedgehog and Super Mario. These types can be mostly found in computer games. However, avatars created as part of an open identification process have the potential to become more immersive. Users develop a personal and intimate relationship with their digital representations as they can decide on their looks and characteristics. This process takes more time, but ultimately this is the type of avatar "that to the highest degree allows its player to construct a virtual self-image and create a personalized protagonist."[29] These avatar types appear in role-playing games like *The Sims* and *World of Warcraft*, but also on social media and social VR. For instance, curating a Facebook profile is an open identification process. Users create their avatars by choosing a specific name and several images and videos that will show themselves in various contexts (the "ideal self"). By combining their real names with mediated content, users create fictionalized versions of themselves that incorporate information from both their physical and mediated selfhood (the "ideal self" perceived as the "actual self"). Kwan Min Lee calls this "self-presence," "a psychological state in which virtual (para-authentic or artificial) self/selves are experienced as the actual self in either sensory or nonsensory ways."[30] Additional information such as listing favorite films and music as well as joining groups and adding friends further contributes to the open-identification process on social media.

Research has shown that an individual's behavior is influenced by the self-presentation as digital avatar. The "Proteus effect" suggests that the photos on a dating app that are perceived as attractive lead to more intimate contacts with others than profiles with photos that are perceived as less attractive.[31] None of that gives an accurate indication of a person's attractiveness in physical reality. However, there are several guidelines online on how to create a dating profile with an ideal headline, description, and photos to be perceived as attractive in the mediation.

At the same time, avatars refer to the moment of escapism in times of distress. Users turn to their avatars as their essential creations in digital spaces. For instance, more than 40 percent of Tinder users are on the platform for "confidence-boosting procrastination."[32] Social activities and online relations seem to be easier to establish and to maintain and allow instant gratification. For instance, with online dating apps, it is much easier to search and match with a large number of people. Also, users do not have to deal with the pressure of rejection online. The whole process is able to satisfy the basic human need to be recognized, valued, and appreciated by others. Marie-Laure Ryan describes this in relation to her concept of possible worlds. She uses the example of Cinderella being locked in her room, which creates two possible worlds. In one possible outcome Cinderella makes it to the ball and marries the prince in the end. In the other version, she is hindered from getting away and one of her stepsisters marries the prince instead. Most readers might choose the first option, especially if they identify with Cinderella as the protagonist. In digital media, the identification with digital avatars allows users to believe in different possible outcomes embodied through their digital selves. They might get to the ball in time, and even marry the prince, but more likely online than in physical reality.

> Profiles (avatars) on social media are created in an *open identification process*. Users are influenced by their self-representation as it allows *instant gratification* to satisfy basic human needs.

Within the concept of the 360° gaze framework, the self is a mediation. Our ideas of self are immersive, as they matter beyond the computer screen, the tablets, and mobile phones we use to access and curate them. Self-expression and self-promotion are ways to organize our digital autobiographies, the ones that are not just private but shared with others. We use strategies to constantly mediate and interpret our own existence in front of others. We arrange and rearrange bits and pieces of our avatars, and the resulting narrative might be a construction in hindsight, a retroactive ordering, but secretly we wish the ideal self to become the actual self.

We are strongly influenced by our mediated selves, and we are using our avatars as the benchmarks for our identity. The self is defined by the social activities and semiotic codes we implement to inscribe ourselves in mediated experiences—everything from posting, sharing, liking, and tweeting to swiping and swapping. We perform our selves in stagelike processes: we measure and optimize our roles by having different avatars, profiles, and personas. The way we showcase our faces and bodies on social media becomes a process of masking and unmasking, reverting the order of perceptual belief by introducing an ideal version of ourselves in the mediation. The psychology behind it is complex. It has to do with the idea of an "ideal self" we wish to portray in front of others (and ourselves), which at the same time is heavily influenced and determined by our intrinsic, extrinsic, and meta-images. Platform providers like Facebook feed the various needs for expressing the self through their interfaces, different presentation styles, and the use of algorithms. Users are only able to curate their avatars under the regulations of the platform. For instance, they have to give a full name on Facebook, preferably their own, which merges the mediated with the physical being. These tactics reveal the deeper ideological and economic interests in creating online identities, those that are ultimately balanced with the company's goal of increasing revenue.

With more social VR world-building platforms such as Horizon and XRSPACE, and free XR-based meeting services such as Spatial and Mozilla Hubs that are also available on non-headset devices, the future of digital avatars is about to become three-dimensional. Social XR experiences will foster new ways to design and curate the digital self, and its appearance will become more realistic with the time. In fact, Facebook was already teasing photorealistic avatars in May 2018 for interactive VR experiences.[33] These three-dimensional lifelike avatars would mirror users' physical appearances by mapping their faces and facial characteristics to synchronize facial movements and expressions in the virtual environment. Seeing photorealistic images of friends in the VR space would make virtual experiences feel more convincing, and ultimately increase the significance of digital encounters. At the same time, three-dimensional avatars will also redefine the performative process of using roles and masks. In fact, the three-dimensional "ideal self" could introduce an idolized version that would appear superior to our identity in physical reality. What would be more "real" then: the physical or the mediated? Ultimately, the answer to this question can be already found in several mediated practices of digital culture that aestheticize and market the self. This will also determine how immersive the mediation of the digital self will be in the future of social XR.

## *Moments of Immersion: Katie Kelly*

Katie Kelly is the head of engagement at AltspaceVR, one of the leading social VR platforms. AltspaceVR was founded in 2013 and was acquired by Microsoft in October 2017. The company offers several social VR experiences, including live events with celebrities, politicians, comedians, and DJs.

**What is your vision for social VR?**

As long as we are not in the *Ready Player One* kind of world, there is no substitute for reality, or meeting people in real life. People were excited when we got the phone, and later video chat. But in both of these cases, we are still tied to a device and we are interacting with another person in an—I would argue—unnatural way. I was actually talking to my father the other day, and we were discussing what happened to him that day. It was all verbal cues about his experience in life. With video chat, you have a screen in front of you and you see somebody's face, but still it is verbal and the expressions attached to it are very minimal. But with social VR, you feel like you are actually there with somebody else. The experience is similar to going to a restaurant and having dinner together. At the moment, there are not enough headsets out there to connect in VR with the people you care about in real life. So we are focusing on how to meet strangers in VR, and ultimately become friends with them.

**But in this kind of experience, I would not be meeting real people but only digital avatars, correct?**

The reason we use avatars is because we wanted to avoid the "uncanny valley" phenomenon. We wanted our avatars to be as simple as possible, so as a user, you know these are not human beings in front of you, and your brain does not have to do the magic trick to say, "Yeah, I am still ok with it." We have four different robots and a humanoid. It is not a big variety yet. But we are creating an avatar system right now that will give people the chance to choose how they want to appear in VR.

**Once I have an avatar, how will I experience communication on your platform?**

As a user, you will be having conversations that bring you out of your echo chamber. You will be learning from other people and growing with them. We make sure that at any time we have AltspaceVR employees present in VR locations to help you if anything is wrong, if you have any questions, or if you want to report anything you think we should know about. And there are also ways to have ownership of your experience—for instance, you can "mute" someone you do not want to talk to anymore, or erase that person from your experience. We also use a so-called "space bubble." This is a bubble that gets turned on once you are on our platform, so people cannot simply invade your space and you have a certain privacy.

**Which sorts of locations do you use for social VR?**

We are constantly adapting this. We want to introduce locations that people know, but that are just better in VR. We have this world and can let anything happen in it, so we want to make sure to bring delight and experiences to people that they could not necessarily experience in the real world. We experiment with locations a lot. Since the beginning of our platform we have been collecting data to understand which experiences people want to have in VR, and in which locations they want to be. Our environments want to mainly foster communication, encourage people to be inside them, and interact with others. We have a desert, and a maze. We have a space station you can go to, and also a cool DJ dance studio that is floating on a planet.

In addition to that, we let our users create their own locations. We had this amazing woman designing a karaoke studio, and she put all those posters of her favorite movies on the wall. We had someone creating a church in VR for others to join where they could pray. These locations are very personal, therefore you can invite only those you really want to be there.

**There are a lot of ethical implications to social VR experiences. Basically you are changing the way in which way we perceive digital communication.**

I do not think we change so much the way people are communicating, but we are bridging the gap between locations. With time, the technology will get better and you will have a more natural way to communicate with each other in VR. A while ago, I walked in one of our main VR lobby spaces and heard a person from Jerusalem and one from Palestine having a conversation. They were clearly arguing, but it was a conversation. They had access to each other that they would not have normally. Interestingly, I think the Internet has taken away ownership to be your real self. People do not have ownership of their own picture anymore. There is a long way to go, but we give people the chance to be their authentic selves. I think we need to be aware of how we perceive reality and emotions in VR, but this is true with any kind of technology. Immersive technologies can bring us back to the roots of how to communicate as a community.

> But that also means we will be depending on the corporations who will provide us with those tools.

> Yes, but I think the world itself is going to be an interconnected web of people making this technology, and as long as we fill it up with the life of people that do care and we keep corporations accountable, I think we can avoid a dystopian future. At the end of the day, it is still made up of people.

**Immersive Parasocial**

As of September 2017, Spotify users could listen to their own personalized playlists, using Time Capsule. The feature is dedicated to music that transports listeners back to their teenage years. By analyzing a user's age and preferences, Spotify pulls more than 50 iconic songs from the user's teens and early twenties and compiles a playlist out of them. That means listeners must be at least over 18 years old to provide enough data and use the feature. I was born in the '80s and grew up in the '90s with Ray Cokes and MTV, Nirvana and Grunge, Take That and Backstreet Boys, and a lot of Eurotrash dance music. In every shop they played Meat Loaf's ballad "I'd Do Anything for Love (But I Won't Do That)"; in every tabloid there was a paparazzi shot of George Michael or the Spice Girls. Time Capsule is surprisingly accurate for me: Oasis, Pearl Jam, Smashing Pumpkins, Björk, Beastie Boys, Hanson, TLC, Red Hot Chili Peppers, even "Breakfast at Tiffany's," the one-hit wonder by American rock band Deep Blue Something. I am not alone in this. Seth Stephens-Davidowitz analyzed data provided by Spotify and measured every Billboard chart-topping song released between 1969 and 2000 in combination with the age of their biggest fans when the songs first came out.[34] The study reveals that the average man's musical taste develops between the ages of 13 and 16, while a woman's takes shape between the ages of 11 and 14. We usually develop our musical tastes in our early teens, around the same time we are exploring ourselves physically, emotionally, and sexually. Spotify's Time Capsule playlist is just one of many playlists that the streaming service provides. Many of them, though, like the Year in Music playlists, offer a deep and personal look into past listening experiences from recent years. These nostalgic playlists draw from a variety of data to encourage users to immerse themselves, with the help of music, in memories of the past as an all-encompassing process of yearning beyond the mediation.

Nostalgia is another sociocultural immersive experience. It can be found in a variety of mediated representations and clearly affects audiences beyond any viewing or listening experience. Our obsession with dwelling in the past, the habit of living in memory rather than the present and idolizing events of former times, is a powerful drug. A song is able to evoke memories of certain life periods—from being in high school to dating someone special decades ago. It might underscore our first kiss and become a permanent resident in our brain. It literally transports us to a mediated space in our mind, one that recreates memories based on feelings that we associate with mediated content. Nostalgia is a powerful emotion.

For several centuries, nostalgia has been considered a psychological disorder and a problem better avoided: do not immerse yourself in the past to escape the present. Lost lovers, happiness, rose-tinted times of peace: nostalgia has been described as the illness of soldiers, who are far away from their homes; a state of mind that made life in the here and now easier to accept while hiding in a foxhole. Literally, the Greek words *nóstos* (to return home) and *álgos* (pain, ache) refer to the conditions of painfully returning home after war, yearning for the past after traumatic experiences, such as the pain, loss, and fear on the battlefield. Psychologists use the German word *Sehnsucht* (longing, pining, craving) to describe the individual's never-satisfied search for lifelong happiness and the struggle to cope with loss and unrealizable wishes.[35] Alternative experiences—that happened in both physical and mediated realities—are romanticized to such an extent that they become an escape from the haunting present. As we are mortal souls, nostalgia intermingles the sadness of loss with the joy of knowing the loss is not complete. In the same way a single song on Time Capsule makes me feel like I am young again, while simultaneously reminding me that I am substantially older.

In postmodern culture, nostalgia becomes a commercialized product to satisfy society's need for the longing for better times. In postmodern media culture, especially, nostalgia is another immersive experience for the self, a way to shape identities, and to curate the individual on digital platforms. Nostalgia is industrialized in almost every area of consumer goods you can imagine, from remakes and reboots of popular films (*Star Wars, Star Trek, Total Recall, Blade Runner,* etc.) and TV shows (*The X-Files, Twin Peaks, Gilmore Girls, Roseanne, Will & Grace,* etc.) to fashion, vehicles, food, vinyl, home decoration, and the relaunch of the classic Nintendo Entertainment System (NES) with more than 30 vintage games in 2017[36] which were originally released in the mid-1980s. Pop stars like Backstreet Boys, Take That, a-ha, and ABBA are celebrating comeback records and tours, while long-established artists such as the Rolling Stones, Bob Dylan, U2, and R.E.M. are re-releasing anniversary editions of their most popular records (including bonus tracks, demo versions, interviews, new photos, documentaries, etc.). Very often these artists even go on tour and play the full album's tracklist—often in correct order—as part of every night's setlist (such as U2's thirtieth-anniversary tour to celebrate their landmark album *The Joshua Tree*). Other artists release albums loaded with emotional nostalgia for legendary genres and singers, such as the Rolling Stones' blues record *Blue & Lonesome* and Bob Dylan's reinterpretation of classic American songs (often performed by Frank Sinatra) on *Shadows in the Night*, *Fallen Angels*, and *Triplicate*.

In their book *Understanding Media Industries*, Timothy Havens and Amanda Lotz have analyzed the industrialization of cultural products.[37] Their framework is crucial for understanding the role of cultural industries and digital ecosystems within the 360° gaze. The authors argue that economic, social, and cultural aspects within the media and cultural industries are influencing our idea of culture. If we see culture as a concept of evolution, adaptation, and reconfiguration—in other words, as a phenomenon under the umbrella of postmodernism—we recognize both nostalgia and immersion as results of cultural processes. The framework, which includes the parameters of (1) mandates, (2) conditions and practices, (3) media texts, (4) public, and (5) social trends, tastes, and traditions, suggests that any cultural idea, trend, tradition, and myth can ultimately become part of a process of industrialization. Nostalgia might be an internal process of the human mind, but the feeling can be evoked by any content production that satisfies the audience's demand for returning to the general recollection of the past. That is why it is a key element for immersion as well.

These days, mass media are widely commercialized, and our human desire to escape the realism of the present is satisfied by cultural industries. Such shows as Netflix's *Stranger Things* activate feelings of nostalgia for a generation of 30-somethings who might feel lost as part of the radical changes that generally occur during adolescence. The show is a pastiche of an '80s memorabilia playbook and sci-fi/horror inspirations by Stephen King, Joe Dante, John Carpenter, and Steven Spielberg. The show's producers, the Duffer Brothers, use the theme of nostalgia as an established semiotic system throughout the series: classic films (*Jaws*, *Gremlins*, *Ghostbusters*, *The Terminator*, *Alien*, *The Exorcist*, *E.T.*, *The Goonies*, *Stand by Me*, *Halloween*, *It*), '80s hits (the Police's "Every Breath You Take," Cyndi Lauper's "Time after Time," Bon

Jovi's "Runaway," Ray Parker Jr.'s "Ghostbusters Theme," Limahl's "NeverEnding Story," and the show's instrumental score by experimental synth quartet S U R V I V E), as well as other cultural references (Farrah Fawcett's hairspray for luscious '80s hairstyles, and the medieval arcade game *Dragon's Lair*). They provide a range of cultural touchstones for a journey back to the audience's past.[38] In fact, streaming providers are using nostalgia programming to revive their audience's treasured memories. With show revivals such as *Gilmore Girls* on Netflix, and a one-off special of *Friends* on HBO Max, streaming services are targeting the segment of the population that looks back fondly on the '80s, '90s, and early 2000s.

Immersive experiences linked to nostalgia are strongly influenced by our own past, or more specifically, how the past has been mediated to us. Records, games, films, and TV shows are remediating the past not as an individual experience but as a general recollection of mediated representation. Certain historic periods are associated with specific cultural products that can be relaunched and rebranded at any given time within the process of industrialization. They open up a playground of nostalgia for audiences to communicate semiotic codes and performative acts through immersive practices: rewatching favorite shows, buying vintage merchandising, playing old records, watching favorite clips on YouTube, and meeting fellow

**Figure 4.2**
Netflix's *Stranger Things*: the boys wearing *Ghostbusters* costumes for Halloween. Reprinted with the permission of Netflix. © 2017 Netflix.

fans who are going through the same process of transition. The past, however, is mediated. It is not always the past that audiences have physically experienced themselves, but one they feel strangely familiar with through the music they listen to, the books they read, and the films they watch. That way, audiences are able to appreciate their former ideas of self, exclude unpleasant memories or find positive reinterpretations to deal with them, and establish important biographical benchmarks to shape their mediated identities.[39] The recollection of the past is a security mechanism for humans to remain optimistic and hopeful. It is the craving for a place that no longer exists, or indeed that never existed at all. It is an easy rabbit hole to fall into.

In her book on nostalgia as a strategy for entertainment industries, Kathrin Natterer argues that nostalgia is a form of escapism, strongly related to the phenomenon of immersion. She situates immersion as a characteristic of nostalgic experiences and its sociopsychological implications, besides mental stability, improved social relations, and increasingly addictive behavior.[40] What she calls "personal nostalgia" are individual long-term memories that can be evoked at any given time, for instance, by music or photographs. However, these experiences do not always necessarily activate specific memories of our past (like our first kiss), but rather general recollections of that moment (a romanticized version of our kiss inspired by several mediated representations around it). This is the liquid space of immersion: we create a mediated version of our selfhood and even a fictionalized timeline of past events that is underscored by various mediations. Nostalgia becomes an immersive experience as the physical experience merges with mediated memories.

What Natterer describes as "historic nostalgia" refers to how we remember and romanticize whole periods of time through mediation. Songs by Elvis Presley, films about the Doors, or documentaries about events like Woodstock stimulate certain collective memories of historic periods, which we idealize in the face of present times. Ultimately, romanticizing the past leads to the design of hyperrealities. Films such as *Back to the Future*, *Forrest Gump*, and *Midnight in Paris*, and TV shows like Netflix's *Stranger Things* and AMC's *Mad Men*, are drawing on the desire to reexperience historic periods, which are only based on mediated recollections.

But why are we fascinated by the past to such an extent that we want to *be there*? Why do we choose to immerse ourselves in the surroundings of former times rather than live in the here and now? In his *Condition of Postmodernity*,[41] David Harvey describes a time-space compression, which defines the terms postmodernity and post-Fordism. Because we are able to overcome the limits of time and space with the help of new technologies (for instance, digital communication tools), our perception of time and place have also changed. Time is a cultural construction: Monday-Sunday, summer-winter, Friday the 13th. For instance, how long does it take you to reply to someone's message? A few decades ago, it would have been fine to not call someone back for several days. With messenger apps, caller identification, read receipts, and constant availability, we now tend to reply much faster. Time is never stable but is instead a carrier of sense and meaning. In postmodern societies, the sense-making of former times, in particular, is accomplished with the help of nostalgia.

Robert Zemeckis's *Back to the Future* trilogy might be the ultimate serial narrative about time and about being immersed in the nostalgia of the past. Time travel is the utopian wish to make the transition of time visible and tangible, and that is why the four "time zones" of *Back to the Future* are not coincidental: 1885, 1955, 1985, and 2015. While 1985 represents the present time of the story (and of its viewers at the time the films were released), 1885 and 1955 represent important moments in the history of American culture. In the year 1885, in the wild, wild West, new technologies were developed and the use of locomotives and trains had increased travel and production. The year 1955 is the post-world-war era, an age of economic growth, television, and car manufacture. And of course, the perception of the 1950s is associated with pure nostalgia that glorifies the past (not exclusively but especially American culture): the times were easy, jobs could be found everywhere, coffee was cheap, a dishwasher could become a millionaire. Finally, 2015 represents the future. And even though some of the futuristic gadgets in the films such as hoverboards and flying cars never came to be in physical reality, the utopia worked in the context of the story back in the 1980s (and still does post-2015). In fact, decades later, the fictional year 2015 in *Back to the Future Part II* (1989) was the inspiration for actual inventions like smart clothing, biometric devices, and wearable technology.[42]

> Nostalgia is an immersive experience. Audiences stimulate *memories of individual and historic past* with the help of mediated content. In postmodern culture, nostalgia is a *strategy for entertainment industries* to satisfy the demand for idealized mediations of former times.

We tend to glamorize periods that are now out of reach. Millennials are looking back to the baby boomers for social stability, just as the hippies of the '60s looked back to the folky pastoralism of the preindustrial revolution in the face of increasing environmental anxiety. Looking back can have a positive effect on our state of mind. Constantine Sedikides et al. even perceive nostalgia as an internal defense mechanism to foster self-continuity: "Nostalgia strengthens a sense of belongingness or acceptance, which, in turn, elevates self-continuity. . . . For example, nostalgia instills perceptions of life as meaningful . . . and renders the past self more relevant or vivid."[43] While it has long been considered a debilitating process of yearning and a psychological disorder, nostalgia could also be a neurological defense mechanism to overcome anxiety and depression. It "generates positive affect, increases self-esteem, fosters social connectedness, and alleviates existential threat."[44] The feeling of nostalgic immersion regulates aversive states such as meaninglessness or loneliness. It often remains a symptom of escapism, but one that could have a restorative effect on the self.

Being immersed in nostalgic mediations is often linked to experiences of fandom. The feeling of nostalgia opens many doors to becoming emotionally attached to a mediated object, but really, fandom is also an immersive experience on its own terms. Fandom means you care deeply about something that is not always (or often never) physically present in your life. You develop a one-sided, nonreciprocal emotional bond with media figures you have never met. Being a fan means you accept the reality of a mediated construction. You identify with mediated personas and objects and feel associated with them through specific rituals and traditions, something you want to belong to, which gives you comfort, and feels as real as anything else in your life.

I previously detailed my immersive experience of being a fan of the band R.E.M., but really, it could apply to any mediated experience. Imagine visiting a football game. You support one of the teams. In the framework of the 360° gaze, there are several semiotic codes and performative practices attached to that experience: you might wear the official team shirt, carry banners and flags with the team's logo, and color your face with the team's colors. You might sing along fan chants and hymns and shout out slogans of encouragement or frustration, depending on whether the team wins or loses. There are also several psychological and receptive implications: you might develop a feeling of belonging and identity. You want to purchase more merchandising and even season tickets and take part in fan-related events such as fan conventions and behind-the-scenes tours. All of that allows you to perform your fandom as an immersive experience. But most importantly, you will speak of *your* team. Its success will have a direct impact on your personal well-being. Fandom is an immersive experience as it connects audiences deeply with a diverse range of mediated representations that become crucial to defining their identities.

In a 2017 essay, *Wired* editor Kevin Kelly nailed the description of a true fan: "A True Fan is defined as someone who will purchase anything and everything you produce. They will drive 200 miles to see you sing. They will buy the super deluxe reissued hi-res box set of your stuff even though they have the low-res version. They have a Google Alert set for your name. They bookmark the eBay page where your out-of-print editions show up. They come to your openings. They have you sign their copies. They buy the t-shirt, and the mug, and the hat. They can't wait till you issue your next work. They are true fans."[45] Fans develop an intensive emotional relationship to a fan object and reach an emotional level of admiration for celebrities, stars, athletes, and musicians, among others. These relationships are crucial to the understanding of selfhood. Even though fans might never physically meet their objects of worship, these *parasocial relationships* can last for decades, often for a lifetime. For instance, when Disney released the highly anticipated first trailer of the *Star Wars* relaunch *The Force Awakens* (2015), fans reacted with emotional intensity on social media (17,000 tweets per minute, according to ABC News),[46] including joy and tears as they saw their favorite characters Han Solo and Chewbacca returning to the nearly 40-year-old franchise. Those characters are fictional, but for their fans they are as real as anything else. Celebrity culture is defined by these mechanisms, and celebrities are aware of them. They use them to establish collective experiences of fandom (for instance, Justin Bieber fans are called Beliebers, and Lady Gaga's followers are named Little Monsters).

Fandom builds upon interpersonal, emotional connections to mediated experiences, be they songs, films, TV shows, sports, or media personalities. Fandom is a performative process as it involves rituals, traditions, and stories, which define our identities as much as any social interaction. In fact, we might know more about Kim Kardashian than about our colleagues at work. This naturally stems from our tendency to latch onto the lives of people who we feel most influenced by and who seem to appear out of reach. However, it is also the result of reaching the level of extended presence, when mediated representation is perceived as meaningful. Listening to songs of your favorite band might make you *feel better*, you *deeply care* about the results of your favorite football team, or you *wait for hours* in the cold to get a glimpse of your favorite actress, and make a selfie with her as *a mediated memory* to hold onto in the future. As fans, we not only want to step into the mediated experience, *we want to belong to it*. All the selfies and videos with celebrities serve as proof and help to construct our identities within the logics and rules of digital ecosystems such as social media.

> Fandom involves several *rituals, traditions, and stories* as performative practices to feel immersed in mediated experiences.

The *parasocial* is a specific form of sociocultural immersion. It can be found in both nostalgic and fandom experiences. The parasocial satisfies fundamental human needs for personal identity and companionship. It can be categorized in three tiers of intensity that are strongly related to the concepts of presence and immersion: parasocial interactions, parasocial relationships, and parasocial attachments. Parasocial interactions occur when audiences develop a feeling of presence while consuming mediated content (cheering at football teams while watching a game on TV; worrying about the fate of television characters while binge-watching a show, etc.). Parasocial relationships on the other hand are a matter of immersion, when the mediation has ended and you still think and deeply care about the characters and actors (very often perceived then as one and the same person). Parasocial attachments, however, go a bit deeper. Fans develop a desire to be with a mediated personality and experience heartbreak and a lack of joy when that person is not around. These parasocial attachments can be quite intense, but can also be easily disturbed by the confusion between physical and mediated realities. For instance, in 2017 as part of the #MeToo movement, actor Kevin Spacey was accused of sexual assault and consequently was replaced by Christopher Plummer in the role of J. Paul Getty in Ridley Scott's *All the Money in the World* (2017). Most of the scenes with Spacey were redone; others combined new footage with shots from the original that initially did not include Plummer. Scott said of the decision that "[i]t would have been a pity if the film were completely neglected because of what happened."[47] He was afraid that audiences would shy away from watching a movie with Spacey in it, because they might project the failings of the physical persona onto his mediated representation of a character in the film. Netflix also decided to fire Spacey from *House of Cards* and produced the last season of the show without the character of Frank Underwood.[48]

In the process of identity formation, fans often transform into collectors. While going on a hunt for memorabilia of their favorite artists, they are developing a parasocial relationship with mediated social entities. These entities (often referred to as celebrities or stars) do not always exist in the physical world of the fan. You might worship Harry Potter, but you surely will never get to meet him in physical reality. You only know him as a mediated representation in a series of books and films. However, you might catch a glimpse of Daniel Radcliffe, who portrayed Harry Potter in the films, but even then you only witness the actor and not the role, the public persona instead of the mediated figure. Even if you admire Radcliffe as an actor, you would merely come close to his mediated persona, the performance he allows the public to see of himself. The psychological processes to establish intimate experiences with mediated characters have been widely explored (mostly in relation to television personalities and the phenomenon of "affect TV"). Research has shown that emotional attachment to media personalities could lead to problematic fan behavior and an extreme experience of immersion. The idea of getting attached to mass media figures has been discussed since Donald Horton and R. Richard Wohl described parasocial interactions between media users and on-screen characters. They argued that the

> crucial difference in experience obviously lies in the lack of effective reciprocity, and this the audience cannot normally conceal from itself. To be sure, the audience is free to choose among the relationships offered, but it cannot create new ones. The interaction, characteristically, is one-sided, nondialectical, controlled by the performer, and not susceptible of mutual development. There are, of course, ways in which the spectators can make their feelings known to the performers and the technicians who design the programmes, but these lie outside the para-social interaction itself. Whoever finds the experience unsatisfying has only the option to withdraw.[49]

Every purchase, every concert, every signed copy brings audiences closer to their object of admiration—the one they only know from television, films, and music. Often they even secretly hope to merge their own physical reality with the mediation, to the extent that they completely disregard the mediated setting, as the example of Madame Tussauds, mentioned earlier. Even though visitors are standing next to wax figures of celebrities, their actions do not refer to the figure itself but to the mediated representation of celebrities. In that particular mediated setting, audiences "meet" George Clooney, but he will probably wear a tuxedo and hold a cup of cappuccino in one hand—familiar scenarios audiences will recognize from Clooney's role in *Ocean's Eleven* and his work endorsing a popular coffee brand. The fact that visitors do not see beyond the mediation is more evidence for how immersive these relationships are for them.

Shalin Hai-Jew describes these kinds of experiences as the *immersive parasocial*, suggesting a wide spectrum of effects of immersive parasocial relationships.[50] She designs a theoretical framework, which shows the range of mild to extreme parasocial effects. Mild effects could be sensory arousal, attention-getting, and increased entertainment value. On the side of extreme effects, she finds symptoms such as illusion-forming, vulnerability to manipulation, mediated voyeurism, obsession, addictiveness, and stalking. What Hai-Jew describes here is an extreme, unhealthy state of immersive experiences—the kind in which audiences forget about what is real to such an extent that it becomes a psychotic illness. In this hyperreality, they prefer the simulacrum to the real, and as Ryan suggested earlier, audiences then begin to create parallel worlds in their minds, which allow them to escape physical reality.

> Immersive fandom evokes a spectrum of *parasocial phenomena and psychological sensations*. They range from mild sensory arousal to extreme effects like illusion-forming, addictiveness, and obsession.

The lines between mild immersive reactions and extreme, unhealthy behavior are often blurred. Being in the liquid space of immersion involves an uncertainty in relation to social participation as audiences are in absolute defiance of objective reality. Visual displays of these relationships are enhanced through social media. Fans now have the opportunity to get in touch with celebrities on Twitter and Facebook and create a different kind of intimacy in the public's eye.[51] These relationships often remain one-sided, and the lack of replies and control might result in frustration and anger. This might lead to losing touch with the standards of physical reality. *Ultrafans* then begin to envision a utopia in which they are side by side with their object of worship, as seen in numerous examples of celebrity stalkers in recent years (those of Taylor Swift, Kate Beckinsale, Selena Gomez, and others). These fans are willing to do anything to immerse themselves in their vision of a parallel world. In 1981, John Hinkley Jr. believed he was in a relationship with actress Jodie Foster and thought he might be able get her attention by shooting President Ronald Reagan. Margaret Mary Ray, who suffered from schizophrenia, believed she was married to David Letterman. These individuals—often suffering from an underlying mental illness—were obsessed with the mediated representation of "their" star, while they were developing an unhealthy sense of intimacy with the mediation. They were living under the impression that a merge between their reality and the mediated image could actually happen—to the point that they were prepared to invade the lives of others.

Attachment theory gives us insights into what is going on in the minds of such obsessed individuals. Extensive attachment to media personalities often mirrors extensive attachment to people in physical reality.[52] First, ultrafans attempt to reduce the distance between themselves and their object of worship. Research has documented that people rearrange their daily schedules to stay in touch with the mediation. Ultrafans collect all sorts of trivia, tape television broadcasts, and contact celebrities on social media. Second, being around the object of worship provides these fans a sense of security and companionship. This form of the immersive parasocial could even have positive effects as it fulfills a need for interaction for those who might be isolated in physical reality or still need to develop a level of emotional autonomy (such as teenagers who are experiencing a transition from parental attachment to peer attachment). Tim Cole and Laura Leets argue that insecurely attached individuals might seek parasocial relationships more often, most likely as a compensatory response to their relational anxiety and incompetence.[53] Attachment to media personalities could provide comfort and companionship. Their presence is perceived as being as real as anything else. Third, detachment from the fan object leads to a form of protest. The individual experiences sadness, distress, and a drastic lack of life quality in times of being isolated from the fan object, which may lead to psychotic illnesses and dangerous obsessions, such as stalking.[54] Brian H. Spitzberg and William R. Cupach add: "A study of threatening fan letters revealed a potentially darker side to such motives, in which many clearly expressed a belief that they had some kind of personal relationship."[55] Detachment threatens the feeling of extended presence and the process of sense-making. The result is often a pathological reaction. The individual tries to merge the mediation with its own physical reality, but fails miserably. It continues to imagine fantasized interactions and scenarios in parallel worlds that it wishes to carry out in physical reality as well. At that point, these individuals no longer accept the mediation. They often no longer even recognize it. The practices of crossing borders and overcoming the frames of the mediation now lead to a dysfunctional and borderline-pathological behavior, which could turn into obsessive relational intrusion. The result might be an addictive attachment, which prevents them from initiating and forming relationships in their lives. These individuals get literally *lost in limbo*, often as part of an underlying mental illness.

These extreme forms of parasocial encounters are surely the exception within the broad spectrum of fandom experiences. However, parasocial relationships are as real as anything else. They can have the same effect on our lives as physical relationships. To become immersed means to be deeply, emotionally invested while establishing an intense attachment to mediated realities. Our affective and cognitive proclivities affect the formation of our identities and the curation of selfhood. Whatever we use as a stimulus to experience the immersive parasocial, the line between mild and extreme effects is quite narrow. Usually, the experience is able to provide safe and healthy relationships with media personalities, which potentially could have positive effects in the development of self-esteem. But it could also turn out to become an extreme form of addiction that influences every aspect of a person's life, an immersion beyond the meditation with strong social implications.

**Binge-Watching and Algorithmic Flow**

You might have had one of these unpleasant conversations with yourself in the past. The clock approaches one in the morning, and you have to wake up early to get to the office. But that tiny whisper in your head assures you: "Only one more episode. Really, I promise." You are already sold. To hell with the default countdown between episodes. You are clicking "next." You *want* this. You *need* this. Eventually, some hours later, one episode becomes three or more. For some reason, the pathetic struggles of Jimmy McGill in *Better Call Saul,* and the rise and fall of whole empires in *Game of Thrones* are more important than whatever you have to face in the morning. You are caught in binge-watching mode, taking in episode after episode due to the irresistible strength of the unfolding plot. Once you have finally made it to bed, the storylines remain in your mind—after all, you were not able to finish the season. Tomorrow, this will follow you around for some time.

Some may call it a "first world problem," but you are not alone with this. A 2017 survey by Nielsen[56] of more than 30,000 VOD (video-on-demand) consumers in 61 countries indicates that a large majority of TV viewers "binge" their shows. Netflix defines binge-watching as "completing at least one season of a show within seven days of starting" and claims it takes just twelve days after joining the platform until most users start their first binge.[57] Two-thirds of all responses indicate that viewers like to catch up on multiple episodes in a single sitting, making it a universal and global phenomenon (particularly in North America, with 73 percent, and in Africa/the Middle East, with 70 percent). Binge-watching redefines the way TV content is consumed and "changes the stakes of narrative engagement by reframing the temporality of viewing experiences to optimize emotional intensity and story immersion."[58]

But "to binge" is more than a way for audiences to get their daily fix of their favorite shows. It is a globally accepted pleasurable experience and an immersive practice of entertainment culture. Social life is segmented into different *media binge activities*,[59] from watching endless loops of videos on YouTube and well-stocked viewing libraries on VOD platforms to listening to individualized and mood-based playlists on Spotify. Binging transports audiences into a process of flow with the help of mediation. For instance, the phrase "Tinder binge" refers to "when a person spends an immense amount of time swiping right."[60] And like any cultural practice that is embedded in everyday life, "to binge" is used in the context of sense-making. Binge-watching is the manifestation of media personalization in which streaming profiles (or better: mediated selves) are designed to suit individual tastes and needs. Entertainment consumption on streaming platforms has been personalized and individualized through the use of machine learning algorithms, which "occupy a substantially wide array of spaces within social life, affecting a broad range of particularly individual choices."[61]

Media binge activities are the result of a process, which in psychology is often defined as "flow." Hungarian-American psychologist Mihály Csíkszentmihályi used the term in the early 1990s to describe a state of mind in which we feel the emotional and cognitive intensity of an activity. While studying the creative process of artists, Csíkszentmihályi was fascinated by the fact that they were able to "lose themselves" in their work and even disregard hunger, fatigue, and discomfort. At the same time, he noticed a certain kind of enjoyment while they were performing their activities, a key for him to understand how to set up and successfully complete goals. He was interested in how we could put ourselves in a mental state to achieve specific goals and "live a good life."[62] Flow activities could be almost anything—sports, games, even ironing clothes, or driving a car. Csíkszentmihályi did not focus on media consumption, but his findings certainly apply to it as well. He argued that the influence of flow on the perception of selfhood is underappreciated and that "[e]*ntering flow* is largely a function of how attention has been focused in the past and how it is focused in the present by the activity's structural conditions."[63] What strikes me here is the idea of structure within flow experiences. In other words, the formal aspects that configure, shape, and modulate an experience have an effect on our psychological reaction to it. That means that media binge activities—which arguably put audiences in a flowlike state of mind, namely, a feeling of enjoyment, distortion of temporal experiences, and a loss of self-consciousness—are strongly influenced by their structural conditions. As soon as those flow activities are enriched with meaning and purpose—for instance, a compelling story or curated playlists—we transition into immersion after the mediation has ended. Whatever that particular experience might mean to us, and how it is linked to mediated content, is based on the programmability of flow.

To uncover how media binge activities transport audiences into a state of immersion, let us look by way of example at binge-watching experiences on streaming services such as Netflix.

Binge-watching as an immersive experience in postmodern culture can be categorized in three different phases. The first one involves the implementation of binge-watching technologies. Networks have always devoted longer periods of their programs to multiple episodes of a single show (from a view hours to multiple days in so-called "marathons"). In addition, video recorders allowed audiences to record episodes and watch them later back-to-back. The introduction of DVD box sets added another distribution channel to the business of multinational entertainment conglomerates and satisfied the desire of audiences to watch serial content as a whole narrative unit. But as Casey J. McCormick points out, the term "binge-watching" has become increasingly popular since 2013, with Netflix expanding across the globe and introducing quality serial content like *House of Cards* and *Orange Is the New Black*.[64] Even before its expansion, Netflix represented innovative television more than anything else and made European audiences such as Germans crave becoming "Netflixed": "[T]he success of Netflix in the United States and elsewhere made it appear as a sort of faraway utopia that, once available, would provide German audiences with content that was otherwise impossible to see on local television."[65] In early 2020, Netflix had 182 million subscribers in over 190 countries.[66] In addition, the company has established itself as a major player in the production of television content and feature films.

Quality TV,[67] a term often discussed in relation to shows like *The Sopranos*, *Lost*, *The Wire*, and *Six Feet Under* in the early 2000s, has moved to digital platforms that privilege algorithmic mechanisms of programming, nonmaterial acquisition, and practices of mobile and individualized consumption. Platform owners implement architectures that set out the parameters of data and algorithmic governance. These mechanisms allow them to govern platform activity and content and enhance user experience. Algorithms favor the categorization in genres, tastes, and recommendations. The more audiences watch and rate, the more the platform learns about their preferences and tastes, and the more it will customize its portfolio accordingly. With a dizzying array of streaming services such as Netflix, Amazon Prime, Hulu, Sky Go, HBO Now, HBO Max, Apple TV+, and Disney+, audiences face countless hours of addictive, high-quality programming. But could immersion—spending multiple hours in their ecosystems—be bad for our health?

This leads directly to the second phase of binge-watching culture in which "to binge" is perceived as a tempting, addictive force for immersing audiences. The addictiveness of the interplay between content and platform has been defining Netflix from the first season of *House of Cards* in 2013. The show's initial success was strongly intertwined with Netflix releasing all 13 episodes on one day. With showrunners such as David Fincher and leading roles assigned to Kevin Spacey and Robin Wright, it had the potential to become a hit. But even though the Machiavellian viciousness of Frank Underwood was based on globally known Shakespearian character roles, the complex conditions of American politics offered hardly any universal appeal beyond the US market. It was the distinctive distribution strategy that made the difference. It introduced narrative parameters linked to the temporality of its consumption. Audiences were already familiar with similar interrelations such as "real-time temporality" in shows like *24* (every episode was 60 minutes long and took place during one hour in the life of its characters). In fact, one day before the first season of *House of Cards* was released, co-showrunner Beau Willimon announced in an interview: "Our goal is to shut down a portion of America for a whole day."[68] His hypothesis was based on the causal correlation between availability and temporality. The more episodes available at once, the more episodes viewers would watch in a single sitting.

Time proved him right. In fact, the increasing amount of high-quality serial content on multiple platforms has resulted in addictive behavior similar to many other binge activities such as eating and drinking, as a much-debated study by the University of Texas suggests.[69] The researchers conducted a survey of 18- to 29-year-olds on how often they binge-watch television and how often they had feelings of loneliness, depression, and self-regulation deficiency. The findings showed that those who lacked the ability to control themselves were more likely to binge-watch. It is not the first time excessive media consumption has been linked to health problems. Studies have shown that prolonged television viewing is linked to an increase in obesity, diabetes, social isolation, and a suffering sex life. Excessive gaming has been linked to physical and mental problems. All of these studies underline a crucial argument: excessive media usage is related to feelings of escapism and the devaluation of physical existence. Peter Vorderer argues, regarding the dangers of interactive-media consumption, that every individual has the potential to feel part of a virtual community. This feeling of belonging would depend on the illusion that mediated experiences veridically reflect and depict physical reality.[70] To binge means to remove yourself from physical reality. These forms of escapism and the neglect of physical reality are often linked to irresponsible and harmful behavior, which ultimately could lead to negative long-term effects.

> Media binge activities transport audiences into a state of flow, a *process of emotional and cognitive intensity*. Excessive media usage has been often linked to physical and mental illnesses.

There is an ongoing debate as to whether excessive use of media is to blame for various health issues, including a serious lack of exercise, social interaction, and self-regulation.[71] Regardless, immersion is now associated with addiction and problematic behavior. The growing number of shows to watch escalates the problem. As addicts often blame the dealer for their addiction, media conglomerates are blamed for providing too many shows at the same time on too many different platforms. In an article with the title "Please No More Brilliant TV. I'm at Breaking Point," Jack Bernhardt worries: "In the UK, if you want to watch all the 'must-see' TV being produced, you have to sign up to Netflix . . . , Amazon Prime . . . and Sky TV. . . . And even if you can afford it, when will you watch it all? Netflix apparently has 34,739 hours of content on it. That's nearly four years."[72]

As traditional, unidirectional structures of programming do not work on VOD platforms, the production cycles of seasons have adapted as well, which results in multiple shows running at the same time, leaving audiences in the dark about how long the break between shows might last. The struggle is real and provokes new strategies of consumption. For instance, "speed watching" (streaming content at accelerated speed, sometimes two times as fast as normal) is a trend to cope with the huge amount of serialized content and to stay on top of it. Netflix's "post-play feature" (the default countdown between episodes) accelerates the immediacy of the experience. Netflix gives audiences exactly five seconds between episodes to decide whether they would rather do something else. If they refuse to make a decision, the platform will make it for them. It will not only autoplay the next episode, it also skips the intro and "previously on" segment to resume the narrative without any interruption.[73] Viewers might not even realize that they are watching a new episode at all. In addition, Netflix allows them to download selected titles to their mobile devices to ensure they can access them without being online.

In the third phase of binge-watching culture, binge activities are embedded in daily routines. Media binge activities are now becoming a form of sociocultural immersion, with effects on social and cultural conditions. Whatever has been described as addictive and problematic before is now a defining element of contemporary media—a lifestyle that celebrates individual freedom by allowing yourself to "get lost" in a binge activity. Audiences are consuming entertainment for multiple hours, watching them simultaneously, and planning other activities around them. Watching episodes during a long commute on the train is a common activity for many people. Getting lost in one of your favorite shows helps to kill time and blocks out the daunting commute for a while. For instance, British telecommunications provider Three uses the binge-watching trend to promote unlimited data allowances with the slogan "Go Binge."[74] In late 2017, Three posted several

Sociocultural Immersions 195

ads with images from Netflix shows with corresponding taglines on the public transport system in London, such as a threatening still of *Stranger Things'* character Eleven with the line "When you've got a series to finish and all the seats are taken," or a smiling Pablo Escobar of Netflix's *Narcos* with the message "When your commute means one more episode."

Of all the binge-watching platforms, Netflix became synonymous with accelerated and intensified media consumption. The platform owners successfully created a brand out of Netflix's cultural significance. Social activities on a Saturday night might now involve enjoying a night in with a glass of wine and Netflix, and statements like, "I have had a hard day and all I want to do this evening is put on Netflix, curl up into the fetal position and consume four straight hours of visual entertainment." The *Urban Dictionary* lists more than 40 different euphemisms related to the platform,[75] including "Netflix and chill" (having casual sex), "Netflixaholic" (dedicating every spare moment to binge-watching), "Netflix and deal" (performing job-related tasks during binge-watching activities), and "Netflix and die" (enjoying a film and falling asleep). In addition, viewers tend to create a specific atmosphere that now romanticizes the binge experience.[76]

Netflix established a new form of mediated immediacy by becoming an influential factor in very personal, intimate, and apparently unmediated experiences. How we think of partners, friends, and colleagues has been modified

Figure 4.3
"Go Binge": British telecommunication provider Three uses Netflix and binge-watching to attract customers on their commute. © 2018 the author.

and governed by the common practices of the platform. For instance, Netflix allows users to share their account details within a household, and Netflix CEO Reed Hastings has been very supportive of costumers sharing their account among friends and family, stating: "We love people sharing Netflix. . . . That's a positive thing, not a negative thing."[77] This allows for the creation of many different individualized profiles in one single account and reveals individual tastes and interests to everyone sharing the account. In other words, if I am only watching rom-coms, I am either very confident anyone else with the password will not check out my list of recently watched titles or that they will not judge me for it. Maybe I am simply good at covering my tracks by including some must-sees and critics' favorites. In any case, my mediated self becomes visible to others (if only through the platform itself).

Binge-watching on Netflix also influences our understanding of social relations and sexuality. Over two-thirds of all listed activities in the *Urban Dictionary* related to Netflix refer to sexual encounters, mostly dominated by the male gaze. The phrase "Netflix and chill" has been innuendo-free for a long time, but is now a euphemism for casual sex. The *Urban Dictionary* added it in April 2015 with the definition: "code for two people going to each other's houses and fucking or doing other sexual related acts."[78] Searching for the phrase online results in hundreds of memes such as images of condoms with slogans printed on them. For example, "Guys tweeted pictures of smug faces alongside captions such as, 'When she says Netflix and chill,' while girls tweeted pictures of shocked, dismayed faces with captions such as, 'When you find out what Netflix and chill means.'"[79] Variations of "Netflix and chill" are "Netflix and chill alone" (masturbation), "Netflix and cuddle" (no sex), "Netflix and freeze" (only one party wants or expects casual sex), "Netflix and Bill" (inviting someone to watch Netflix but luring them into sex instead, a reference to former US president Bill Clinton), and "Netflix and eat food" (oral sex during binge-watching).

As the immersive practice of binge-watching becomes more immanent in our lives, it also has an effect on how we behave in relationships. For instance, for Valentine's Day 2018, Netflix reminded its users of their first binge-watching experience as a nostalgic (and romantic) event: "You never forget your first." According to Netflix, *Breaking Bad*, *Orange Is the New Black*, and *The Walking Dead* are "the shows that people fall in love with first" and 35 percent of its users have rewatched their first binges.[80] "Netflix cheating," on the other hand, is a phenomenon related to watching a show on your own instead of waiting to watch it together with your partner. According to Netflix, nearly half of its users are "cheaters," a threefold increase since Netflix first conducted a study on it in 2013. In a more recent study from 2017 of roughly 30,000 users, 66 percent of them said, "the shows are just so good we can't stop binging" while 81 percent of all cheaters said they were repeat offenders.[81] The temptation to move forward within a show has the potential to cause an argument between partners. To avoid this, viewers are able to delete whatever they have been watching on their viewing-activity page.

With Netflix becoming a defining part of social interactions, there are now several ways to enhance the experience. Netflix recommends a list of smart TV sets with faster access to the app, a designated "Netflix button" on the remote control, a high-resolution interface, and the latest update of the Netflix app with the newest features.[82] But there are also hundreds of apps and browser extensions helping to sort, filter, and rank titles, such as *Netflix Rate* and *Nenhancer Chrome* to add critical aggregates from *Metacritic* and *Rotten Tomatoes*, or *Netflix Roulette*, an app offering a random choice of titles for audiences to watch next. *Rabbit* is a video chat service that supports the sociality of binge-watching by running a chat while watching episodes together with others. *Flix Plus* helps users to customize the Netflix platform (e.g., move the "My List" section to the top), adds notes to their favorite titles, and hides potential episode spoilers. *Flix Assistant* is specifically designed to enhance the immediacy of the experience—the browser extension eliminates the default countdown between episodes to get to the next one more quickly.

Netflix is embracing the fact that binge-watching activities have gained in importance and become part of daily routines. Similar to Nintendo's Labo platform Toy-Con Garage, Netflix is asking its users to provide ideas to "make the Netflix-watching experience even more spectacular."[83] In several manuals and descriptions on their website *makeit.netflix.com*, Netflix takes on suggestions from its audiences and explains how to interrelate binge-watching with other activities, or how to avoid any disruption of the experience. Projects include The Switch, a button to dim lights and to silence incoming calls during binge-watching sessions. Once the button has been pushed, The Switch automatically orders takeout to avoid audiences moving away from the screen to prepare food. A similar project involves knitted socks with a built-in accelerometer, a sensor that detects when viewers have stopped moving for a prolonged period of time. In case the viewer dozes off while watching a show, the sensor triggers a signal to the TV, which then pauses Netflix. Another project is called *Netflix Personal Trainer* that lets users transform their favorite characters from *House of Cards, Narcos, Jessica Jones, BoJack Horseman,* and *Orange Is the New Black* into fitness coaches. The gadget provides words of motivation with predefined phrases in three different coaching modes to enhance the training experience.

> Binge-watching influences the knowledge about *social relations and cultural conditions*. For example, common *platform-related practices* govern the understanding of relationships and sexuality.

Binge-watching is determined by both the narrative structure of the content and the design of the platform. In other words, any psychological sensation emerging from binge experiences relies on how content is conceptualized and how it is positioned within the technological and commercial logic of the platform. That also implies though that the story itself is not more important than its formal aspects. Umberto Eco famously noted that serial content is defined by repetition, iteration, and duplication to achieve an intrusive result.[84] Audiences reach a level of security by noticing recurring patterns within the structure of serial content—a comfort zone, which assures them they know what is going in. Characters, for instance, are defined by specific characteristics, internal and external conflicts, and those hardly change throughout the whole series. Temporal structures and dramaturgical elements like cliffhangers are also recurring themes that can be found in every episode.

In this regard, Casey J. McCormick's analysis focuses on the temporal structure of *House of Cards* and how the formal representation of the narrative is embedded in the overall logic of Netflix's platform mechanism. *House of Cards* has all the recurring elements previously discussed to establish a state of flow: clearly defined characters and storylines, repeating themes and conflicts, and aesthetic iterations in the fields of cinematography and sound design. Some commentators even noticed a particular habit in the color palette (dominantly blue and yellow) of the series and ask: "Why does every scene in 'House of Cards' look the same?"[85] By introducing paratextual framing, *House of Cards* also has no specific titles for episodes but rather "chapters," as in literary storytelling, which link "the show to a history of serial fiction, it separates it from the dominant way

## Sociocultural Immersions

of organizing TV, and it creates continuity across seasons."[86] McCormick adds: "The fact that the show's creators have placed such an emphasis on binging as an ideal mode of consumption ... sets up a preferred reading that is reliant upon accelerated temporality and the play between story time and viewing time."[87] Other formal aspects, therefore, include the play with narrative temporality and temporal play—the overlap of story time and real-world time—as well as characters breaking the fourth wall (Frank Underwood addresses viewers several times to make them accomplices). In 2017, Netflix started to experiment with interactive animated formats,[88] in which audiences could decide, out of a set of options, how the story progresses (*Puss in Book: Trapped in an Epic Tale*, *Buddy Thunderstruck: The Maybe Pile*, *Stretch Armstrong: The Breakout*). The platform's experiments with the structure of interactive storytelling culminated in the success of the interactive film *Black Mirror: Bandersnatch* (2018), which had five different endings and several hidden Easter eggs. *Unbreakable Kimmy Schmidt: Kimmy vs. the Reverend* (2020), an interactive movie spun off from the successful Netflix sitcom, applied the concept of different choices to the comedy genre. However, interactive formats remain the exception, as traditional structures of serial storytelling are still most popular among audiences.

**Figure 4.4**
*Unbreakable Kimmy Schmidt: Kimmy vs. the Reverend*: Netflix experiments with the structure of interactive storytelling. Reprinted with the permission of Netflix. © 2020 Netflix.

At the same time, the structural elements of storytelling are intertwined with the design of the binge platform. For instance, scrolling down on Netflix's interface magically reveals more and more titles, giving the impression of infinite options to watch. In late 2017, Netflix also introduced personalized visuals and artworks on its homepage that represent the viewer's taste and recommendations. For instance, if you are interested in specific actors, future recommendations on your personalized Netflix homepage would highlight them: "A member who watches many movies featuring Uma Thurman would likely respond positively to the artwork of Pulp Fiction that contains Uma. Meanwhile, a fan of John Travolta may be more interested in watching Pulp Fiction if the artwork features John."[89] That way, Netflix creates a personalized and intimate setting for its viewers. Taste in that regard is not necessarily a quality of reason or aesthetic judgment, as Kant once put it.[90] It is, instead, a result of relativism and individual experiences. Through the use of algorithms, including audience composition and constitution (demographic profile, interests, lifestyle, etc.), and consumption preferences (program choice, amount of viewing, duration of viewing, etc.), audiences become predictable. Netflix analyzes an overwhelming amount of information to create more than 70,000 hidden microgenres[91] and produce predominantly content which is able to provoke binge activities.

What I would like to call *algorithmic flow* is based on what Antoinette Rouvroy describes as "data behaviorism," meaning "producing knowledge about future preferences, attitudes, behaviors or events without considering the subject's psychological motivations, speeches or narratives, [instead] relying on *data*."[92] She notices the installation of *digital regimes*, in which the practices of big data analytics and data mining govern the ways audiences behave and watch and predict what they want to see in the future. To cluster data from time series analysis, GPS, genres, classifications, and ratings allows platform providers to come up with causal relations between specific parameters. For instance, audiences who have watched (a) films by director David Fincher, (b) films with actor Kevin Spacey, and (c) the original UK version of *House of Cards*, and who (d) are interested in the genres "crime," "thriller," and "suspense," will most likely watch the US remake of *House of Cards*.

Actions and behavior become predictable currencies. Personalization, recommendation, and search (PRS) are the keys to understanding binge platforms. Once you start a video on YouTube, the platform chooses similar videos to continue your experience. Once you search for an artist on Spotify, the platform will give you "similar artists" (mostly connected through band members, genre, compilations, and your search history) to make you stay in the flow of the experience. Netflix's premium content is actually very limited. But its titles are carefully planned and researched to maximize their appeal to a targeted section of the audience, exemplified in the relaunch of *Gilmore Girls* (a popular and missed TV show with a strong fan base), and the science fiction show *Stranger Things*, which exploits nostalgia from the 1980s among today's 30- to 40-year-olds, who grew up with films such as *E.T.*, *The Goonies*, *Gremlins*, and *Stand by Me*. All of these titles are based on data and preferences provided by the platform's users through ratings, searches, and algorithms. The phrase "You might also like . . ." at the end of an episode is more than just a recommendation. It is a way to collect data from the audience. Netflix knows what its users want—and what they *will* want—long before Netflix actually produces it. For a long time, Amazon even offered audiences pilot episodes for planned shows, greenlighting whichever pilots got the best ratings.

Biometric data, in particular, becomes a valuable asset for platforms to monetize. There are behavioral biometrics, such as gestures, handwritten text, and vocal recognition that are used in several devices by Google, Amazon, and Apple. There are also physical biometrics, comprised of fingerprint, vein, iris, and facial recognition used as security mechanisms on smartphones and tablets. There is also DNA, the most valuable human data. In 2018, Spotify teamed up with the world's largest for-profit genealogy company, Ancestry, to determine listeners' "musical DNA" based on DNA test results. Ancestry promises "a unique Spotify mix of music, inspired by your origins."[93] Critics warn about the risks of misuse and express concerns that valuable data assets could be hacked, as well as the potential for mass surveillance: "Once collected, it points to you and you alone; once lost or stolen, it is open to permanent abuse. You can always rethink a password but you can't rewrite your DNA."[94]

Despite all the risks for users, algorithmic mechanisms are established modes of audience research on digital platforms. As Cédric Courtois and Elisabeth Timmermans notice in regard to Tinder: "algorithms are practically invisible to users. Users are seldom informed on how their data are processed, nor are they able to opt out without abandoning these services altogether. . . . Due to algorithms' proprietary and opaque nature, users tend to remain oblivious to their precise mechanics and the impact they have in producing the outcomes of their online activities."[95] Tinder, for instance, is known for merging a gamification approach (similar to a narrative structure) with machine-learning algorithms. Gamification is the use of games design in nongame contexts: "There are certain controls and maneuvers that hold specific meaning within the game, like swiping left and right to pass or like users, and a magic circle is established that allows players to behave inside the realm of the app."[96] From the user's perspective, the process seems very simple, but there is more to Tinder than meets the eye. The logics of the platform draw upon algorithmic filtering, which controls who sees whom: "we assume Tinder carefully doses matches, meaning that its governing algorithm monitors activity and intervenes in its outcomes to keep the user experience in check."[97] Built-in restrictions to push users to paid services such as Tinder Boost and Tinder Gold and Platinum subscriptions install a "digital regime" that influences not only match-making on the platform but the dating culture in general.

> The *mechanics of platform-governed algorithms* provoke *social and psychological sensations*. Data and algorithmic governance are hardly ever transparent but are strongly linked to user behavior and actions.

The logics and tactics of these platforms transform media binge activities to immersive experiences—both on a narrative and structural level. These activities can be found in a variety of different areas, as they often become culturally significant *beyond the use of media*. They are defined by the interplay of user behavior and the algorithmic curation of big-data assets. Technology supports this process by questioning the limitations of consumption and by modifying our understanding of the human-technology relationship.

### Posthuman Hybridity and Artificial Intelligence

How is immersion influenced by the growing importance of human-machine hybridization, artificial intelligence, and smart and wearable technologies? Over the course of this book, I have defined immersion as the liquid space we are entering, a threshold to cross, which ultimately epitomizes all the conventional boundaries between the physical and the mediated, the analog and the digital, the natural and the artificial, the animate and inanimate, the human and the machine, the conscious and the seemingly insentient. Our understanding of these dichotomies has always been based on the assumption that there are borders and thresholds between them. Initially, the medium has always been something else than us.

Within the last few years, though, we have noticed tendencies that reframe the process of crossing borders to mediations. The pull of technologies on every aspect of our lives, the influence of machines on how we perceive our selfhood, free will, data, and surveillance has distorted our relationship to the mediated. Machine entities, often in the shape of wearable and smart technologies, become attached to our bodies and create a new version of human-machine hybrids. We are confronted with intelligent machines questioning the "realness" of what we have always perceived as part of the human condition: work, leisure, friends, and lovers. Their influence reframes the human species through permeable and communicative technologies and introduces new ways of conscious sense-making. I have previously compared immersion to the act of swimming and the experience in which we are protecting our bodies against the fluid mass around us. We are determined to maintain the separation between body and water. We are aware that once we tried to breath underwater, we would be in serious danger. However, what if we were *required* to open our mouths? What if the act of swimming entailed that our bodies *become one with the water* by letting it enter our organisms? In the process of *human-machine hybridization*, it now becomes essential that *body and mind merge with the medium*.

One cannot think of humanity as distinct from technology anymore. New technologies are designed to improve certain aspects of our lives, but by doing so they frequently move closer to our bodies. They seemingly "improve" our physical beings and enhance them through technical prostheses to restore and modify their functionalities. This process is reminiscent of the image of Jude Law as PR nerd Ted Pikul in David Cronenberg's film *eXistenZ* (1999). In order to play a VR game, Law's character needs to connect to a synthesized amphibian game port through an opening in the human's lower spine. At first it is a painful procedure, but quite soon the game port becomes part of the human body, it connects to an alternative reality. Playing the game creates a "cyborg body"[98] in which the physical appearance is considered to be something put together, artificial, and mutated. This new organism is not merely a technoid mixture of human and machine, or the extension of the body through media prostheses in the McLuhanian sense.

Instead, it is a mutation of organic data and mediated features. The result is the formation of a biotechnically mutated organism that further dissolves the thin line between the physical and the mediated until it entirely disappears. It is the end of the body image as we know it.

The Cypriot-Australian performance artist Stelarc once noted that "the body becomes obsolete." Stelarc uses various mechanical and digitally controlled devices attached to his body to construct a collaborative physiology between human and machine. The conventional body is replaced by a cyborg creature, which uses the practices of immersion to reposition it "from the psycho world of the biological to the cyber zone of the interface and extension—from genetic containment to electronic extrusion."[99] Stelarc is a bodyhacking pioneer, a kind of DIY experimenter, who seeks to enhance his own flesh and blood with biometric implants and cognitive enablers. Like the artists Guillermo Gómez-Peña and Eduardo Kac, whose cyborg visions include transplanting computer chips into their bodies, Stelarc searches for aesthetic strategies to challenge the parameters of embodiment. For instance, he questions the notions of physiology by using VR systems to break out of the vessel of his own corpus. Skin is no longer the interface that separates the body from the world outside. It is now replaced by a technological surface, which is exploited by the globalized techno-industries to establish new power relations. The result is a *posthuman organism* in which body and technology each condition the other. Within this *mediated hybrid* it becomes difficult to distinguish where the materiality of the physical ends and the medium begins. Marwan M. Kraidy adds: "Hybridity is a risky notion. It comes without guarantees."[100]

Hybridity is both a marginal and an emergent phenomenon. Its role in contemporary media culture implies a profound transformation of the human consciousness to separate between mediated and physical worlds. The closer the medium is attached to the body, the more it possesses autonomy and materiality and oscillates from an external technocratic object to an internalized posthuman form. The critical fractions and personal frictions between the human and the machine, boundaries that arbitrarily include or exclude whatever is mediated and physical, become obsolete. There is only one agent in all of this: the *posthuman*. What does that mean for immersion? How are we able to assess immersion beyond any mediated experience, when we are constantly part of the mediation, as the medium needs us and we need the medium to exist?

> Posthuman hybridity is the *merging of organic data and mediated features* such as technology. The result is an artificial, mutated *posthuman organism*.

Posthuman hybridity can be categorized in two phases. The first phase describes the convergence of media technology and the human body. It is a creeping process of assimilation as technologies slowly and almost unnoticeably move closer to our bodies, mainly through the improvement of design, size, handling, and user experience. The most cited mainstream example in postmodern media culture in that regard is the smartphone. It involves the convergence of communication channels, which are now streamlined side by side to each other within one single, wireless, mobile device. The smartphone is the advancement of the mobile phone, which was envisioned already in the mid-1970s. However, smartphones in the twenty-first century are much more than mobile phones. They are comprehensive personal computers that play a significant role in people's lives. If you have ever lost your phone, or even, for that matter, were confronted with a dying battery, you will know about the feelings of fear, anxiety, and disorientation that might result from it. Early smartphones were bulky, but fairly exclusive brick phones such as the IBM Simon Personal Communicator released in 1994. Despite being relatively expensive, the fact that it was mobile changed how users perceived communication and ultimately reality. With the Simon you could not only place and receive calls. You were able to send emails, facsimiles, and cellular pages. It had a calendar, address book, calculator, scheduler and notepad, and it was able to display maps, news, and stocks. These functionalities made everyday life easier, and users started to develop closeness and intimacy with a device like they never had experienced before.

Fast-forward to today, the smartphone is a multimedia powerhouse. It organizes not only the lives of business people and enthusiasts, it connects us to endless possibilities that we continue to explore: watching movies, playing games, chatting, sharing content. For most people, it is the gateway to the world of social media, supported by several apps, which help to organize social media accounts, platforms, and avatars. Some find it difficult to function without their phones by their sides: most people are rarely more than five feet away from their phones. Nomophobia is considered to be a disorder in postmodern societies; the term is used to describe the discomfort caused by the nonavailability of communication devices. Its symptoms have a lot in common with "smartphone addiction," a dependence syndrome that could possibly lead to behavioral changes, obsessive-compulsive disorders, sleeplessness, stress, and anxiety.[101]

Smartphones have changed how we perceive and engage with digital communication.[102] We feel the urge to constantly check our phones for new messages and become obsessed with real-time messaging and personal feedback. This "fear of missing out" (FOMO) is defined by the desire to stay continually connected with others.[103] We wonder whether and when someone reads our messages (and why there is no instant reply). For instance, the introduction of the blue ticks in WhatsApp chats is Facebook's response to the "always on" mentality in digital communication. However, the read receipts have been criticized for allowing users to spy on others and enabling passive-aggressive behavior. Only after a public backlash did Facebook include an option to disable the feature entirely.

The smartphone is still an external feature separated from the human body, but we keep it close, almost too close. We keep it in our pockets, put in on a table during a meeting, no matter if professional or private. We are connected with it through Bluetooth earbuds and gesture controls. Biometric features such as fingerprints, as well as iris and facial recognition, are used to familiarize the machine with the human organism, and the other way round. For instance, Apple introduced fingerprint authentication in 2013 with the release of the iPhone 5s, and a number of other providers like Amazon, Starbucks, PayPal, Dropbox, and eHarmony added support for biometric authentication features to log into their services without a password. In 2017, and with the release of the iPhone X, Apple replaced the fingerprint with the Face ID facial recognition. The built-in phone camera analyzes more than 30,000 invisible dots and creates a map of the face while you look at it. As soon as the facial map is encrypted, the phone unlocks. At the same time, Apple saves exact biometric data; therefore, the feature is limited to only one face per device, as it recognizes slight changes in facial appearances. Apple explains, "Face ID automatically adapts to changes in your appearance, such as wearing cosmetic makeup or growing facial hair. If there is a more significant change in your appearance, like shaving a full beard, Face ID confirms your identity by using your passcode before it updates your face data. Face ID is designed to work with hats, scarves, glasses, contact lenses, and many sunglasses. Furthermore, it's designed to work indoors, outdoors, and even in total darkness."[104]

In addition, Apple uses the distinctive features of the human face to create new forms of emoji for the iPhone X, called Animoji.[105] The camera of the phone captures more than 50 facial muscle movements and transfers them to AR-animated emoji: "You can become a dragon, a bear, a skull, or a lion."[106] Animoji are the link to semiotic codes and performativity as outlined in the 360° gaze framework. The AR-animated emoji are globally used icons as well as performative practices. They transform individual facial expressions of physical faces to semantic units within a semiotic system and establish another form of human-machine hybridization.

Through headphone and earbud cables, we are plugged into our machines, or rather, the machines are plugged into us. While listening to music, watching movies, and speaking on the phone, we are isolating ourselves from our physical surrounding. This expands our notion of the innumerable and intertwined layers of reality. The perception of our physical reality changes while we are connected to our phones.

Machines are creating a distorted version of physical reality by connecting us to devices in various ways, often without a visible separation. For instance, Apple's AirPods are wireless earbuds and connect through Bluetooth with the phone, making users look as if they have mechanical sound extensions in their ears. Traditional earbud cables had always been a thin but still visible

**Figure 4.5**
iPhone X users can send AR-animated emoji that mirror their facial expressions. Reprinted with the permission of Apple. © 2017 Apple.

dividing line between the human and the device. Wireless earbuds, however, merge with the ears and become socially accepted and widely used. They tend to have minimal builds, not a lot of padding, and not only become hidden, invisible connectors between humans and sound-playing devices but also reframe, deconstruct, and shatter body aesthetics. Technology, which was once an external feature, is now an integral part of both the human body and consciousness through its applications and accessories. Future generations of earbuds will be solely wireless and users will be required to use them, mainly because phones without headphone jacks will become the industry norm. Users will depend on wireless solutions to access the intimacy of the soundscape of their favorite music, films, and TV shows. That offers the opportunity for technology providers and platform manufacturers to literally "put a flea in their ear," and surround them with content and advertising. In addition, gesture control technology promises that in the future, users will not need to touch their phones at all. They can simply wave their hands or pinch their fingers in the air to unlock their phones and control music, videos, phone calls, and alarms.

Eliminating the cables on high-end VR headsets is a similar case. The tether "keeps us connected to reality (the one we're trying to escape with VR)"[107] and distracts audiences from feeling present in a digitally created environment. The mind is concerned with the tether of the system as "the cable keeps us from completely detaching from the physical space we're in; somewhere in the back of your head your brain is tracking the (virtually invisible) cable and deciding when you need to step over it, twist a different direction, or avoid hitting it with your arms."[108] A variety of technology solutions such as the new standalone headset generations by Oculus and HTC try to overcome the problems related to bandwidth, latency, and price. Ultimately, every visible separation between the human and the machine will disappear like the frames from a painting. And while this happens, the medium merges with us, hidden, invisible, and unnoticed—and often out of convenience, trend, or fashion.

This leads us to the second phase of posthuman hybridity: the evolutionary fusion of human and media technology to the extent that one cannot exist without the other. This creates a posthuman version of organic life that needs to go back to the very basic questions of humanity such as: Why are we? What are we? What is "we"? How do we understand the concept of humanity and our awareness of it?

Examples of technological gravitation and artificial intelligence, which arguably accelerate this development, can be noticed everywhere around us. Recall when Google considered introducing Glass to the consumer market, and that Apple plans to do the same with its own AR eyewear. Wearing lightweight glasses with integrated displays, which connect the carrier to the online world, was too risky to pursue at the time. However, as everyone who wears conventional glasses will know, after a while you forget that you are wearing them. At the beginning you might see them as an uncomfortable external object, but at some point you hardly notice them anymore. They become part of your physical appearance. We need to assume that Apple would like to achieve the same level of comfort with its AR glasses, while they would obviously be much more than just a vision aid. The device would become a permanent rabbit hole to a world of mediated experiences, governed by a powerful media conglomerate. Wireless earbuds influence our perception on the same level (even if they only impact our sense of hearing), especially because future generations promise us more benefits from wearing them.

For instance, Google's Pixel Buds are wireless earbuds, which offer easier access to Google Assistant and Google Translate, and therefore real-time translation. Users merely need to touch the right earbud to activate the Translate app on their Pixel phone to translate whatever has been said in another language: "You hold your finger down on your ear and say, 'help me speak French,' and speak a phrase. When you lift your finger, the Translate app speaks and displays your translation. Then the person you're speaking to holds a button down on your phone and says their reply, which you hear in your ear."[109] In an ideal situation, both participants in a conversation would wear the earbuds and could talk in their native language, while the Pixel translates in real time through each other's ears. The gadget, which reminds many of the Babel Fish in Douglas Adams's novel *The Hitchhiker's Guide to the Galaxy*, positions the wireless earbud as a social tool, necessary to avoid unpleasant conversations and encounters associated with a lack of proficiency in speaking in another language.

> As *media technology merges with the human body*, it introduces *new semiotics of communication* and *different performative practices* now *linked to devices, machines, and digital platforms*.

As soon as technology merges with the human organism, it is also linked to the formation of the subject. Devices become agential practices and uncover the indeterminacy of boundaries and closure. They become essential to mediate our access to the material world, as Judith Butler suggested in her account of "materialization"[110] and Donna Haraway in her notion of "materialized refiguration."[111] As a result of that, mediation moves beyond representation to the field of performativity. In that sense, performativity determines what is real. Through performances, we constitute our bodies with the help of technology and create an organism that did not exist before. This trans-/posthuman hacking process redefines the preconstituted physical limitations of the human body through performativity. We are only able to comprehend the functionalities of our bodies through the performative features of technology.

This complex shift in our understanding can be seen in various examples. Neurobiology and brain research, especially, in recent years have introduced the decade of wearable technologies, smart computing, deep learning, intelligent sensor networks, and the development of automated infrastructures, the latter also known as the Internet of things. American feminist theorist Karen Barad calls this the "thingification" of our times—"the turning of relations [also between humans and technologies] into 'things,' 'entities' and 'relata,'" while she asks in the same breath, "[w]hy do we think that the existence of relations requires relata?"[112]

For instance, wearable technology is used in various fields in our daily lives, as Kashif Saleem et al. point out:[113] security and safety, medical, wellness, sports and fitness, glamour, communication, and lifestyle computing. Our bodies merge with different kinds of tech products for measuring and tracking activities to monitor, manage, and optimize our physical and mental performances. Besides smartwatches there are smartglasses, contact lenses, smart tracking shoes and activity trackers in jackets, masks, bracelets, headsets, and wristbands like BodyMedia Fit Core (worn on the upper arm), Nike+ FuelBand (worn on the wrist), Jawbone, and Fitbit Flex that constantly process, cluster and profile our body data. In combination with several apps like Fitbit, rolling live data is able to visualize and remodel the way we perceive our physical condition.[114] As part of trends such as "lifelogging" and "self-tracking,"[115] users are measuring their physical activities, health, mood, memory, and sleep to optimize their performances through mediation. Mental performance can be influenced, for instance, with a PIP device, a tiny gadget coupled with a smartphone app to give immediate feedback about stress levels. Other examples are the Muse headband, a brain-sensing headband that gives real-time biofeedback during meditation practice, and several VR apps such as Visitations that create virtual meditation environments.[116]

Monitoring physical and mental performances is an immersive practice, as the perception is based on mediation. In other words, we are only able to comprehend how we feel and how we could improve by the rules and conditions of media. In this mediated perception, the motivational tagline is: "There is a better version of you out there." To find that "ideal self," wearable and smart technologies will monitor the heart, log exercises and diet, and estimate how many calories have been burned. They will compare the results with those of peers and encourage users to set goals to find further motivation to reach for, for instance, the prominently advertised 10,000 steps a day. Whatever the goal is, it is measured and visualized as mediated models, graphs, concepts, particle tracks, and photographic images. After achieving their goals, users might receive digital gold medals or bonus points that might unlock additional features in the app. All of this is part of an immersive experience mediated by the representation of human data.

As technology is intertwined with the human body, it materializes data which otherwise would never get materialized without the human organism attached to it. These materializations are performances positioned between the human and the machine. In fact, machines ("devices"/"apparatuses") are not framed anymore but merge with the organic. As Karen Barad notes, they dynamically set so-called "intra-actions" and "agential practices": "Apparatuses are not inscription devices, scientific instruments set in place before the action happens, or machines that mediate the dialectic of resistance and accommodation. They are neither neutral probes of the natural world nor structures that deterministically impose some particular outcome . . . but rather *apparatuses are dynamic (re)configurings of the world, specific agential practices/intra-actions/performances through which specific exclusionary boundaries are enacted.* Apparatuses have no inherent 'outside' boundary."[117] According to Barad's theory of "agential realism,"[118] "phenomena" or "objects" do not precede their interaction through representation, but rather emerge through what she calls "intra-actions" as part of performative practices. That means, for instance, that wearable devices would not be able to actually monitor the heart rate, but only visualize the interaction between the body and the device. However, as users are only able to rely on the data provided by the device and have no other reference point, their understanding of "feeling well" is conditioned solely by the medium. If an app tells me, I have walked 10,000 steps today, should I feel "good" now? It is a vicious cycle, as it leads to only one inevitable result: we begin to develop an understanding of our bodies through the mediation presented by machines that are now an integral part of our existential agency.

Barad's theory stands in opposition to Bruno Latour's actor-network theory (ANT) in which different actors constitute networks.[119] Barad rejects the clear distinction between human and nonhuman forms of agency. She argues that both human and nonhuman forms contribute to the production of knowledge and meaning. That way, she acknowledges the fluidity of different organisms merging with one another and (through a postmodern lens) points out that "[a]pparatuses are constituted through particular practices that are perpetually open to rearrangements, rearticulations, and other reworkings."[120]

In the age of posthuman hybridity, there is no clear distinction between humans and media. We are part of the mediation that only exists to mediate us. It is all part of mediated realities or "phenomena": "Reality is not composed of things-in-themselves or things-behind-phenomena but 'things'-in-phenomena."[121] For Barad, phenomena are "produced through agential intra-actions of multiple apparatuses of bodily production."[122] The posthuman organism transforms into an immersive experience. Our consciousness relies on both organic and inorganic material, but the differentiation is no longer unequivocal. The "things-in-phenomena" are *we*, who are immersed in dynamic reconfigurations of the human body, the mediation and the physical, the organic and the data, the agencies of the self and the medium combined. The apparatuses, devices, and machines disappear—in our bodies, hands, clothes, and up in the cloud(s). That way the human body can always be immersed within media.

> The posthuman is under the *permanent influence of immersion,* as it is positioned between the frictions of the physical and the mediated. The posthuman organism transforms into an immersive experience.

With that in mind we can now ask how the posthuman organism is influenced by the mechanics of artificial intelligence. For many, AI is the innovation of the future, the one that connects the machine with the human mind. It has progressed rapidly in recent years in widely used technologies such as spam filters, search engines, speech recognition, and image classification. AI systems can now produce "deepfake" content that is almost indistinguishable from photographic material and makes it nearly impossible for audiences to recognize whether the mediation has a reference point to physical reality or is artificially created. Silicon Valley tech industries, in particular, celebrate AI and robotics as a way to enhance and ease people's consciousness with, for instance, self-driving cars and robot carers for the elderly. Soon, machines will make ethical decisions for us. They will be able to create, reflect, and learn on their own. Google's AI division promises "[o]ur mission is to organize the world's information and make it universally accessible and useful, and AI is enabling us to do that in incredible new ways—solving problems for our users, our customers, and the world. AI makes it easier for you to do things every day, whether it's searching for photos of people you love, breaking down language barriers, or helping you get things done with your own personal digital assistant."[123] In its labs, Google works on solutions for deep learning, data mining, and custom-built units for cloud machine learning.

Lev Manovich has pointed out that ideas and methods used today for data analysis were developed in the nineteenth and twentieth centuries.[124] In the late 1950s, early pioneers in the fields of AI (including network analysis, neural networks, and deep learning), such as Herbert Simon, Allen Newell, John McCarthy, and telepresence guru Marvin Minsky, investigated cerebral processes associated with pathfinding in games, such as player affective modeling, coevolution, and automatically generated procedural environments. Their work pointed out that machines could perform various tasks that looked like human thinking, but that an artificial intelligence that thinks like a human would be impossible. In his book *Granular Society*,[125] Christoph Kucklick compares AI to playing chess against a computer. The human mind might fail to win almost every time in the competition as the machine learns and builds up a repertoire of moves and strategies. However, the mind is able to combine experience with empathy and creativity, which the computer is not able to do yet. A more updated, interdisciplinary understanding of AI comes directly out of the AI research divisions from Google, Apple, and Microsoft[126] and is influenced by the methods of neuroinformatics. This research field investigates the complex structures and functions of the nervous system by looking at neuroscience data and knowledge bases together with computational models for sharing, integrating, and analyzing data. Today's AI researchers build machines as complete autonomous agents that are able to show self-regulatory learning behavior. This machine-based intelligence employs self-learning algorithms and predictive models and is capable of mimicking human behavior through advances in robotics, sensor technology, semantics and language, and cognitive science.[127] John Johnston points out that, similar to human behavior that "entails a fluid linking of body, brain, and environment," a new understanding of AI needs to map behavior "as a dynamical system with varying 'trajectories,' 'attractors,' and 'bifurcations.'"[128] A new movement in AI research has abandoned the cognitivist perspective and instead believes that intelligent behavior of nonhuman agencies should be analyzed using synthetically produced equipment and control architectures.[129]

AI has evolved in fields like health, the automobile industry, education, aviation, transport, and telecommunication. Machine-based intelligence is able to support and even replace human capabilities. In many heavy industries, robots are already given jobs that are considered dangerous to humans. A 2017 study reveals that "[r]oughly three-quarters of Americans who have heard a lot about this concept (76 %) express some level of worry about a future in which machines do many jobs currently done by humans."[130] Machines are creating and processing data at a rate that is impossible for the human mind to comprehend. They shape our understanding of a post-digital and posthuman environment.

Intelligent machines are increasingly becoming intertwined with people's lives, and more importantly, are able to trigger human emotions. Media industries, especially, are about to become completely transformed by AI. When 5G becomes the new standard for mobile broadband, it will be completely designed, monitored, managed, and optimized by AI. AI machine-learning algorithms are embedded in apps such as TikTok to provide each viewer a different series of clips. Machine learning is also used by companies like Cinelytic, ScriptBook, and Vault to predict the eventual box-office success of movies. Postproduction processes, in particular, benefit from AI to improve the "realness" of characters in computer-generated imagery. In 2020, Creative Artists Agency, one of Hollywood's top agencies representing A-listers such as Steven Spielberg, Will Smith, and Scarlett Johannson, signed up "Miquela," a 19-year-old Brazilian model who is actually a virtual creation. Miquela is already a celebrity and has over 2.2 million Instagram followers, and she may be be less difficult to deal with than her physical competition when it comes to salary demands.[131] But it also works the other way round. For *Gemini Man* (2019), the VFX team of director Ang Lee made lead actor Will Smith appear younger by scanning his face, building a database of his expressions, and merging it with earlier performances of his career, such as in *The Fresh Prince of Bel-Air*. For his film *The Irishman* (2019), Martin Scorsese used AI technology to "de-age" Robert De Niro, Al Pacino, and Joe Pesci. In two years of postproduction, hours of footage and gigabytes of data captured by a new type of camera rig were used to mask the actors with imagery of their younger selves.[132] But even nonprofessionals can use AI technology to, for instance, upscale old video games for display on higher-resolution monitors.

AI systems are part of numerous devices in our daily lives. Over 100 million people every day talk to Amazon's Alexa,[133] and with other voice-driven AI assistants like Apple's Siri, the number grows into many hundreds of millions. Improved smart TVs with integrated Google and Amazon assistants allow audiences to talk to their televisions instead of using remote controls. AI-governed programming and fully responsive adbots will soon become the industry norm. Millions of people every day interact with chatbots via messaging apps like Kik and WeChat. Ray Kurzweil even predicted that AI chatbots would eventually pass the Turing test, meaning they would be indistinguishable from a physical person: "If you think you can have a meaningful conversation with a human, you'll be able to have a meaningful conversation with an AI in 2029. But you'll be able to have interesting conversations before that."[134]

AI is also redefining our understanding of creativity and art. Deep learning, in particular generative adversarial networks (GANs), have produced highly convincing simulations of aesthetics and styles. Artists such as Mario Klingemann and Helena Sarin are generating novel visual aesthetics with the help of AI, choreographers like Wayne McGregor collaborate with neural networks, and sculptors such as Scott Eaton and Ben Snell create new aesthetic forms with machine learning mechanisms. Music is already strongly influenced by AI technology. For instance, in 2019 Warner Music Group became the first major record label to sign a deal with an algorithm. The German mood music app Endel creates personalized soundscapes for users and is expected to produce 20 records a year.[135] In another example, avant-pop musician Holly Herndon used the machine-learning software Spawn to create her 2019 album *PROTO*.[136] Herndon's experiment in treating AI as another ensemble member stretches back at least as far as the 1990s, when David Cope, a music professor, created a computer program called Emily Howell, an interactive interface that composed music based on a database of compositions and feedback from the listeners. The machine produced classical music that was directed to the parts of the human brain that link the semantic dimensions of music to meaning and emotions. If immersive experiences are currently mediations at least partly created by humans, how will immersion change if the mediation is entirely produced by AI systems and autonomous machines? Ahmed Elgammal is a pioneer in the field of AI, and is interested in how AI can be creative without human intervention. His main argument is that creativity produces something novel, and the result is always embedded in social, economic, and cultural contexts. In one of his experiments, he used a range of data on various art forms, styles, and aesthetics from different significant periods in human history to let an AI system create new versions of them. While the results produced by the machine were certainly remarkable, they lacked any social and cultural meaning, suggesting that AI systems are currently able to enhance but not autonomously create immersive experiences.[137]

In Spike Jonze's *Her* (2014), a quirky, lonely man falls in love with the soft, witty, sensual human voice of a hypersophisticated AI, which gives him personal advice and intimate suggestions. The AI system is voiced by Hollywood actress Scarlett Johansson to make it sound more appealing, and that is already very close to how AI is currently influencing our understanding of immersion. In fact, in 2019, Amazon's Alexa introduced its celebrity voice program with Hollywood star Samuel L. Jackson: "Instead of relying entirely on prerecorded phrases, the Samuel L. Jackson voice is powered in part by Amazon's neural text-to-speech model. It's like a lightweight deepfake, but the actor obviously gave his permission to stand in for Alexa's standard voice."[138] The goal is to perceive AI devices as natural parts of the users' daily lives, maybe even to build up a personal relationship with them. The partial human resemblance through appearance (skin, face, hair) or human communication (language, voice, semantics) allows the human mind to connect with the machine much easier. Biological-information processing is already strongly embedded in the Google Assistant, Amazon's Alexa, Apple's speech interpretation and recognition interface Siri, and Facebook's facial-recognition algorithm DeepFace.[139] Siri, for instance, is embodied by a human, expressive voice (by choice either male or female) in different languages to act as a task manager of all sorts, while its contextual-learning methodology syncs all of the user's devices to extend Amazon's digital ecosystem. That might seem like a distant future, but who says you would not consider Siri and Alexa as "friends," "companions," or even "lovers," if they offered you advice, stability, and support in your life, maybe even more than any physical person around you, as these devices could draw from all the data and knowledge ever produced? In a time in which communication and social activities are already defined by digital realities, it does not seem like a far stretch. There is certainly no shortage of predictions about how AI will influence our lives in the future.[140]

> Artificial intelligence is already embedded in several mediated experiences such as *AI-governed communication and programming*. In the future, AI will merge with immersive technologies to create *all-surrounding media experiences governed with the help of machine learning*.

At one point, AI systems will be able to extend their knowledge independently, even beyond the knowledge level of their creators. That is when artificial intelligence will surpass human intelligence. AI might then be able to cure diseases, solve political conflicts, and establish lasting solutions for a sustainable environment. But machine-driven governmentality might also have extreme risks—what is our role as humans then compared to machines that lack human values and emotions? What happens when AI creates social partners that appear to be more ideal than other humans? How could we protect our biometric data from these kinds of machines? Ray Kurzweil is convinced that with the help of nanobot-based technology "there won't be a clear distinction between real and virtual people. . . . We will all become virtual humans."[141]

British contemporary artist Ed Atkins investigates humanoids and bodies in digital spaces. In his video art pieces such as *Ribbons* (2014), he creates uncanny CGI vessels that appear as bodies unable to claim their authenticity. Everything in Atkins's pseudo-historic world is understood as fake, including virtual humans. In our interview for this book, Atkins reflected on the idea of digital vessels replacing physical bodies entirely: "It is pretty much entirely figurative. It is a shift regarding the location of metaphor or literalism. As in, those things that might traditionally be thought of as figurative are possible to approach as literal. Like a cartoon, I suppose. Only with an emphasis on verisimilitude, which is peculiar and creates a kind of viscerality. I would also add that in the idea of this gratuitous cartoonified realism is the idea of creating a surrogacy. A body that can go through something in my stead, so that I do not have to." He later adds, in relation to the ethical implications of artificial intelligence and posthuman bodies: "The cocktail of anonymity, vast faceless communities, populism, and Randian individualism makes for a permissive culture surrounding tech. But when you describe it like that, it sounds pretty terrifying. Like drowning."

With major tech companies setting the standards for AI to generate revenue, "[h]ow do we protect our communities—and particularly already vulnerable and marginalized groups—from the potential uses of these systems for surveillance, harassment, detainment or deportation?"[142] How can we prevent AI systems from discriminating against certain populations and exacerbating financial inequalities? As soon as immersive technologies merge with AI—and they already do to an extent as, for instance, humanoids are implemented in social VR environments—immersion would become a permanent condition for mediated experiences in the age of technocapitalism. There would be nothing beyond the mediation anymore. Audiences would be plugged into their machines, constantly feeding their data, both organic and inorganic, into devices that are inseparably connected to their bodies. They would "evolve thereby into a hybrid of biological and nonbiological thinking."[143] In the future of the posthuman, how will we be able to protect our privacy, our values, and our ethical standards from powerful corporations?

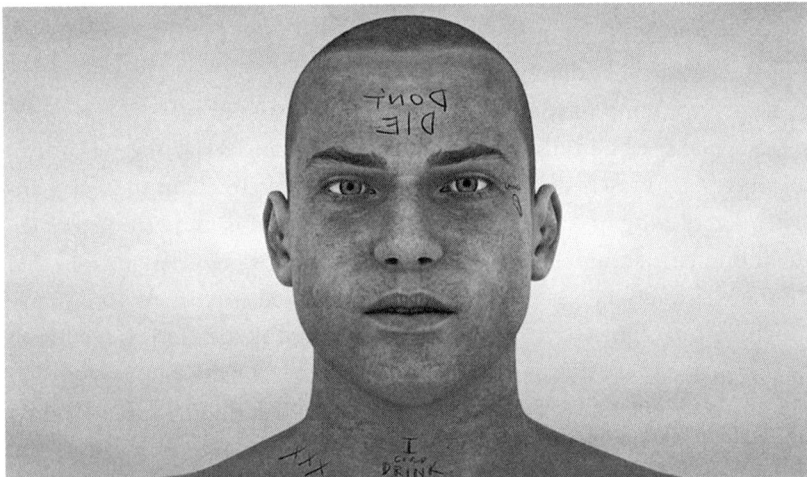

**Figure 4.6**
Ed Atkins: *Ribbons*: the creation of digital vessels. The body is no longer the bastion of authenticity. Reprinted with the permission of the artist. © 2014 Ed Atkins.

The posthuman knows no mediated entities anymore. It does not differ between human and artificial intelligence. The posthuman transforms into an immersive experience that will be potentially governed by intelligent machines using ever-growing amounts of data to control the semiotic systems, performative practices, and psychological dimensions of immersion.

## Moments of Immersion: Ed Atkins

Ed Atkins is one of the most important video and digital artists of our time. His works deal with the ambivalent relationship between the virtual and the real and the submersive environment of avatars and bodies in digital spaces. Atkins exploits the definition of the hyperreal with the help of technologies and virtual characters. His works include *Us Dead Talk Love* (2012), *DEPRESSION* (2012), *Ribbons* (2014), and *Old Food* (2017), and they have been exhibited at the Museum of Modern Art in New York, the Serpentine Gallery in London, the Kunsthalle Zürich, Berliner Festspiele, and the Tate Britain.

> **I argue that when we enter any immersive experience, we establish an understanding of selfhood defined by the conditions of the experience we are positioned in. That ultimately means we have many different phenomenological selves and those are mediated as well. In your opinion, how is selfhood in the physical world influenced by the creation of digital avatars?**

The figures I create are psychic avatars, and sub- or unconscious ones at that. I mean, the intentional creation of avatars for particular use—social media, gaming—is one thing. An avatar created inadvertently is another. By this I mean the idea that you are being yourself in any given context, but that "yourself" is a kind of iterative performance, there being no authentic self or means by which one interacts or represents oneself. Performance, and particularly Butlerian ideas of performativity and psychoanalysis, contains within it this idea of different selves, an indeterminate number, created in order to survive or thrive or whatever in particular environments. I guess that is usually meant socially, and not phenomenologically. My figures are not referred to as avatars by me. I prefer surrogate. Or crash test dummy is better: not alive, simply allegorical vessels. Whether the allegory is understood or not—by me or the viewer—is another thing altogether.

**How does our understanding of the body change when we use such an allegorical vessel?**

It is pretty much entirely figurative. It is a shift regarding the location of metaphor or literalism. As in, those things that might traditionally be thought of as figurative are possible to approach as literal. Like a cartoon, I suppose. Only with an emphasis on verisimilitude, which is peculiar and creates a kind of viscerality. I would also add that in the idea of this gratuitous cartoonified realism is the idea of creating a surrogacy. A body that can go through something in my stead, so that I do not have to. Something like that. I mean a sort of accelerationism regarding modeling, speculation—but the speculative confused with desire (which is probably always the case, but conspicuously so here). It's also got an ethical lock around it, a kind of ambivalence that pushes around the realism and the overt digitality.

**What is your opinion on terms such as "the empathy machine"?**

Maybe that says more about a lack that we might need something literally shown to us before we can feel a sufficient empathy for it? I am certainly interested in this, but it cuts down on the freakiness of something being both literal—kind of real—and figurative. Like an allegory. Or like dragons that look very convincing. I guess if it works?

**Would you say it works?**

There are technologies that claim a more immersive experience, but that tends to be precisely the opposite case: VR headsets are heavy and cumbersome; resolutions and frame rates far inferior to standard or "normal" moving imagery. It has the ring of sloganeering, to me; a corporate radicality to sell speculative developments, when the current reality is far from convincing. I suppose, technically, it is true that those technologies loaded with that term are sensorially more immersive—but this seems at the expense of fidelity. Or rather, the stakes are higher when you make a claim like "immersive": it has to be entirely convincing, or it fails completely. Immersive, of course, could just be an engrossing book, music, cinema.

**Multinational corporations are strongly influencing what we perceive as real and authentic in digital spaces. It is all about access, but not ethics. Where do you see the ethical implications of artificial intelligence and posthuman bodies?**

That is huge. Everywhere? Of course, the disappearing of direct encounters with ethically pointed situations is to the benefit of each of those corporations. Or it is of benefit to their unchecked operation. I think there is a certain duty or responsibility to research what is going on, to read the terms and conditions, to read the pieces about sweatshop work, mines, foreign bodies going through hell. Somehow, the onus has shifted to the consumer—which I guess is a prime move of biopolitical power—so that it is up to us to agree or not, to buy into what we almost always know is a lie, or not. The cocktail of anonymity, vast faceless communities, populism, and Randian individualism makes for a permissive culture surrounding tech. But when you describe it like that, it sounds pretty terrifying. Like drowning. But you could probably equally describe it in fun, optimistic tones.

# 5  In Limbo

Not bad for a race of demented monkeys
From a cave to a city to a permanent party
—Father John Misty, "Total Entertainment Forever," *Pure Comedy*, Sub-Pop, 2017

## Virtual Reality: The End Is the Beginning Is the End

Twenty years before Justin Timberlake apparently considered using a projected hologram of the late Prince at the Super Bowl LII halftime show in Minneapolis,[1] the Purple One did an interview with the magazine *Guitar World* discussing the mutations of digital re-creation. In fact, he had been asked if he would ever consider using digital editing to perform with an artist from the past. Prince found the whole idea of the mediated being inseparably interwoven with the physical quite disturbing: "That's the most demonic thing imaginable. Everything is as it is, and it should be. If I was meant to jam with Duke Ellington, we would have lived in the same age. That whole virtual reality thing . . . it really is demonic. And I am not a demon. Also, what they did with that Beatles song ['Free as a Bird'], manipulating John Lennon's voice to have him singing from across the grave . . . that'll never happen to me. To prevent that kind of thing from happening is another reason why I want artistic control."[2] Although according to the *Minneapolis Star Tribune*, a Friday night rehearsal of Timberlake's halftime show did indeed include a Prince hologram,[3] fans and members of Prince's former band immediately condemned the plans to project the late artist as a holographic image to create the illusion that he would be performing live with Timberlake on stage. To show his respect for Prince, Timberlake backed out, and the hologram ended up being merely a video projection on a massive sheet. However, Prince would not have been the first musician to become recreated on stage after his death. It has been done with Tupac Shakur at Coachella 2012, Michael Jackson in 2014, Elvis in 2016, Buddy Holly, Roy Orbison, and Frank Zappa in 2019, and Whitney Houston in 2020. ABBA performed as their "digital selves" as part of an "avatar tour project" in 2018.[4] Gorillaz, one of the most successful bands of our times, even made it from holograms to global headliners.

The whole issue of holographic images raises significant ethical questions about immersive technologies that are able to bring human beings back to physical reality after they have left it forever—for one-off stunts to potentially lucrative tours.[5] What are the moral and legal implications? Who will be in charge and for what purpose? Who owns digital re-creations in general? What exactly are the implications of submitting to a dominant mediated reality, which creates an illusionary setting that, depending on the creator, could range between a pleasurable experience and a violent, sadomasochistic, and authoritarian-technocratic reality? Where do we draw the ethical boundaries between the physical and the digital, especially if we perceive the latter as an integral, defining part of our lives?

This future is right at our doorstep, at least according to Facebook. In early 2016, Mark Zuckerberg introduced the company's ten-year road map for the first time, and revisited it again at the F8 developer conference in 2018. Unveiling his vision for the next decade,[6] Zuckerberg imagines a future in which his company is in control of a very powerful ecosystem that spans all areas of our lives—and immersive technologies are a major part of that strategy. Obviously, Facebook's social networking platform would still be the foundation for the growth of the company's business, with additional products Instagram, Video, Search, Groups, and communication tools WhatsApp (including the newly acquired GIF search engine Giphy) and Messenger as the most influential assets to reach audiences through mobile devices. The road map also highlights lesser-known services such as Workplace (a tool to assess how Facebook's own employees use the platform to communicate at work) and Marketplace, Facebook's own version of Craigslist.

This is a future in which holograms of late artists, AI characters in cinematic immersive story worlds, and in-depth psychological explorations of ourselves in virtual live-action role plays merge with our understanding of "reality." Currently, the digital space is mostly flat, but it is about to become fully three-dimensional. Essential for Facebook's strategy to achieve this are several connectivity services (satellite and high-altitude internet, telecom and shared infrastructure projects, experimental wireless and fiber technology, and affordable access to faster, open-source telecom hardware). AI solutions allowing communication through vision and language are about to offer new ways to talk to the machine. Human-computer interaction beyond touch—beyond the keyboard, mouse, and touchscreen—will dominate the ways we are connected and will be using our voices, gestures, and contextual awareness. Facebook plans to build generative network AI machines that get better the more they are exposed to our data. In addition, VR and AR technologies are part of Facebook's strategy, such as AR tech (mainly AR glasses), standalone VR (Oculus Quest 2), high-end PC VR experiences, and social VR apps (Horizon, Parties, and Venues) to grasp every aspect of three-dimensional digital communication.

In this ecosystem, Facebook envisions Horizon becoming one gigantic, OASIS-like virtual reality, a holodeck experience that offers everything, a "world beyond our world."[7] We would be communicating with powerful AI devices that appear as digital characters on the same level as avatars on social media, or friends in our WhatsApp contact list. We would meet them at three-dimensional virtual locations that would be impossible to visit in physical reality, partly because we are not able to do so (e.g., Mars); partly because they might not even exist anymore (e.g., polar regions). AI would be able to adapt to our needs and tell us what we want to hear through powerful brain-computer interfaces (BCI)—an echo chamber that is only there for us and is constantly feeding from our biometric and digital data. We would not need to worry about life-changing decisions anymore. Our AI friends would know us based on data clustering and profiling, probably better than the people around us.[8] They would give us guidance on which job to apply for, which people to date, and when to invest in cryptocurrencies. Every aspect of our daily lives could potentially be incorporated into that ecosystem. We would not just visit. We would be *living in it*.

Would that not be the ultimate form of immersion, while at the same ending what we now define as the difference between the physical and the mediated?

Facebook is not the only company expanding its digital ecosystem with immersive technologies. As part of the multinational conglomerate Alphabet Inc., Google is just one of Alphabet's subsidiaries dealing with AI. Others include X (e.g., self-driving cars), Nest Labs (self-learning WiFi-enabled

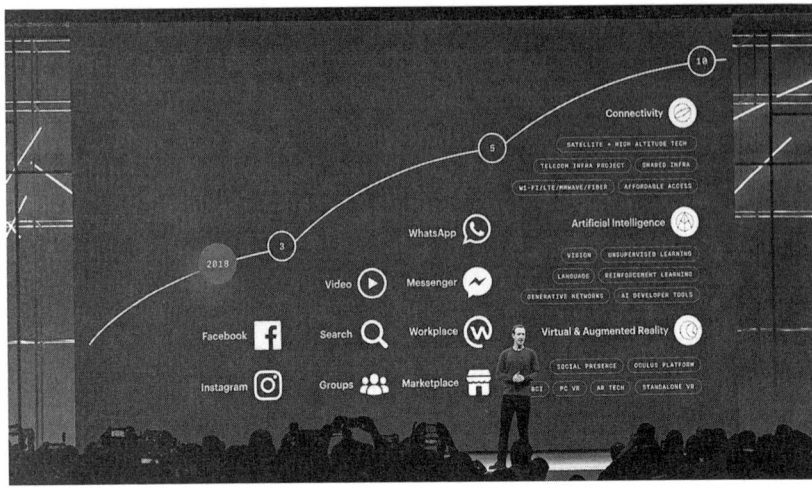

**Figure 5.1**
Mark Zuckerberg revisits Facebook's 10-year roadmap at the F8 developer conference in 2018. Reprinted with the permission of Facebook. © 2018 Facebook.

thermostats, smoke detectors, security systems), and DeepMind (machine learning and AI in healthcare). At the same time, Google has the largest database for human data and constantly learns from our behavior in digital spaces through several mobile apps like YouTube and Google Maps. Google knows where we have been and where we are likely to go (Google Maps), what we are interested in (Google Search), which music we listen to or films we watch (Google Play), which emails we send (Gmail), even how our voices sound (Google Assistant), and what our fingerprints looks like (fingerprint screen lock). Apple uses multiple devices to let users move between them smoothly and seamlessly, uses facial recognition, and installed its own streaming platform (Apple+). Amazon, on the other hand, extends its e-commerce business to the physical environment of its customers by using so-called Dash Buttons[9]—shortcuts in the physical space to order more than ten million of its most popular brands online. In addition, Amazon built up its own entertainment powerhouse with in-house film productions and the streaming platform Prime Video. I have already previously discussed how Netflix enhances binge-watching experiences, from algorithmic mechanisms to personalized recommendations and devices like The Switch.

We can hardly avoid getting trapped in one or more digital ecosystems at the same time. Since these companies constantly aim to increase their

# In Limbo

revenue streams, they put forward strategies and tactics to give us the reality we seem to want, or make us want it in the first place. The installment of digital regimes introduces powerful mechanisms of behavioral manipulation through immersive experiences that deeply affect us, especially when there are commercial, governmental, political, or religious interests behind their creation and maintenance.

But why should we fall in their rabbit holes? Why should we immerse ourselves in their mediation of the real? There is not a simple or straightforward answer, but it has to do with the configurations of postmodernism, the construction of mediated realities, and what sorts of mediation we are willing to accept. In late 2017, I watched a fascinating and bleak documentary called *The Sex Robots Are Coming*[10], produced by Raw TV for British broadcaster Channel 4. It explored the success of the Californian company RealDoll, which sells realistic life-size sex robots that appear like women to their mostly male customers. One of them was James, an engineer from Atlanta, who had been married to his wife, Tine, for quite some time, but is also attracted to a collection of sex dolls. He treats them like real people. He talks to them, dresses them, and applies their makeup. All of these dolls have a prosthetic vagina, and at one point James is asked if he would choose his dolls over his wife to provide sexual pleasure ("I honestly could not say"). Later in the documentary, James meets a prototype of RealDoll's newest creation: Harmony, an AI-enabled silicone sex doll. Harmony looks like a plastic version of Pamela Anderson. After the doll politely begins a conversation ("I am delighted to sweeten your life") and compliments James on his looks, much to the man's amusement, it becomes evident that this is a man's image of a passive, submissive, and obedient porn star ("Shoot your load for me baby, I want it so bad"). Nevertheless, James almost naturally begins to develop a feeling of presence while being confronted with the mediation. He starts to flirt with the doll. Many viewers were confused by James's attraction to the machine. Some commentators even called it "awkward, unsettling, and horrifying."[11] The prospect of sexual relationships with humanoid robots certainly poses moral dilemmas, and the ethical, legal, and social implications of AI robotics,[12] as well as sexual activities with AI machines, are currently being explored.[13]

But I would like to point out how immersion is used here to construct a conceptualized, controlled version of reality. Three-dimensional AI sex dolls are the most recent example of conditioning audiences in regard to digital pornography.[14] They follow up on the mechanics of porn-streaming sites such as Pornhub and VR porn, which appear to put the viewer in the middle of a sexual encounter.[15] Both examples reframe the idea of visual pleasure through the relationship between embodied observer and mediated images. Humanoid sex robots, however, are different. They are embodied mediations planted inside physical reality. They are full-sensory, three-dimensional mediated experiences. Feona Attwood speaks about a crisis of representation, a side effect caused by immersive experiences: "It marks what seems to be a disappearing gap between reality and representation. . . . It marks the pervasiveness of media and of mediation itself in which technologies appear to have closed the gap between real and imagined worlds or destroyed our ability to tell the difference between them."[16]

The medium no longer represents, it performs. In this scenario, technological agencies exist on the same level as human agencies, and the human consciousness takes them for real, as much as James believes in the realness and attractiveness of his sex doll Harmony.

With the framework of the *360° gaze*, I have tried to point out that this is not a new phenomenon. In fact, the "realness" of immersive experiences has always been determined by their semiotic codes, performative practices, psychological implications, and the cultural industries and digital ecosystems by which they are governed. But all of that has been based on the assumption that media are indeed *mediating* between physical reality and representation. But what if the mediation replaces the physical indefinitely?

The 360° gaze framework makes us look into the plasticity of the human mind, one that is easily affected by immersive experiences. There are several reasons why humanity might turn to complete immersion: escapism, identity formation, sociality, and connectivity, among many others. When we turn on the news, we get the impression that the world is falling apart. Everyone seems to feel that there is a disruption of traditional and established structures. Climate change and global warming, contagious diseases and viruses, global migration crises, the rise of the far right, Brexit, Trump, Le Pen, and the production of deliberate misinformation in the form of fake news and deepfakes, are just a few examples from recent years. The "fake-news nightmarescape"[17] created by the coronavirus (COVID-19) outbreak, fake celebrity porn videos, fake political videos, or the question of how many people actually attended Trump's inauguration ceremony compared to Obama's,[18] made it clear how immersive experiences feed into the perception of our understanding of truth and reality. Yi-Fu Tuan adds: "Facing

**Figure 5.2**
*The Sex Robots Are Coming*: James and the AI-enabled sex robot Harmony. Reprinted with the permission of Raw TV and Channel 4. © 2017 Raw TV and Channel 4.

reality, then, implies accepting one's essential powerlessness, yielding or adjusting to circumambient forces, taking solace in some local pattern or order that one has created and to which one has become habituated."[19]

At the same time, digital media offer us several rabbit holes into experiences that apparently seem to make our lives easier and more comfortable, from shopping to meeting other people. This version of reality might be disruptive and overwhelming as well, but not to the degree that physical reality seems to be. The level of our control—at least in our perception—appears to be much higher in the mediation. We believe we are in charge of our profiles, email accounts, and bookmarks. And more importantly, our human desires seem to get satisfied by mediated experiences because immersive media are designed to do exactly that, while in physical reality, our desires seem to be very often ignored. Participation, engagement, involvement, identity formation, and our need for recognition—all of that appears to be so much easier once we are immersed in mediated experiences. Would you not turn to an animatronic sex robot, if it provided the best (and maybe only) opportunity to satisfy your sexual needs? Especially if the sensory and perceptual experience is as real as you can imagine? Would you not prefer to meet up with your best friends at a virtual birthday party in one of Facebook Horizon's digital environments rather than not meeting them at all, since they are spread across the globe? Would you not rather see a larger-than-life, crisp, and high-definition hologram of Elvis than never seeing him perform in concert at all? Postmodern media culture is full of these examples, and they are already part of our lives. If you were ever concerned about the destiny of Harry Potter, or have ever developed feelings for someone based on their online-dating profile, you have already chosen immersion over physical reality on many occasions.

But our control in immersive experiences is very limited, often nonexistent. The deeper we immerse ourselves in these kind of mediations, the more likely we are to get lost *in limbo*, an endless ocean of uncontrolled desires. For Feona Attwood this is "a state of immersion whereby the viewer becomes horribly animated, possessed and perverse"[20] as his autonomy and sense of agency become permanently violated. The ethical implications of immersive media stress the affective nature of these experiences. For instance, South Korean TV broadcaster MBC aired a documentary called *Meeting You* that centered on a family's loss of their seven-year-old daughter Nayeon to a severe illness. Using VR in addition to photogrammetry and motion capture technology, the team "recreated" Nayeon as a virtual simulacrum for one last goodbye with her mother and to reexperience some happy memories, such as Nayeon's birthday party. Recreating a deceased loved one in VR—let alone using this in a television show—raises some immanent ethical concerns regarding the human implications of immersive media.[21] In addition, VR content might put audiences directly into risky environments through perceptual and sensual engagement. They might be confronted with disturbing sexual and violent experiences, which would be crossing red lines on many levels. Content could encourage and reinforce undesirable mental problems and have lasting effects on the audience's psyche. This could lead to deep experiences that become a playground for the subconscious mind. Much like Leonardo DiCaprio's character in *Inception* (2010), audiences then would need a totem, an object to test whether they are still in physical reality, a spinning top that might topple over, or continue to spin forever.

> The *end of immersion* refers to the *absoluteness of full-sensory three-dimensional mediation* and therefore the *denial of physical reality*. In this scenario, technological agencies exist on the same level as humans and are governed by powerful digital regimes to manipulate human behavior.

For many years we have assumed that people value physical reality over mediated and virtual experiences. This barely questioned assumption goes back to Robert Nozick's well-known thought experiment Experience Machine. In this experiment, we are asked to imagine that we are able to plug ourselves into a virtual reality for a very long time. Nozick argued that we might not wish to "plug in" the machine indefinitely, as the mediation would always lack something specific, a kind of meaningful authenticity compared to physical environments, "something [that] matters to us in addition to the experience."[22] Now, what if this experience were able to add exactly that "something," maybe even more so, as outlined in several examples in this book. Immersive experiences claim to define reality, even co-create it, and therefore also establish an authentic framework that could possibly become the foundation of our understanding of reality. Recent works, such as one by Felipe De Brigard, suggest exactly that and provide a twist to Nozick's experiment. In De Brigard's version, he asked his students to do the same experiment with him, but this time he asked them to imagine that they were already plugged into the mediated experience. In order to return to their "real" lives, they would need to unplug from what they currently perceive as "reality." Naturally, many of his students wished to not be unplugged and wanted to remain in the status quo. Their reaction suggests that our understanding of reality can be manipulated, almost to the point when we would need a reminder that another version of reality still exists, even if that "other version" was our physical reality. That would be the *end of immersion* in mediated experiences. We would then need to consider strategies for how to develop feelings of extended presence in physical environments.

I have outlined earlier that we simply do not know the effects of long-term immersion yet, especially if we look at it as a sociocultural phenomenon that can be found in various cultural and media practices. While we are still figuring out the future of immersive technologies, we need to evaluate immersive practices as part of already established mediated experiences to learn from them. We have been made aware of addictive behavior and tendencies of escapism as a result of excessive media consumption, and we know about the structural and narrative persuasion of content. We know about the performative strategies of the self, and the roles we play in theater and games. We should also be more sensible about data protection and security, platform logics, and the implications of human-machine hybridization. Still, there is a deep underlying concern about the impact of virtual worlds, especially if they are introduced as three-dimensional experiences cutting us off from our physical surroundings.

It is important to note that discussing ethics in relation to immersive technologies is not fundamentally different to other ethical dilemmas in relation to media. In general, immersive experiences and virtual worlds are able to benefit individuals and society, but they also confront us with fundamental fears we already know too well. One of the biggest concerns is that we might isolate ourselves completely from society in favor of an artificial social world created by corporate organizations. We have witnessed these developments now for several years as digital media becomes deeply intertwined with us moving further into digital spaces. We are not yet abandoning our physical reality or bodies completely, but digitalization in many areas of our lives support or even require us to choose the mediated before the physical, from online shopping and online dating to digital communication. That means that we, more than ever, need to reflect on what media we are consuming and in which ways. With immersive technologies, our perception of mediated realities and the world around us will be changing drastically, and the barriers for us to develop a sense of presence in mediated spaces will soon be practically nonexistent. Ultimately, this will impact the rules and conditions of social interaction, social activism, politics, and so forth. Immersive media have the ability to replace the "real" and (re)create it in digitally mediated environments. That raises questions such as: How do immersive technologies modify and influence our understanding of the world and reality? Are users really able to act autonomously once they are fully surrounded by mediated realities governed by someone else? In which ways are privacy issues affected if biometric data is passed on to third parties through VR devices? How do we reimagine our understanding of the "real" if presence, involvement, and engagement exist predominantly in virtual spaces? What happens to our memories, stories, and ideas if the mediated becomes the new real? And will our digital avatars and holograms take on our existence once our physical bodies have passed away?

Answering these questions requires an ongoing discourse of immersion across disciplines such as sociology, computational science, media studies, drama and performance studies, film studies, game studies and games design, semiotics, interactive arts, cognitive science, human-computer interaction, business and economics, and transmedia studies. But more importantly, it demands a different approach to media literacy to make sure we gain enough knowledge to reflect on immersive experiences. That means we have to get rid of the common understanding that so-called "digital natives" know about the impact and foreseeable risks of emergent technologies just because they are growing up with it. We need to implement issues of digital media and emerging technologies in the curricula and schoolbooks of our educational facilities. We have to make sure technological developments are not only in the hands of those that want to increase sales and revenue, or have specific political or religious agendas. We have to ensure that we are shaping emerging technologies, such as XR, AI, and robotics, in such a way that they are ethical, transparent, open, and inclusive to all our communities for a responsible and sustainable technological future. In fact, the 360° gaze framework supports the critical reflection of these technologies as it highlights their impact on selfhood, mediated realities, and media usage in general.

The developments in the area of immersive media are rapidly moving forward with news about innovative headsets, controllers, platforms, and content emerging every day. While finishing this book, I came across an advertisement on TV about a metal band that users wear around their necks in order to manipulate brain activity while playing a VR game. AzanaBand[23] promised to make players feel physical pain in VR through a combination of electric pulsation to create an intense sensation of deep fear and mimic the heart rate of excitement. Advertisements spread across London and Manchester, and YouTube influencers promoted the sensory gaming device as well. However, AzanaBand was nothing but a hoax. Many viewers (including me) were duped by a viral-marketing campaign promoting *Kiss Me First* (2018),[24] a sci-fi series about loneliness, paranoia, and identities in the digital age. The sensory gaming device merely exists in a mediated representation. But to be honest, it sounded quite plausible.

Shows like *Kiss Me First*, *Black Mirror*, *Electric Dreams*, *Westworld*, *Years and Years*, *Upload*, and *Devs*, channel our deepest fears of a dystopian future ruled by technology. They also make us aware that just because something can be done, does not mean it is inevitable. Among other things, our desire for immersive experiences is a direct result of how we perceive the uncertainty of the world around us. As long as we find our times unstable and complex and perceive things to be falling apart, we will consider other (mediated) options. Existential crises in physical reality, such as the coronavirus pandemic, also present us with opportunities to recalibrate our relationship with mass media and establish a healthier, more sophisticated and balanced use of technology.

In the year 2045 of *Ready Player One* "there is nowhere else to go, except the OASIS, a whole virtual universe."[25] That would be, of course, the ultimate showdown. But it refers to the complex intersections between the human mind and immersive experiences that influence each other in ways we are just beginning to comprehend. The only way to do this is to develop institutionalized and strategic mechanisms to stay in control while being exposed to liquid spaces and immersion.

It is never too late to learn how to swim.

# Notes

**Ready Player One**

1. Carolin Gorlitz and Anne Helmond, "The Like Economy: Social Buttons and the Data-Intensive Web," *New Media and Society* 15, no. 8 (2013): 1348–1365.

2. Donna Haraway, "A Cyborg Manifesto: Science, Technology and Socialist-Feminism in the Late Twentieth Century," in *Simians, Cyborgs and Women: The Reinvention of Nature* (New York: Routledge, 1991), 149.

3. Mark Andrejevic et al., "Participations: Dialogues on the Participatory Promise of Contemporary Culture and Politics. Part 2: Labor," *International Journal of Communication* 8 (2015): 1089–1106. See also Mark Andrejevic, "Facebook als neue Produktionsweise," in *Generation Facebook: Über das Leben im Social Net*, ed. Oliver Leistert and Theo Röhle (Bielefeld: Transcript, 2011), 31–49.

4. Haraway, "A Cyborg Manifesto," 149.

5. The tech support of the lucid dream company in *Vanilla Sky* explains: "You sculpted your lucid dream out of the iconography of your youth. An album cover that once moved you . . . a movie you once saw that showed you what a father could be like—or what love could be like." (Cameron Crowe, *Vanilla Sky*, directed by Cameron Crowe [Los Angeles: Paramount, 2001], 01:50:45–01:52:45.) See also Philipp Schmerheim, *Skepticism Films: Knowing and Doubting the World in Contemporary Cinema* (New York: Bloomsbury, 2015), 272–278.

6. Isra Daraiseh and M. Keith Booker, "Unreal City: Nostalgia, Authenticity, and Posthumanity in 'San Junipero,'" in *Through the Black Mirror: Deconstructing the Side Effects of the Digital Age*, ed. Terence McSweeney and Stuart Joy (London: Palgrave Macmillan, 2019), 151–164.

7. *Black Mirror's* episode "Striking Vipers" (S05E01, 2019) explores sexual gratification and fetishism in the VR space. In a *Mortal Kombat*-style fighting game, two longtime friends develop an attraction to one another's digital avatar that leads to in-game sex in VR.

8. Despite significant security issues, the free web-based video conferencing application Zoom has exploded onto the scene and became a popular choice for many people and businesses to connect with others virtually. In addition, other remote-working tools have surged in popularity as well, such as Microsoft's collaborative apps Teams and Skype, and Google's Meet and Hangouts. The coronavirus pandemic also resulted in the launch of several XR-based meeting services for businesses, such as HTC's VR platform Vive Sync (part of Vive XR Suite, HTC's bundle of VR apps for various remote experiences), and the free XR meeting apps Spatial and Mozilla Hubs, that are accessible via a variety of devices and do not solely rely on headset technology. Facebook even teased a new feature using the outward-facing cameras of its VR headsets Rift and Quest, in addition to pass-through technology, resulting in the creation of XR workspaces unbounded by the limits of physical monitors. See also Nick Statt, "Facebook Teases a Vision of Remote Work Using Augmented and Virtual Reality," *TheVerge.com*, accessed May 23, 2020, https://www.theverge.com/2020/5/21/21266945/facebook-ar-vr-remote-work-oculus-passthrough-future-tech.

9. Edward Castronova, *Exodus to the Virtual World: How Online Fun Is Changing Reality* (London: Palgrave Macmillan, 2007).

10. Deborah Tannen, "The Personal Becomes Dangerous," *Politico.com*, accessed March 23, 2020, https://www.politico.com/news/magazine/2020/03/19/coronavirus-effect-economy-life-society-analysis-covid-135579.

11. In the *Black Mirror* episode "USS Callister" (S04E01, 2017), software designer Robert Daily abuses his colleagues by creating digital clones with their DNA and forcing them to enact his favorite science-fiction TV show in an immersive VR universe called "Infinity." Inside the universe—that strongly resembles *Star Trek: The Original Series*—he is able to torture them without any consequences.

12. This study will not focus exclusively on visual media, mainly because I understand immersion as a sociocultural phenomenon that spans across different forms of immersive and pervasive media, but also because the relation between image-based art and immersion has been already covered extensively in Oliver Grau, *Virtual Art: From Illusion to Immersion* (Cambridge, MA: MIT Press, 2003).

13. ARTE, "Digital Productions," *Arte.tv*, accessed April 8, 2020, https://www.arte.tv/sites/webproductions/en/.

14. Berliner Festspiele, "Immersion," *Berlinerfestspiele.de*, accessed April 8, 2020, https://www.berlinerfestspiele.de/en/immersion/start.html.

**Chapter 1**

1. Neil Postman, *Amusing Ourselves to Death: Public Discourse in the Age of Show Business* (New York: Penguin, 1986), viii.

2. Neil Postman, *Technopoly: The Surrender of Culture to Technology* (New York: Vintage Books, 1993).

3. Ibid., 3.

4. José van Dijck, "Facebook as a Tool for Producing Sociality and Connectivity," *Television and Media* 13, no. 2 (2012): 161.

5. Martin Warnke, "On the Spot: The Double Immersion of Virtual Reality," in *Immersion in the Visual Arts and Media*, ed. Fabienne Liptay and Burcu Dogramaci (Leiden: Brill Rodopi, 2016), 205.

6. Facebook, "Facebook Spaces," *Facebook.com*, accessed June 30, 2020, https://www.facebook.com/spaces

7. Mark Zuckerberg, "Oculus Connect Demo, October 6, 2016," *Facebook.com*, accessed July 24, 2020, https://www.facebook.com/zuck/posts/10103154542263811.

8. I understand "rabbit hole" as a gateway, loophole, door, window, passage, or portal—any sort of friction and border between physical and mediated environments that audiences have to pass. On one hand, rabbit holes can be understood as defining elements of media technologies (cinema and television screens, computer interfaces, web browsers, mobile phones, VR glasses). On the other hand, rabbit holes are embedded as conceptual elements within the structure of content to engage and stimulate audiences to feel present in a mediated environment. Rabbit holes lead the way to a different mode of reality.

9. José van Dijck, "Facebook and the Engineering of Connectivity: A Multi-layered Approach to Social Media Platforms," *Convergence* 19, no. 2 (2012): 141–155.

10. Nick Wingfield and Mike Isaac, "Mark Zuckerberg, in Suit, Testifies in Oculus Intellectual Property Trial," *NYTimes.com*, accessed April 15, 2020, https://www.nytimes.com/2017/01/17/technology/mark-zuckerberg-oculus-trial-virtual-reality-facebook.html.

11. In May 2020, NextVR was acquired by Apple to support the development of Apple's own VR and AR headsets. See Nick Statt, "Apple Confirms It Bought Virtual Reality Event Startup NextVR," *TheVerge.com*, accessed May 14, 2020, https://www.theverge.com/2020/5/14/21211254/apple-confirms-nextvr-acquisition-purchase-vr-virtual-reality-company.

12. Jose Fermoso, "Facebook Invests $250m More in VR as Zuckerberg Shows Off Wireless Oculus," *TheGuardian.com*, accessed July 20, 2020, https://www.theguardian.com/technology/2016/oct/06/zuckerberg-facebook-virtual-reality-wireless-oculus-connect-3.

13. Facebook, "Facebook Spaces: This App Is No Longer Available," *Facebook.com*, accessed February 4, 2020, https://www.facebook.com/spaces.

14. Oculus, "Introducing 'Facebook Horizon,' a New Social VR World, Coming to Oculus Quest and the Rift Platform in 2020," *Oculus.com*, accessed February 4, 2020, https://www.oculus.com/blog/introducing-facebook-horizon-a-new-social-vr-world-coming-to-oculus-quest-and-the-rift-platform-in-2020/.

15. Marie-Laure Ryan, *Narrative as Virtual Reality: Immersion and Interactivity in Literature and Electronic Media* (Baltimore: Johns Hopkins University Press, 2001), 1.

16. Ben Gilbert, "Facebook Is Changing Its Logo to Make Sure Users Know It Owns Instagram and WhatsApp," *Inc.com*, accessed February 12, 2020, https://www.inc.com/business-insider/facebook-new-logo-instagram-whatsapp-parent-company.html.

17. Adi Robertson, "Facebook Is Making Oculus' Worst Feature Unavoidable," *The Verge.com*, accessed August 19, 2020, https://www.theverge.com/2020/8/19/21375118/oculus-facebook-account-login-data-privacy-controversy-developers-competition.

18. Facebook, "Reports Fourth Quarter and Full Year 2019 Results," *Facebook.com*, accessed February 15, 2020, https://investor.fb.com/investor-news/press-release-details/2020/Facebook-Reports-Fourth-Quarter-and-Full-Year-2019-Results/default.aspx.

19. See Rebecca Hills-Duty, "Apple Team Up with IKEA to Create AR Shopping App," *VR Focus*, accessed July 6, 2020, https://www.vrfocus.com/2017/06/apple-team-up-with-ikea-to-create-ar-shopping-app/.

20. Glass has not been introduced to the consumer market for various reasons. The device has been described as "overpriced and socially awkward," but has also raised ethical concerns because of the use of its incorporated camera. See Samuel Gibbs, "Google Glass Review: Useful—but Overpriced and Socially Awkward," *TheGuardian.com*, accessed July 6, 2020, https://www.theguardian.com/technology/2014/dec/03/google-glass-review-curiously-useful-overpriced-socially-awkward. See also Nicky Woolf, "Google Glass Ceases Production 'in Present Form,'" *TheGuardian.com*, accessed July 6, 2020, https://www.theguardian.com/technology/2015/jan/15/google-glass-ceases-production-for-now.

21. Pearl Jam "Super Blood Wolf Moon," *Moon.pearljam.com*, accessed February 4, 2020, https://moon.pearljam.com.

22. The term "mixed reality" is also confusingly used by Microsoft for its blend of VR and AR applications accessible through four Microsoft-branded, lower-priced headsets and its Windows Mixed Reality platform. The latter is a standard feature of the Windows 10 operating system. See Microsoft, "Introducing Windows Mixed Reality," *Microsoft.com*, accessed January 30, 2020, https://www.microsoft.com/en-gb/windows/windows-mixed-reality.

23. Joel Hruska, "Microsoft Wants to Bring HoloLens to the Consumer Market Once the Technology Matures," *Extremetech.com*, accessed January 30, 2020, https://www.extremetech.com/gaming/243643-microsoft-wants-bring-hololens-consumer-market-technology-matures.

**Notes** 237

24. Michael R. Heim, "The Paradox of Virtuality," in *The Oxford Handbook on Virtuality*, ed. Mark Grimshaw (Oxford: Oxford University Press, 2015), 117.

25. PwC, "Perspectives from The Global Entertainment and Media Outlook 2017–2021," *Pwc.com*, accessed February 12, 2020, https://www.pwc.com/gx/en/entertainment-media/pdf/outlook-2017-curtain-up.pdf.

26. ResearchAndMarkets, "Virtual Reality—Global Market Outlook (2017–2026)," *ResearchAndMarkets.com*, accessed February 12, 2020, https://www.researchandmarkets.com/reports/5017503/virtual-reality-global-market-outlook-2018?w=4.

27. Paul Moody, "An 'Amuse-Bouche at Best': 360 Degree VR Storytelling in Full Perspective," *International Journal of E-Politics* 8, no. 3 (2017): 42–50.

28. Road to VR, "2019 was a Major Inflection Point for VR—Here's the Proof," *RoadtoVR.com*, accessed February 12, 2020, https://www.roadtovr.com/2019-major-inflection-point-vr-heres-proof/.

29. PwC, "Perspectives from The Global Entertainment and Media Outlook 2017–2021," *Pwc.com*, accessed February 12, 2020, https://www.pwc.com/gx/en/entertainment-media/pdf/outlook-2017-curtain-up.pdf.

30. Ray Kurzweil, "Foreword to Virtual Humans," *Kurzweilai.net*, accessed February 14, 2020, https://www.kurzweilai.net/foreword-to-virtual-humans.

31. Kevin Kelly, "AR Will Spark the Next Big Tech Platform—Call It Mirrorworld," *Wired.com*, accessed February 14, 2020, https://www.wired.com/story/mirrorworld-ar-next-big-tech-platform/.

32. Philippe Fuchs, *Virtual Reality Headsets: A Theoretical and Pragmatic Approach* (London: CRC Press, 2017), 9–10.

33. Ibid., 11.

34. Nicholas Negroponte, *Being Digital* (New York: Knopf, 1995).

35. John Perry Barlow, "A Declaration of the Independence of Cyberspace," *Eff.org*, accessed February 18, 2020, *https://www.eff.org/cyberspace-independence*.

36. Marshall McLuhan, *Understanding Media* (New York: McGraw-Hill, 1964).

37. Stanley Grauman Weinbaum, "Pygmalion's Spectacles," *Gutenberg.org*, accessed July 6, 2020, https://www.gutenberg.org/files/22893/22893-h/22893-h.htm.

38. Ivan Sutherland, "The Ultimate Display," in *Information Processing 1965: Proceedings of International Federation for Information Processing Congress 65*, ed. Wayne A. Kalenich (Washington, DC: Spartan Books, 1965), 506.

39. William R. Sherman and Alan B. Craig, *Understanding Virtual Reality: Interface, Application, and Design* (San Francisco: Morgan Kaufmann, 2003), 141.

40. Visually induced motion sickness, a condition that might occur during or after viewing dynamic images while being physically still, has been reported in a variety of media with three-dimensional images, especially 3D films. See Angelo G. Solimini, "Are There Side Effects to Watching 3D movies? A Prospective Crossover Observational Study on Visually Induced Motion Sickness," *PloS ONE* 8, no. 2 (2009): 1–8.

41. Sutherland, "The Ultimate Display," 508.

42. Gordon Calleja, "Immersion in Virtual Worlds," in Grimshaw, *The Oxford Handbook on Virtuality*, 222.

43. Jaron Lanier, "Virtual Reality: The Promise of the Future," *Interactive Learning International* 8, no. 4 (1992): 275–279.

44. Quoted in Steve Ditlea, "Inside Virtual Reality," *PC/Computing* (1998): 97.

45. In one scene in *The Matrix*, lead character Neo picks up a book in which he keeps illegal software and reveals its title *Simulacra and Simulation* to the camera. Baudrillard himself was not so fond of the reference. He felt the directors have misapplied his theory as they designed the virtual as a concrete reality. *The Matrix* references him but does not apply his understanding of image, mediation, and the hyperreal. See also Seyda Öztürk, "Simulation Reloaded," *Cinetext.philo.at*, accessed June 29, 2020, http://cinetext.philo.at/magazine/ozturk/seyda_ozturk-simulation_reloaded.pdf.

46. David Hesmondhalgh, *The Cultural Industries*, 3rd ed. (London: Sage, 2013).

47. Henry Jenkins, *Convergence Culture: Where Old and New Media Collide* (New York: New York University Press, 2006), 18–19.

48. Jaron Lanier, *Dawn of the New Everything: A Journey through Virtual Reality* (London: Bodley Head, 2017).

49. Fuchs, *Virtual Reality Headsets*, 17.

50. Oculus Link allows tethering the Quest 2 to a computer to play high-end VR games, but the feature still requires a VR-ready PC. See "Oculus Link-Compatibility," *Support.Oculus.com*, accessed February 20, 2020, https://support.oculus.com/444256562873335/.

51. Ben Bloch, "Volumetric Photogrammetry—Big Words, Bigger Impact on VR," *TechCrunch.com*, accessed January 19, 2020, https://techcrunch.com/2018/01/17/volumetric-photogrammetry-big-words-bigger-impact-on-vr/.

52. Nicholas Sutrich, "Nearly Half of All VR Headsets Sold in 2019 was an Oculus Quest," *Androidcentral.com*, accessed February 15, 2020, https://www.androidcentral.com/nearly-half-all-vr-headsets-sold-2019-were-oculus-quest. In 2020, the Quest 2 replaced the Quest as Facebook's fastest-growing headset. See Scott Hayden, "Oculus Quest 2 Surpasses Original Quest in Monthly Active Users," *RoadtoVR.com*, accessed January 7, 2021, https://www.roadtovr.com/oculus-quest-2-monthly-active-users/.

53. With the Rift, its successor Rift S, and the Oculus Quest being discontinued in Spring 2021, Facebook is entirely focusing on the Quest 2 as the single way forward for the company's VR strategy. See Adi Robertson, "Facebook Is Discontinuing the Oculus Rift S," *TheVerge.com*, accessed October 13, 2020, https://www.theverge.com/2020/9/16/21422717/facebook-oculus-rift-s-discontinued-quest-2-vr-connect/.

54. Ben Lang, "HTC Unveils Three New Vive Cosmos Headsets: Elite, Play, and XR," *RoadtoVR.com*, accessed February 29, 2020, https://www.roadtovr.com/htc-vive-cosmos-series-cosmos-elite-cosmos-play-cosmos-xr/; Ben Lang, "HTC Reveals Vive Proton, a Compact Standalone VR Headset Prototype," *RoadtoVR.com*, accessed February 29, 2020, https://www.roadtovr.com/htc-vive-proton-prototype-reveal/.

55. Ben Lang, "5 Million VR Units Sold, Sony Announces," *RoadtoVR.com*, accessed February 10, 2020, https://www.roadtovr.com/playstation-vr-sales-5-million-milestone-psvr-units-sold/.

56. Hesmondhalgh, *The Cultural Industries*, 195–197.

57. Daydream, "Phones Built for Virtual Reality," *VR.google.com*, accessed February 12, 2020, https://vr.google.com/daydream/smartphonevr/phones/.

58. Janko Roettgers, "Google Ships Pixel 4 without Daydream VR Support, Stops Selling Daydream Viewer," *Variety.com*, accessed February 12, 2020, https://variety.com/2019/digital/news/pixel-4-google-daydream-vr-1203371182/.

59. Ben Lang, "Daydream and VR Nowhere to Be Seen at Google I/O," *RoadtoVR.com*, accessed February 10, 2020, https://www.roadtovr.com/google-io-2019-daydream-vr/.

60. James Vincent, "Copy and Paste the Real World with Your Phone Using Augmenting Reality," *TheVerge.com*, accessed June 10, 2020, https://www.theverge.com/2020/5/4/21246386/augmented-reality-ar-copy-cut-paste-real-world-photoshop-demo.

61. Apple, "The Future Is Here: iPhone X," *Apple.com*, accessed February 10, 2020, https://www.apple.com/newsroom/2017/09/the-future-is-here-iphone-x/.

62. Scott Hayden, "Facebook Is Using Target Tracking to Unlock AR Experiences from Movie Posters and More," *RoadtoVR.com*, accessed April 10, 2020, https://www.roadtovr.com/facebook-using-target-tracking-unlock-ar-experiences-movie-posters/.

63. Scott Hayden, "Oculus Go Headsets Are Now Shipping to Developers," *RoadtoVR.com*, accessed January 6, 2020, https://www.roadtovr.com/oculus-go-headsets-now-arriving-developers/.

64. Ben Lang, "Oculus Quest Review—the First Great Standalone VR Headset," *RoadtoVR.com*, accessed January 6, 2020, https://www.roadtovr.com/oculus-quest-review-the-first-great-standalone-vr-headset/, and Ben Lang, "Oculus Quest 2 Review—the Best Standalone Headset Gets Better in (Almost) Every Way," *RoadtoVR.com*, accessed October 13, 2020, https://www.roadtovr.com/oculus-quest-2-review-better-in-almost-every-way/.

65. Scott Hayden, "HTC Unveils Vive Focus Plus with 6DOF Controllers, Built for Enterprise," *RoadtoVR.com*, accessed January 6, 2020, https://www.roadtovr.com/enterprise-vive-focus-plus-6dof-controllers/.

66. Ellen Daniel, "Apple AR Headset Rumours Suggest That a Launch Could Be Immanent," *Verdict.co.uk*, accessed May 13, 2020, https://www.verdict.co.uk/apple-glasses/.

67. As the device vaporizes aromatic liquids, it needs to pass FDA approval before it could get launched onto the market. Scott Hayden, "Feelreal VR Scent Masks Hits Roadblock amidst Crackdown on Flavored Vaping Products," *RoadtoVR.com*, accessed February 20, 2020, https://www.roadtovr.com/feelreal-vr-scent-mask-vaping-fda-ban/.

68. The Void, "Ghostbusters," *TheVoid.com*, accessed January 31, 2020, https://www.thevoid.com/dimensions/ghostbusters/.

69. The Void, "Star Wars: Secrets of the Empire," *TheVoid.com*, accessed January 31, 2020, https://www.thevoid.com/dimensions/star-wars-vr/.

70. Jeremy White, "I Was a Stormtrooper for 15 Minutes and It Was Awesome," *Wired.co.uk*, accessed January 31, 2020, http://www.wired.co.uk/article/star-wars-vr-london-secrets-of-empire-void-experience.

71. Ibid.

72. Global Times, "China Releases New Policy Guideline to Master Key VR Technologies by 2025," *Globaltimes.cn*, accessed January 31, 2020, http://www.globaltimes.cn/content/1133605.shtml.

73. Meredith Bricken, "Virtual Worlds: No Interface to Design," in *Cyberspace: First Steps*, ed. Meredith Bricken (Cambridge, MA: MIT Press, 1991), 372.

74. Steve Mann coined the term "mediated reality" to describe the ability to manipulate the perception of reality through the use of wearable technologies. I will extend the term to describe any form of reality created through the use of a medium. See Steve Mann and Woodrow Barfield, "Introduction to Mediated Realities," *International Journal of Human-Computer Interaction* 15, no. 2 (2003): 205–208.

75. In the 1990s, Paul Milgram and Fumio Kishino created a diagram of the "virtuality continuum." At one end they put what they called the "real environment" (the physical world) and a complete "virtual environment" at the other (VR). See Paul Milgram and Fumio Kishino, "A Taxonomy of Mixed Reality Visual Displays," *IEICE TRANSACTIONS on Information and Systems* E77-D, no. 12 (1994): 1321–1329.

76. For Plato, an idea is a "form, original, true entity, the real." These "universals" are mind-independent, abstract objects. They are "ideal," and no representation could ever capture them entirely. See also Gilles Deleuze, "Plato and the Simulacrum," in *The Logic of Sense,* trans. Mark Lester and Charles Stivale, ed. Constantin V. Boundas (New York: Columbia University Press, 1990), 253–266.

77. Guy-Ernest Debord, *La société du spectacle* (Paris: Buchet-Chastel, 1967).

78. Guy-Ernest Debord, "Society of the Spectacle," *Library.nothingness.org*, accessed July 2, 2020, http://library.nothingness.org/articles/SI/en/display/16.

79. Oliver Grau, *Virtual Art: From Illusion to Immersion* (Cambridge, MA: MIT Press, 2003), 15.

80. Slavoj Žižek, "The Reality of the Virtual," *Openculture.com*, accessed July 2, 2020, http://www.openculture.com/2014/07/the-reality-of-the-virtual-zizek.html.

81. Stefan Weber, "Media and the Construction of Reality," *Mediamanual.at*, accessed July 6, 2020, https://www.mediamanual.at/en/pdf/Weber_etrans.pdf.

82. Grau, *Virtual Art*, 17.

83. According to Fichte, the essence of the self consists of its own awareness. The self ("I") differs from its surrounding, the other or the non-self ("not-I"), while at the same time it has to develop its self-consciousness. Fichte's theories of the self and representation lead to a postmodern understanding of reality construction through the influence of mediation. See Johann Gottlieb Fichte, *The Science of Knowledge*, ed. and trans. Peter Heath and John Lachs (Cambridge: Cambridge University Press, 1982).

84. Weber, "Media and the Construction of Reality," 2.

85. Ibid., 6.

86. A similar meaning can be uncovered in his painting *Ceci n'est pas une pomme* (This is not an apple), 1964.

87. For instance, reality TV appears to mimic physical reality, although it is often scripted and edited to meet the requirements of a TV format. The characters might refer to individuals in physical reality, however, their appearance is mediated and conditioned for entertainment purposes, e.g., the Kardashians or the Osbournes.

88. The Globe at London's Bankside is not an actual replication of the Elizabethan playhouse, which was destroyed in a fire in 1613, rebuilt a year later, and then again completely demolished in 1644. The modern Globe is partly based on several descriptions and drawings of the first Globe. However, it is mainly based on the original Rose Theatre, one of the few playhouses of that era of which full design plans still exist. The 1997 replication is a hybrid version of the Globe and Rose theaters.

89. Gordon Calleja, "Digital Games and Escapism," *Games and Culture* 5, no. 4 (2010): 339.

90. Jean Baudrillard, *Simulations*, trans. Paul Foss, Paul Patton, and Philip Beichtman (New York: Semiotext[e], 1983), 150.

91. See also Gary Genosko, *Baudrillard and Signs: Signification Ablaze* (London: Routledge, 1994); William Merrin, "To Play with Phantoms: Jean Baudrillard and the Evil Demon of the Simulacrum," *Economy and Society* 30, no. 1 (2001): 85–111; Michael

Camille, "Simulacrum," in *Critical Terms for Art History*, ed. Robert S. Nelson and Richard Shiff (Chicago: University of Chicago Press, 2003), 35–50.

92. Jean Baudrillard, *Simulacra and Simulation*, trans. Sheila Faria Glaser (Ann Arbor: University of Michigan Press, 1994), 12–13.

93. Werner Wolf, "Aesthetic Illusion," in *Immersion and Distance: Aesthetic Illusion in Literature and Other Media*, ed. Werner Wolf, Walter Bernhart, and Andreas Mahler (Amsterdam: Rodopi, 2013), 24.

94. Baudrillard, *Simulacra and Simulation*, 23.

95. Langdon Winner, *Autonomous Technology: Technics-Out-of-Control as a Theme in Political Thought* (Cambridge, MA: MIT Press, 1977).

96. Stephen K. Reed, *Cognition: Theory and Applications*, 6th ed. (Belmont, CA: Wadsworth, 2004), 179.

97. Vilém Flusser, *Kommunikologie, Schriften 4*, ed. Vera Eckstein and Stefan Bollmann (Mannheim: Bollmann, 1996).

98. Vilém Flusser, "The Codified World," in *Writings*, ed. Andreas Ströhl, trans. Erik Eisel (Minneapolis: University of Minnesota Press, 2002), 40.

99. Gordon Calleja, *In-Game: From Immersion to Incorporation* (Cambridge, MA: MIT Press, 2011), 18.

100. Ibid.

101. For instance, Cheryl Campanella Bracken and Renée A. Botta, "Telepresence and Television," in *Immersed in Media: Telepresence in Everyday Life*, ed. Cheryl Campanella Bracken and Paul Skalski (London: Routledge, 2009), 39–62.

102. For instance, Frank Biocca and Ben Delaney, "Immersive Virtual Reality Technology," in *Communication in the Age of Virtual Reality*, ed. Frank Biocca and Mark R. Levy (Hillsdale, NJ: Lawrence Erlbaum, 1995), 57–124; Matthew Lombard and Theresa Ditton, "At the Heart of It All: The Concept of Presence," *Journal of Computer-Mediated Communication* 3, no. 2 (1997): unpaginated.

103. Mel Slater and Sylvia Wilbur, "A Framework for Immersive Virtual Environments (Five): Speculations on the Role of Presence in Virtual Environments," *Presence: Teleoperators and Virtual Environments* 6, no. 6 (1997): 607.

104. Bob G. Witmer and Michael J. Singer, "Measuring Presence in Virtual Environments: A Presence Questionnaire," *Presence: Teleoperators and Virtual Environments* 7, no. 3 (1998): 227.

105. Emily Brown and Paul Cairns, "A Grounded Investigation of Game Immersion," in *CHI 2004: Extended Abstracts on Human Factors in Computing Systems* (Vienna, April 24–29, 2004), 1299.

106. Ibid.

107. McLuhan, *Understanding Media*.

108. Wijnand Ijsselstein and Giuseppe Riva, "Being There: The Experience of Presence in Mediated Environments," in *Being There: Concepts, Effects and Measurements of User Presence Synthetic Environments*, ed. Wijnand Ijsselstein and Giuseppe Riva (Amsterdam: Ios Press, 2003), 3–16.

109. Mel Slater, "A Note on Presence Terminology," *Cs.ucl.ac.uk*, accessed July 7, 2020, http://www0.cs.ucl.ac.uk/research/vr/Projects/Presencia/ConsortiumPublications/ucl_cs_papers/presence-terminology.htm.

110. Janet Murray calls this phenomenon "immersion," but she is actually referring to presence, the mediated state of "attending," "participating," "placing yourself" in a mediated environment. See Janet Murray, *Hamlet on the Holodeck: The Future of Narrative in Cyberspace* (New York: Free Press, 1997), 98.

111. Lev Manovich, "To Lie and to Act: Potemkin's Village, Cinema and Telepresence," in *The Robot in the Garden: Telerobotics and Telepistemology in the Age of the Internet*, ed. Ken Goldberg (Cambridge, MA: MIT Press, 2000), 175.

112. Calleja, *In-Game*.

113. International Society for Presence Research, "Presence Defined," *Ispr.info*, accessed July 8, 2020, https://ispr.info/about-presence-2/about-presence/.

114. Calleja, "Immersion in Virtual Worlds," 222.

115. International Society for Presence Research, "Presence Defined," [7a] and [7b].

116. Ibid., [7c]–[7e].

117. Giuseppe Riva and John A. Waterworth, "Being Present in a Virtual World," in Grimshaw, *The Oxford Handbook on Virtuality*, 205–221.

118. Ibid., 207.

119. In contrast, Alan Dix points to the notion of absence and "not being there anymore," in particular to the traces of presence in digital environments. See Alan Dix et al., "absenT Presence," *AlanDix.com*, accessed July 8, 2020, http://alandix.com/academic/papers/absent-presence-2004/absent-presence.pdf.

120. Gabriella Giannachi and Nick Kaye, *Performing Presence: Between the Life and the Simulated* (Manchester: Manchester University Press, 2011), 2.

121. For the etymology of *in-lusio* see Roger Caillois, *Les jeux et les hommes: Le masque et le vertige*, rev. ed. (1958; Paris: Gallimard, 1991).

122. Roland Barthes, "L'effet de réel," *Communications* 11 (1968): 84–89.

123. Annabel J. Cohen, "Music as a Source of Emotion in Film," in *Music and Emotion: Theory and Research*, ed. Patrik N. Juslin and John A. Sloboda (Oxford: Oxford University Press, 2001), 249–271. See also Victor Nell, *Lost in a Book: The Psychology of Reading for Pleasure* (New Haven: Yale University Press, 1988).

124. See Nell, *Lost in a Book*, 199 and 211.

125. Ryan, *Narrative as Virtual Reality*, 21–23.

126. Kendall L. Walton, *Mimesis as Make-Believe* (Cambridge, MA: Harvard University Press, 1993).

127. Rita Felski, *Uses of Literature* (Malden, MA: Blackwell, 2008).

128. Walton, *Mimesis as Make-Believe*, 240–289.

129. Alison McMahan, "Immersion, Engagement, and Presence: A Method for Analyzing 3-D Video Games," in *The Video Game Theory Reader*, ed. Mark J. P. Wolf and Bernard Perron (London: Routledge, 2003), 67–86.

130. Richard J. Gerrig, *Experiencing Narrative Worlds: On the Psychological Activities of Reading* (New Haven: Yale University Press, 1993).

131. Frank Rose, *The Art of Immersion: How the Digital Generation Is Remaking Hollywood, Madison Avenue, and the Way We Tell Stories* (New York: W. W. Norton, 2011).

132. Deepfakes are manipulated videos and images that swap one face for another using machine learning techniques and artificial intelligence, as seen in various examples of fake celebrity videos since 2017.

133. See Ryan, *Narrative as Virtual Reality*, 89–114, and Marie-Laure Ryan, *Narrative as Virtual Reality 2: Revisiting Immersion and Interactivity in Literature and Electronic Media* (Baltimore: Johns Hopkins University Press, 2015), 61–84.

134. See Ryan, *Narrative as Virtual Reality*, 2.

135. Gundolf S. Freyermuth, "From Analog to Digital Image Space: Toward a Historical Theory of Immersion," in Liptay and Dogramaci, *Immersion in the Visual Arts and Media*, 166.

136. Ibid., 181.

137. Ryan, *Narrative as Virtual Reality 2*, 71.

138. Calleja, "Immersion in Virtual Worlds," 232.

139. Niels Christian Nilsson and Rolf Nordahl, "Immersion Revisited: A Review of Existing Definitions of Immersion and Their Relation to Different Theories of Presence," *Human Technology* 12, no. 2 (2016): 108–134.

140. Calleja, *In-Game*, 35f.

141. Laura Emri and Frans Mäyrä, "Fundamental Components of the Gameplay Experience: Analysing Immersion," in *Worlds in Play: International Perspectives on Digital Games Research*, ed. Suzanne de Castell and Jennifer Jenson (New York: Peter Lang, 2005), 15–27.

142. Jan-Noël Thon, "Immersion Revisited: On the Value of a Contested Concept," in *Extended Experiences: Structure, Analysis and Design of Computer Game Player Experience*, ed. Amyris Fernandez, Olli Leino, and Hanna Wirman (Rovaniemi: Lapland University Press, 2008), 29–43.

143. Fabienne Liptay, "Neither Here nor There: The Paradoxes of Immersion," in Liptay and Dogramaci, *Immersion in the Visual Arts and Media*, 88.

144. Roland Barthes, "Leaving the Movie Theater," in *The Art of the Personal Essay: An Anthology from the Classical Era to the Present*, ed. Phillip Lopate, trans. Richard Howard (New York: Anchor Books, 1995), 421.

## Chapter 2

1. Janet Murray uses the metaphor of "being submerged in water" and claims "[w]e seek the same feeling from a psychologically immersive experience that we do from a plunge in the ocean or a swimming pool." See Janet Murray, *Hamlet on the Holodeck: The Future of Narrative in Cyberspace* (New York: Free Press, 1997), 98f.

2. Neil Postman, *Technopoly: The Surrender of Culture to Technology* (New York: Vintage Books, 1993), 3.

3. Boris Groys, "Die Dauer der Bilder," in *Beat Streuli: City*, ed. Boris Groys and Rupert Pfab (Ostfildern: Hatje Cantz, 1999), 13–19.

4. Zygmunt Bauman, *Liquid Modernity* (Cambridge: Polity Press, 2000).

5. In one of the defining scenes of *The Matrix* (1999), Neo's mentor Morpheus is liquefying a mirror to prove that the world Neo is living in is nothing but a digital simulation. In discussing the scene, Bettina Papenburg mentions that the unsettling ambiguity of substances resists the classifications "solid" and "liquid." Neo is now absorbed by a world that is no longer defined by frames and surfaces. See Bettina Papenburg, "Touching the Screen, Striding through the Mirror: The Haptic in Film," in *What Does a Chameleon Look Like? Topographies of Immersion*, ed. Stefanie Kiwi Menrath and Alexander Schwinghammer (Cologne: Halem, 2011), 121.

6. Oliver Grau, *Virtual Art: From Illusion to Immersion* (Cambridge, MA: MIT Press, 2003), 249.

7. Burcu Dogramaci and Fabienne Liptay, "Immersion in the Visual Arts and Media," in *Immersion in the Visual Arts and Media*, ed. Fabienne Liptay and Burcu Dogramaci (Leiden: Brill Rodopi, 2016), 1.

8. Walter Benjamin, *The Arcades Project*, trans. Howard Eiland and Kevin McLaughlin (Cambridge, MA: Belknap Press of Harvard University Press, 1999), 494.

9. Gundolf S. Freyermuth, "From Analog to Digital Image Space: Toward a Historical Theory of Immersion," in Liptay and Dogramaci, *Immersion in the Visual Arts and Media*, 181.

10. Dogramaci and Liptay, "Immersion in the Visual Arts and Media," 3.

11. Peter Sloterdijk, "Architektur als Immersionskunst," *Arch+* 178 (2006): 80.

12. Béla Balázs, *Early Film Theory: Visible Man and the Spirit of Film*, ed. Erica Carter, trans. Rodney Livingstone (New York: Berghahn Books, 2010), 99.

13. Bauman, *Liquid Modernity*, 2.

14. Georg Simmel, "The Picture Frame: An Aesthetic Study," *Theory, Culture and Society* 11, no. 1 (1994): 11.

15. Jay David Bolter and Richard Grusin, *Remediation: Understanding New Media* (Cambridge, MA: MIT Press, 1998), 21.

16. Freyermuth, "From Analog to Digital Image Space," 179.

17. Bolter and Grusin, *Remediation*, 6.

18. Oliver Grau, "Immersion and Interaction: From Circular Frescoes to Interactive Image Spaces," trans. Gloria Custance, *Media Art Net*, accessed July 20, 2020, http://medienkunstnetz.de/themes/overview_of_media_art/immersion/.

19. Grau, *Virtual Art*, 203.

20. "Alles, was über die Welt gewusst, gedacht und gesagt werden kann, ist nur in Abhängigkeit von den Medien wissbar, denkbar und sagbar, die dieses Wissen kommunizieren. Nicht die Sprache, in der wir denken, sondern die Medien, in denen wir kommunizieren, modulieren unsere Welt." Aleida Assman and Jan Assmann, "Schrift—Kognition—Evolution. Eric A. Havelock und die Technologie kultureller Kommunikation," in *Schriftlichkeit. Das griechische Alphabet als kulturelle Revolution*, ed. Eric A. Havelock (Weinheim: VCH, 1990), 2.

21. Werner Wolf, "Introduction: Frames, Framings and Framing Borders in Literature and Other Media," in *Framing Borders in Literature and Other Media*, ed. Werner Wolf and Walter Bernhart (Amsterdam: Rodopi, 2006), 5.

22. Ibid.

23. Theodor W. Adorno, *Ästhetische Theorie* (Frankfurt am Main: Suhrkamp, 1973), 460.

24. Arnold Gehlen, *Zeitbilder* (Frankfurt am Main: Klostermann, 1986), 60.

25. Hartmut Böhme, *Natur und Subjekt* (Frankfurt am Main: Suhrkamp, 1988), 221.

26. Wolf, "Introduction," 4.

27. Bauman, *Liquid Modernity*, 6.

28. Bolter and Grusin, *Remediation*, 24.

29. Ray Kurzweil, "By 2030 We'll Have Full-Immersion, Shared, Virtual-Reality Environments," *Kurzweilai.net*, accessed June 29, 2020, https://www.kurzweilai.net/ray-kurzweil-by-2030-full-immersion-vr.

30. Katja Kwastek, "Immersed in Reflection? Aesthetic Experience of Interactive Media Art," in Liptay and Dogramaci, *Immersion in the Visual Arts and Media*, 73.

31. Jennifer Alsever, "Is Virtual Reality the Ultimate Empathy Machine?," *Wired.com*, accessed June 29, 2020, https://www.wired.com/brandlab/2015/11/is-virtual-reality-the-ultimate-empathy-machine/.

32. Jordan M. Carpenter and Melanie C. Green, "Flying with Icarus: Narrative Transportation and the Persuasiveness of Entertainment," in *The Psychology of Entertainment Media: Blurring the Lines between Entertainment and Persuasion*, ed. L. J. Shrum (London: Routledge, 2012), 170.

33. Werner Wolf, "Aesthetic Illusion," in *Immersion and Distance: Aesthetic Illusion in Literature and Other Media*, ed. Werner Wolf, Walter Bernhart, and Andreas Mahler (Amsterdam: Rodopi, 2013), 29.

34. Kingscross.co.uk, "Platform 9 3/4 at King's Cross Station," *Kingscross.co.uk*, accessed August 2, 2020, https://www.kingscross.co.uk/harry-potters-platform-9-34.

35. Urban Dictionary, "Pics or It Didn't Happen," *Urbandictionary.com*, accessed August 2, 2020, https://www.urbandictionary.com/define.php?term=pics+or+it+did n%27t+happen.

36. American author Henry James coined the term as "a state of vision, of feeling, and of consciousness," as well as a set of aesthetic practices and techniques among media, contexts, and institutions. See Henry James, quoted in Linda Williams, *Playing the Race Card: Melodramas in Black and White from Uncle Tom to O. J. Simpson* (Princeton, NJ: Princeton University Press, 2001), 6.

37. Nigel Thrift, "Afterwords," *Environment and Planning D: Society and Space* 18 (2000): 223.

38. Thomas Metzinger, *Being No One: The Self-Model Theory of Subjectivity* (Cambridge, MA: MIT Press, 2001), 1.

39. See also Thomas Metzinger, *The Ego Tunnel: The Science of the Mind and the Myth of the Self* (New York: Basic Books, 2009).

40. Edward O. Wilson, *Consilience: The Unity of Knowledge* (New York: Vintage Books, 1990), 130.

41. Wolf, "Aesthetic Illusion," 12.

42. Michel Foucault, *Discipline and Punish: The Birth of the Prison* (New York: Random House, 1975).

43. Michel Foucault, "The Confessions of the Flesh," in *Power/Knowledge: Selected Interviews and Other Writings 1972–1977*, ed. Colin Gordon (New York: Harvester Press, 1980), 194.

44. Laura Mulvey, "Visual Pleasure and Narrative Cinema," in *Film Theory and Criticism: Introductory Readings*, ed. Leo Braudy and Marshall Cohen (New York: Oxford University Press, 1999), 833–844.

45. Judith Butler, *Gender Trouble: Feminism and the Subversion of Identity* (New York: Routledge, 1990).

46. Jeremy Bentham, *The Panopticon Writings* (New York: Verso, 2010).

47. Sigmund Freud, *Group Psychology and the Analysis of the Ego*, trans. James Strachey (New York: Boni and Liveright, 1922).

48. Jacques Lacan, "The Mirror Stage," in *Identity: A Reader*, ed. Paul du Gay, Jessica Evans, and Peter Redman (London: Sage, 2000), 44–50.

49. The self is always mediated, even through a mirror. Standing in front of the mirror will not give an accurate simulation of the real, simply because the image will be shown mirror-inverted. Therefore, it is actually pointless to check your look in the mirror before you go out. Nobody will see you as a mirror-inverted image. As a result, an objective evaluation of your own selfhood is impossible.

50. Immanuel Kant, *Kritik der praktischen Vernunft* (Stuttgart: Reclam, 1973).

51. James W. Carey, "A Cultural Approach to Communication," in *Communications as Culture: Essays on Media and Society*, rev. ed., ed. James W. Carey (London: Routledge, 2009), 19–20.

52. Gernot Böhme, *Theorie des Bildes* (Munich: Fink, 1999).

53. Hans Belting, *Bild und Kult. Eine Geschichte des Bildes vor dem Zeitalter der Kunst*, 6th ed. (Munich: C. H. Beck, 2004).

54. Gottfried Boehm, *Wie Bilder Sinn erzeugen. Die Macht des Zeigens* (Berlin: Berlin University Press, 2007).

55. For instance, Hans Ulrich Gumbrecht, *Production of Presence: What Meaning Cannot Convey* (Stanford: Stanford University Press, 2004); Martin Seer, *Aesthetics of Appearing*, trans. John Farell (Stanford: Stanford University Press, 2005).

56. Susan Sontag, *Against Interpretation and Other Essays* (London: Picador, 1966), 14.

57. Harold D. Lasswell, "The Structure and Function of Communication in Society," in *The Communication of Ideas: A Series of Addresses*, ed. Lyman Bryson (New York: Institute for Religious and Social Studies, 1948), 32–52.

58. Denis McQuail, *McQuail's Mass Communication Theory* (London: Sage, 2010), 458.

59. Vilém Flusser, "The Codified World," in *Writings*, ed. Andreas Ströhl, trans. Erik Eisel (Minneapolis: University of Minnesota Press, 2002).

60. Vilém Flusser, *Kommunikologie, Schriften 4*, ed. Vera Eckstein and Stefan Bollmann (Mannheim: Bollmann, 1996), 107.

61. Ibid., 107. "Der Mensch wird aus der Welt verstoßen (Verfremdung 1), versucht, den klaffenden Abgrund durch die Projektion von Bildern zu überbrücken, und dank des Feedback zwischen Existenz und Bild gewinnt er einen Standpunkt zur 'Welt' (magisches Bewußtsein)."

62. Flusser, "The Codified World," 36.

63. Grau, *Virtual Art*, 25f.

64. Juri Lotman first introduced the term in his Russian article *On the Semiosphere* in 1984, describing the semiotic space, outside of which semiosis cannot exist. See Juri Lotman, "On the Semiosphere," trans. Wilma Clark, *Sign System Studies* 33, no. 1 (2005): 205–229. See also Peeter Torop, "Semiosphere and/as the Research Object of Semiotics of Culture," *Sign System Studies* 33, no. 1 (2005): 159–173.

65. Grau, *Virtual Art*, 28.

66. Ibid., 25.

67. Ibid., 27.

68. Carey, "A Cultural Approach to Communication," 21.

69. Alfred North Whitehead, *Process and Reality* (London: Macmillan, 1967), 76.

70. Jonathan Koestlé-Cate, *Art and the Church: A Fractious Embrace. Ecclesiastical Encounters with Contemporary Art* (London: Routledge, 2016), 22.

71. Ibid., 23.

72. Andreas Mahler, "Aesthetic Illusion in Theatre and Drama: An Attempt at Application," in Wolf, Bernhart, and Mahler, *Immersion and Distance*, 162.

73. Lev Manovich, *The Language of New Media* (Cambridge, MA: MIT Press, 2001), 76.

74. Ibid.

75. Ibid.

76. Freyermuth, "From Analog to Digital Image Space," 177.

77. Ibid., 178.

78. The album format is adapting to emerging technologies in order to stay relevant. For instance, such artists as indie singer-songwriter Moses Sumney are releasing new

albums in two or more parts, each spaced out over months. Mark Hogan, "Why Are Artists Breaking Up Their Albums into Separate Releases?," *Pitchfork.com,* accessed February 27, 2020, https://pitchfork.com/thepitch/breaking-up-albums-into-parts-hayley-williams-moses-sumney-bill-callahan/.

79. Experimental streaming services such as the short-form platform Quibi planned to condition viewing habits even further by becoming the home for thousands of chapter-like videos, or "quick bites." However, due to the lack of breakout content and free competition in the form of YouTube and TikTok, the mobile-only streaming service has been shut down after just six months. Chaim Gartenberg, "Quibi Is Shutting Down. The Shortform Video Service Had the Shortest Run of All," *TheVerge.com,* accessed October 21, 2020, https://www.theverge.com/2020/10/21/21527197/quibi-streaming-service-mobile-shutting-down-end-katzenberg.

80. Mark Zuckerberg, "Oculus VR, 25 March, 2014," *Facebook.com,* accessed January 11, 2020, https://www.facebook.com/zuck/posts/10101319050523971.

81. Scott Stein, "I Tried Facebook's Vision for the Social Future of VR, and It's Full of Question Marks," *Cnet.com,* accessed January 20, 2020, https://www.cnet.com/news/i-tried-facebooks-vision-for-the-social-future-of-vr-full-of-question-marks/.

82. For instance, metallic blue is the color of the digital representation of controllers and teleportation techniques in the semiotic system of many VR experiences. Sony's PlayStation VR uses the color blue even as part of the headset design.

83. Rikard Küller and Byron Mikellides, "Simulated Studies of Color, Arousal, and Comfort," in *Environmental Simulation,* ed. Robert W. Marans and Daniel Stokols (Boston, MA: Springer, 1993), 163–190.

84. Fritz Tepper, "Facebook Brings Emoji to VR with 360 Reactions," *Techcrunch.com,* accessed January 11, 2020, https://techcrunch.com/2016/06/22/facebook-brings-emoji-to-vr-with-360-reactions/.

85. Anna Bringas, "Facebook's New VR Emoji to Mimic Facial Expressions," *iTech Post.com,* accessed January 11, 2020, http://www.itechpost.com/articles/37733/20161007/facebooks-new-vr-emoji-mimic-real-life-facial-expressions.htm.

86. Ibid.

87. Josh Constine, "Facebook Invents 'Virtual Reality Emoji' Gestures," *Techcrunch.com,* accessed January 11, 2020, https://techcrunch.com/2016/10/06/vr-emoji/.

88. David-Paul Pertaub, Mel Slater, and Chris Barker, "An Experiment on Public Speaking Anxiety in Response to Three Types of Virtual Audience," *Presence: Teleoperators and Virtual Environments* 11, no. 1 (2002): 68–78.

89. Constine, "Facebook Invents 'Virtual Reality Emoji' Gestures."

90. Thomas Elsaesser, "The 'Return' of 3-D: On Some Logics and Genealogies of the Image in the Twenty-First Century," *Critical Inquiry* 39, no. 2 (2013): 241.

91. See Gabriella Giannachi and Nick Kaye, *Performing Presence: Between the Life and the Simulated* (Manchester: Manchester University Press, 2011), 1–25.

92. Margaret Werry and Bryan Schmidt, "Immersion and the Spectator," *Theatre Journal* 66 (2014): 475.

93. Josephine Machon, *Immersive Theatres: Intimacy and Immediacy in Contemporary Performance* (London: Palgrave Macmillan, 2013), 123f.

94. Pierre Lévy, *Collective Intelligence*, trans. Robert Bononno (New York: Perseus Books, 1999), 149.

95. Grau, *Virtual Art*, 25f. See also Alison Griffiths, *Shivers Down Your Spine: Cinemas, Museums and the Immersive View* (New York: Columbia University Press, 2008), 37f.

96. Michel Foucault, "Of Other Spaces," *Diacritics* 16, no. 1 (1986): 24.

97. Ibid.

98. Laura Bieger, *Ästhetik der Immersion. Raum-Erleben zwischen Welt und Bild. Las Vegas, Washington und die White City* (Bielefeld: Transcript, 2007), 9. ("Immersive Räume sind ein markanter Teil der Ästhetisierung von Lebenswelten, die unsere heutige Kultur so nachhaltig prägt.")

99. Ibid. ("Es sind Räume, in denen Welt und Bild sich überblenden und wir buchstäblich dazu eingeladen sind, uns in die Welt des Bildes zu begeben und in ihr zu bewegen.")

100. Ibid., 224.

101. Murray, *Hamlet on the Holodeck*, 106f.

102. Tony Bennett, "A Thousand and One Troubles: Blackpool Pleasure Beach," in *Formations of Pleasure*, ed. Fredric Jameson, Terry Eagleton, and Cora Kaplan (London: Routledge and Kegan Paul, 1983), 150f.

103. Robin Curtis, "Immersion and Abstraction as Measures of Materiality," in Liptay and Dogramaci, *Immersion in the Visual Arts and Media*, 50f.

104. Werry and Schmidt, "Immersion and the Spectator," 475.

105. Balázs, *Early Film Theory*, 66.

106. Rose Biggin, *Immersive Theatre and Audience Experience: Space, Game and Story in the Work of Punchdrunk* (London: Palgrave Macmillan, 2017).

107. Anne Corlett, "Punchdrunk's Kabeiroi: A Shapeshifting Six-Hour Mystery Tour of London," *TheGuardian.com*, accessed January 30, 2020, https://www.theguardian.com/stage/2017/sep/27/punchdrunk-kabeiroi-immersive-experience-aeschylus.

108. Jesper Juul, *Half-Real: Video Games between Real Rules and Fictional Worlds* (Cambridge, MA: MIT Press, 2005), 164–167.

109. Film terminology uses the definitions "plot" and "story" to define what is chronologically presented on screen and what is part of the film's *diegesis* (the narrative world).

110. Grau, *Virtual Art*, 17.

111. Even though there is an ongoing academic debate on when exactly silent reading gained mass popularity, scholars agree that reading aloud as a performative practice was common since antiquity. Since the eighteenth century, the act of reading is primarily a private, silent practice, and an introspective, intimate performance of the individual. See Mary Carruthers, *The Book of Memory: A Study of Memory in Medieval Culture*, 2nd ed. (Cambridge: Cambridge University Press, 2008).

112. Richard J. Gerrig, *Experiencing Narrative Worlds: On the Psychological Activities of Reading* (New Haven: Yale University Press, 1993., 17–24.

113. Amy Coplan, "Empathic Engagement with Narrative Fictions," *Journal of Aesthetics and Art Criticism* 62, no. 2 (2004): 141–152.

114. David Jagneaux, "Hands-On: Skyrim VR without Teleportation Is Much More Immersive," *UploadVR.com*, accessed January 30, 2020, https://uploadvr.com/preview-skyrim-vr-without-teleportation/.

115. Hans-Georg Gadamer, *Truth and Method*, 2nd ed., ed. and trans. Joel Weinsheimer and Donald G. Marshall (London: Continuum, 2004), 103.

116. Brian Boyd, *On the Origins of Stories: Evolution, Cognition, and Fiction* (Cambridge, MA: Harvard University Press, 2010), 189f.

117. Wolf, "Introduction: Frames, Framings and Framing Borders in Literature and Other Media," 5.

118. Wolfgang Iser, *The Act of Reading: A Theory of Aesthetic Response* (Baltimore: Johns Hopkins University Press, 1978).

119. Shalin Hai-Jew, "Exploring the Immersive Parasocial: Is It You or the Thought of You?," *Merlot Journal of Online Learning and Teaching* 5, no. 3 (2009): 550–561.

120. Edward Castronova, *Synthetic Worlds: The Business and Culture of Online Games* (Chicago: University of Chicago Press, 2005), 238.

121. Grau, *Virtual Art*, 13.

122. Mihály Csíkszentmihályi, *Flow: The Psychology of Optimal Experience* (New York: Harper Perennial, 1990).

123. Bolter and Grusin, *Remediation*, 30.

124. Ibid., 31. It remains a persistent myth that the crowd at the first public screening of the Lumière brothers' film *L'arrive d'un train en gare de La Ciotat* at the Grand

Café in Paris in January 1896 was running away from the train on the screen. Both Tom Gunning and Martin Loiperdinger provide findings that discount such a naïve reaction from the crowd. See Tom Gunning, "An Aesthetic of Astonishment: Early Film and the (In)Credulous Spectator," *Art and Text* 34 (1989): 31–45, and Martin Loiperdinger, "Lumière's Arrival of the Train: Cinema's Founding Myth," *Moving Image* 4, no. 1 (2004): 89–118.

125. Thomas Metzinger, "The Myth of Cognitive Agency: Subpersonal Thinking as a Cyclically Recurring Loss of Mental Autonomy," *Frontiers in Psychology* 4, accessed January 30, 2020, https://www.frontiersin.org/articles/10.3389/fpsyg.2013.00931/full.

126. Metzinger, *Being No One*.

127. Carrie Heeter, "The Subjective Experience of Presence," *Presence: Teleoperators and Virtual Environments* 1, no. 2 (1992): 262–271.

128. Charla Mathwick and Edward Rigdon, "Play, Flow, and the Online Search Experience," *Journal of Consumer Research* 31, no. 2 (2004): 324–332.

129. Yi-Fu Tuan, *Escapism* (Baltimore: Johns Hopkins University Press, 1998), 5–6. See also Andrew Evans, *This Virtual Life: Escapism and Simulation in Our Media World* (London: Vision, 2001).

130. Andrew Kuo, Richard J. Lutz, and Jacob L. Hiller, "Brave New World of Warcraft: A Conceptual Framework for Active Escapism," *Journal of Consumer Marketing* 33, no. 7 (2016): 498–506.

131. Steinicke and Bruder gave a first glimpse into the effects of long-term immersion by exposing a participant of their study to an HMD for 24 hours. See Frank Steinicke and Gerd Bruder, "A Self-Experimentation Report about Long-Term Use of Fully-Immersive Technology," in *The 2nd ACM Symposium*, ed. Andy Wilson, Frank Steinicke, Evan Suma, and Wolfgang Stürzlinger (Honolulu: AMC, 2014), 66–69.

132. Kimberly S. Young, "Internet Addiction: The Emergence of a New Clinical Disorder," *CyberPsychology and Behavior* 1, no. 3 (1998): 237–244, and Hanna O. Price, *Internet Addiction* (Hauppauge, NY: Nova Science Publishers, 2001).

133. Shaun Gallagher, *How the Body Shapes the Mind* (Oxford: Clarendon Press, 2005).

134. Daphne Simeon and Jeffrey Abugel, *Feeling Unreal: Depersonalization Disorder and the Loss of the Self* (Oxford: Oxford University Press, 2006).

135. Curtis, "Immersion and Abstraction as Measures of Materiality." 51.

136. Ibid. See also Robin Curtis, "'Einfühlung' and Abstraction in the Moving Image," *Science in Context* 25, no. 3 (2012): 425–446.

137. Phone Arena, "5 of the Best Virtual Pet Apps That Will Scratch That Tamagotchi Itch," *Phonearena.com*, accessed July 30, 2020, https://www.phonearena.com/news/5-best-Tamagotchi-apps-games-virtual-pet-iPhone-iOS-Android_id95038.

138. Gary Bente and Ansgar Feist, "Affect-Talk and Its Kin," in *Media Entertainment: The Psychology of Its Appeal*, ed. Dolf Zillmann and Peter Vorderer (Mahwah, NJ: Lawrence Erlbaum Associates, 2000), 21–36.

139. Elena Kokkinara and Rachel McDonnell, "Animation Realism Affects Perceived Character Appeal of a Self-Virtual Face," in *Proceedings of the 8th ACM SIGGRAPH Conference on Motion in Games* (New York: AMC, 2015), 221.

140. Victor Nell, *Lost in a Book: The Psychology of Reading for Pleasure* (New Haven: Yale University Press, 1988), 225.

141. AltspaceVR, "A Very Sad Goodbye," *Altvr.com*, accessed July 30, 2020, https://altvr.com/good-bye/.

142. AltspaceVR, "AltspaceVR Joins Microsoft," *Altvr.com*, accessed July 30, 2020, https://altvr.com/joining-microsoft/.

143. Microsoft, "Introducing Windows Mixed Reality," *Microsoft.com*, accessed July 30, 2020, https://www.microsoft.com/en-gb/windows/windows-mixed-reality.

144. Karl Prümm, "From the Unchained to the Ubiquitous Motion-Picture Camera: Camera Innovations and Immersive Effects," in Liptay and Dogramaci, *Immersion in the Visual Arts and Media*, 140.

145. Ibid., 141.

146. Grau, *Virtual Art*, 65–71.

147. Randall Colburn, "Disney Is Planning Enough Star Wars Movies to Last into the 2030s," *Consequenceofsound.net*, accessed July 30, 2020, https://consequenceofsound.net/2017/03/disney-is-planning-enough-star-wars-movies-to-last-into-the-2030s/.

148. Timothy Havens and Amanda Lotz, *Understanding Media Industries* (Oxford: Oxford University Press, 2012).

149. David Hesmondhalgh, *The Cultural Industries*, 3rd ed. (London: Sage, 2013), 4.

150. Ien Ang, *Desperately Seeking the Audience* (London: Routledge, 1991), 62.

151. José van Dijck and Thomas Poell, "Understanding Social Media Logic," *Media and Communication* 1, no. 1 (2013): 2–14.

152. Ibid., 5.

153. Ibid., 7.

154. Viktor Mayer-Schönberger and Kenneth Cukier, *Big Data: A Revolution That Will Transform How We Live, Work and Think* (London: John Murray, 2013).

155. Christian Stiegler, "The Politics of Immersive Storytelling: Virtual Reality and the Logics of Digital Ecosystems," *International Journal of E-Politics* 8, no. 3 (2017): 1–15.

156. Mark Andrejevic, "The Webcam Subculture and the Digital Enclosure," in *MediaSpace: Place, Scale and Culture in a Media Age*, ed. Nick Couldry and Anna McCarthy (London: Routledge, 2004), 197.

## Chapter 3

1. Oliver Grau, *Virtual Art: From Illusion to Immersion* (Cambridge, MA: MIT Press, 2003), 24–139.

2. Ibid., 44.

3. Marie-Laure Ryan, *Narrative as Virtual Reality: Immersion and Interactivity in Literature and Electronic Media* (Baltimore: Johns Hopkins University Press, 2001), 90.

4. See also Dalia Nassar, *The Relevance of Romanticism: Essays on German Romantic Philosophy* (Oxford: Oxford University Press, 2014).

5. Grau, *Virtual Art*, 91f.

6. Novalis (Friedrich von Hardenberg), *Henry of Ofterdingen*, trans. Palmer Hiltey (New York: Ungar, 1964), 195.

7. Ryan, *Narrative as Virtual Reality*, 4.

8. "Werther effect" is still used in media-effect theories to define copycat suicides that are influenced by (real and fictional) suicides in mass media.

9. Wolfgang Iser, *The Act of Reading: A Theory of Aesthetic Response* (Baltimore: Johns Hopkins University Press, 1978), 36.

10. Marisa Bortolussi and Peter Dixon, *Psychonarratology: Foundations for the Empirical Study of Literary Response* (Cambridge: Cambridge University Press, 2003), 37.

11. Ibid.

12. J. K. Rowling, *Harry Potter and the Sorcerer's Stone* (New York: Arthur A. Levine Books, 1997), 20.

13. In their study on immersion in digital literature, Alice Bell et al. propose the distinction between "reader constructions of immersion" and "immersive features." See Alice Bell, Astrid Ensslin, Isabelle van der Bom, and Jen Smith, "Immersion in Digital Fiction," *International Journal of Literary Linguistics* 7, no. 1 (2018): 1–22.

14. Roland Barthes, "The Death of the Author," in *Image / Music / Text*, trans. Stephen Heath (New York: Hill and Wang, 1977), 142–147.

15. Ryan, *Narrative as Virtual Reality*, 5.

16. Andreas Mahler, "Aesthetic Illusion in Theatre and Drama: An Attempt at Application," in *Immersion and Distance: Aesthetic Illusion in Literature and Other Media*, ed. Werner Wolf, Walter Bernhart, and Andreas Mahler (Amsterdam: Rodopi, 2013), 157.

17. Marie-Laure Ryan, "Immersion vs. Interactivity: Virtual Reality and Literary Theory," *SubStance* 28 (1999): 110–137.

18. Frank Biocca, "Virtual Reality Technology: A Tutorial," *Journal of Communication* 42, no. 4 (1992): 25.

19. Ryan, "Immersion vs. Interactivity," 117.

20. Gérard Genette, *Narrative Discourse: An Essay on Method* (Ithaca: Cornell University Press, 1972).

21. Ryan, *Narrative as Virtual Reality*, 4.

22. Ryan, "Immersion vs. Interactivity," 134.

23. Interactive fiction has a long tradition and was quite popular in printed versions in the form of "gamebooks" in the 1970s.

24. Ryan discusses several world-building concepts, most importantly "possible worlds" (literary worlds of imagination) and "impossible worlds" (literary worlds that transgress the basic laws of logic). See Marie-Laure Ryan, *Possible Worlds, Artificial Intelligence and Literary Theory* (Bloomington: Indiana University Press, 1991) and Marie-Laure Ryan, "Impossible Worlds and Aesthetic Illusion," in Wolf, Bernhart, and Mahler, *Immersion and Distance*, 131–148.

25. Henry Jenkins, "Transmedia Storytelling 101," *Henryjenkins.org*, accessed July 20, 2020, http://henryjenkins.org/blog/2007/03/transmedia_storytelling_101.html; Marie-Laure Ryan, *Avatars of Story* (Minneapolis: University of Minneapolis Press, 2006); Samuel Earl Ford, "As the World Turns in Convergence Culture," MIT Comparative Media Studies thesis, 2007, accessed July 20, 2020, https://cmsw.mit.edu/as-the-world-turns-in-a-convergence-culture/; Roberto Simanowski, *Digital Art and Meaning: Reading Kinetic Poetry, Text Machines, Mapping Art, and Interactive Installations* (Minneapolis: University of Minnesota Press, 2011); Bell, Ensslin, van der Bom, and Smith, "Immersion in Digital Fiction," 1–22.

26. Werner Wolf, *Ästhetische Illusion und Illusionsdurchbrechung in der Erzählkunst. Theorie und Geschichte mit Schwerpunkt auf englischem illusionsstörenden Erzählen* (Tübingen: Max Niemeyer Verlag, 1993). For the resistance of lyric poetry to aesthetic illusion, see Werner Wolf, "Aesthetic Illusion as an Effect of Lyric Poetry?," in Wolf, Bernhart, and Mahler, *Immersion and Distance*, 183–233.

27. Mark Z. Danielewski, "Bookworm Podcast: The Familiar," *KCRW.com*, accessed July 20, 2020, https://www.kcrw.com/culture/shows/bookworm/mark-z-danielewski-the-familiar.

28. Jacques Ehrmann, "Homo Ludens Revisited," in *Game, Play, Literature*, ed. Jacques Ehrmann (Boston: Beacon Press, 1971), 55.

29. Andreas Mahler, "Glauben, Nicht-Glauben, Anders-Sagen: Wege des Fingierens in Englands früher Neuzeit," in *Fiktionen des Faktischen in der Renaissance*, ed. Ulrike Schneider and Anita Traninger (Stuttgart: Steiner, 2010), 23–44.

30. Andreas Mahler, "Presented Representation: Intermedial Go-Betweens on the Shakespearean Stage," *Shakespeare Jahrbuch* 143 (2007): 147–158.

31. Mahler, "Aesthetic Illusion in Theatre and Drama," 157.

32. Fernando de Toro and Carole Hubbard, *Theatre Semiotics: Text and Staging in Modern Theatre* (Toronto: University of Toronto Press, 1995), 69.

33. Petr Bogatyrev, "Les signes du théatre," *Poétique* 8 (1971): 529.

34. De Toro and Hubbard, *Theatre Semiotics*, 91.

35. Mahler, "Aesthetic Illusion in Theatre and Drama," 157.

36. Robert Leach, *Theatre Studies: The Basics* (London: Routledge, 2008), 176.

37. Mahler, "Aesthetic Illusion in Theatre and Drama," 166.

38. Ibid., 169.

39. Richard Wagner, "Outlines of the Artwork of the Future," reprinted in *Multimedia: From Wagner to Virtual Reality*, ed. Randall Packer and Ken Jordan (New York: W. W. Norton, 2001), 5–6.

40. Bertolt Brecht, "On Chinese Acting," *Tulane Drama Review* 6, no. 1 (1961): 136.

41. Josephine Machon, *Immersive Theatres: Intimacy and Immediacy in Contemporary Performance* (London: Palgrave Macmillan, 2013), 142.

42. Myrto Koumarianos and Cassandra Silver, "Dashing at a Nightmare: Haunting Macbeth in *Sleep No More*, *The Drama Review* 57, no. 1 (2013): 168.

43. Lauren Rabinovitz, *Electric Dreamland: Amusement Parks, Movies, and American Modernity* (New York: Columbia University Press, 2012), 134.

44. After seveal years of development, *Back to the Future: The Musical* celebrated its world premiere in February 2020 at the Manchester Opera House.

45. See Steve Dixon, *Digital Performance: A History of New Media in Theater, Dance, Performance Art, and Installation* (Cambridge, MA: MIT Press, 2007).

46. See Daniel Paul O'Brian, "The Pervasive and the Digital: Immersive Worlds in Blast Theory's 'A Machine to See With' and Dennis Del Favero's 'Scenario,'" *International Journal of E-Politics* 8, no. 3 (2017): 30–41.

47. Oliver Grau, "Immersion and Interaction: From Circular Frescoes to Interactive Image Spaces," trans. Gloria Custance, *Media Art Net*, accessed July 20, 2017, http://medienkunstnetz.de/themes/overview_of_media_art/immersion/.

48. Frances Dyson, *Sounding New Media: Immersion and Embodiment in the Arts and Culture* (Berkeley: University of California Press, 2009), 138.

49. Institute for Sound and Music Berlin, "ISM Hexadome Artist Selection," *Berlin-ism.com*, accessed February 20, 2020, https://berlin-ism.com/en/news/ism-hexadome-artist-selection.

50. Stuart Grant, "Performing an Aesthetics of Atmospheres," *Literature and Aesthetics* 23, no. 1 (2014): 21.

51. Gerald Lynch, "Dolby Atmos for the People: R.E.M. on the Lost Art of Listening," *Techradar.com*, accessed February 20, 2020, https://www.techradar.com/news/dolby-atmos-for-the-people-rem-on-the-lost-art-of-listening.

52. VR and Fun, "Google Makes It Possible to Explore Each Layer of Music in VR with Inside Music," *VRandfun.com*, accessed February 20, 2020, https://www.vrandfun.com/google-makes-possible-explore-layer-music-vr-inside-music/.

53. Marc Hogan, "Is Secretive Virtual Reality Startup Magic Leap Dreaming Up the Future of Music? A First Look at the Billion-Dollar Company's Potentially Game-Changing Collaboration with Sigur Rós," *Pitchfork.com*, accessed February 20, 2020, https://pitchfork.com/features/article/is-secretive-virtual-reality-startup-magic-leap-dreaming-up-the-future-of-music/.

54. Noah Yoo, "Fields Wants to Be *the* Augmented Reality App for Experimental Music Fans and Creators Alike," *Pitchfork.com*, accessed May 15, 2020, https://pitchfork.com/thepitch/fields-wants-to-be-the-augmented-reality-app-for-experimental-music-fans-and-creators-alike/.

55. Samsung News, "Samsung Announces New 2018 Home Entertainment Lineup," *NewsSamsung.com*, accessed March 14, 2020, https://news.samsung.com/uk/samsung-announces-new-2018-home-entertainment-lineup

56. Paul Sawers, "Beyond Foldable Phones: How Flexible Displays Are Shaping Technology," *Venturebeat.com*, accessed March 30, 2020, https://venturebeat.com/2019/03/01/beyond-foldable-phones-how-flexible-displays-are-shaping-technology/.

57. Casey J. McCormick, "'Forward Is the Battle Cry': Binge-Viewing Netflix's *House of Cards*," in *The Netflix Effect: Technology and Entertainment in the 21st Century*, ed. Kevin McDonald and Daniel Smith-Rowsey (New York: Bloomsbury, 2016), 106.

58. Béla Balázs, *Early Film Theory: Visible Man and the Spirit of Film*, ed. Erica Carter, trans. Rodney Livingstone (New York: Berghahn Books, 2010), 95.

59. See Tom Gunning, "An Aesthetic of Astonishment: Early Film and the (In)Credulous Spectator," *Art and Text* 34 (1989): 31–45.

60. "The Talking Phonograph," *Scientific American* 37, no. 25 (1877): 384–385.

61. Balázs, *Early Film Theory*, 95.

62. See also Yasmin Ibrahim, *Production of the "Self" in the Digital Age* (London: Palgrave Macmillan, 2018).

63. André Bazin, "The Myth of Total Cinema," in *What Is Cinema*, vol. 1, ed. and trans. Hugh Gray (1967; Berkeley: University of California Press, 2005), 20.

64. Scott MacDonald, *The Garden in the Machine: A Field Guide to Independent Films about Place* (Berkeley: University of California Press, 2001), 9.

65. Tom Gunning calls the experience of moving images the "cinema of attractions." See Tom Gunning, "The Cinema of Attractions: Early Film, Its Spectator and the Avant-Garde," in *Early Cinema: Space Frame Narrative*, ed. Thomas Elsaesser (London: British Film Institute, 1990), 56–62.

66. The lack of editing (or the illusion of it) often creates a strong sense of immediacy for audiences. "One-shot films" are feature films filmed in one long take without any editing. Examples are *Victoria* (2014) and *Lost in London* (2017). Films that merely present the illusion of uninterrupted continuity with hidden cuts are Alfred Hitchcock's *Rope* (1948), *Irréversible* (2002), *Enter the Void* (2009), *Birdman or (The Unexpected Virtue of Ignorance)* (2015), and *1917* (2019).

67. Balázs, *Early Film Theory*, 99.

68. Karl Prümm, "From the Unchained to the Ubiquitous Motion-Picture Camera: Camera Innovations and Immersive Effects," in *Immersion in the Visual Arts and Media*, ed. Fabienne Liptay and Burcu Dogramaci (Leiden: Brill Rodopi, 2016), 155.

69. For an extensive list and introduction see Thomas Elsaesser, "The Mind-Game Film," in *Puzzle Films: Complex Storytelling in Contemporary Cinema*, ed. Warren Buckland (London: Wiley-Blackwell, 2009), 13–41.

70. Alison Griffiths, *Shivers Down Your Spine: Cinemas, Museums and the Immersive View* (New York: Columbia University Press, 2008), 87.

71. Ibid., 286.

72. Raymond Williams, *Television: Technology and Cultural Form* (London: Fontana, 1974).

73. Yasmin Ibrahim, "Self-Production through the Banal and the Fictive: Self and the Relationship with the Screen," *International Journal of E-Politics* 7, no. 2 (2016): 55.

74. Ibid., 58.

75. Richard Butsch, *Screen Culture: A Global History* (Cambridge: Polity, 2019).

76. Nintendo, "Nintendo Labo," *Labo.Nintendo.com*, accessed April 30, 2020, https://labo.nintendo.com/

77. Nintendo, "Nintendo Labo VR Kit," *Labo.Nintendo.com*, accessed April 30, 2020, https://labo.nintendo.com/kits/vr-kit/.

78. Dami Lee, "Programming a Song on Nintendo Labo with Toy-Con Garage Took All of My Brainpower," *TheVerge.com*, accessed April 30, 2020, https://www.theverge.com/2018/4/19/17253688/nintendo-switch-labo-cardboard-toy-con-garage.

79. Tuomas Kari, "Pokémon Go 2016: Exploring Situational Contexts of Critical Incidents in Augmented Reality," *Journal of Virtual Worlds Research* 9, no. 3 (2016): 1–12.

80. Mara Faccio and John J. McConnell, "Death by Pokémon Go: The Economic and Human Cost of Using Apps while Driving," *SSRN.com*, accessed February 20, 2020, https://papers.ssrn.com/sol3/Papers.cfm?abstract_id=3073723.

81. Andrea Peterson, "Holocaust Museum to Visitors: Please Stop Catching Pokémon Here," *Washington Post*, accessed January 15, 2020, https://www.washingtonpost.com/news/the-switch/wp/2016/07/12/holocaust-museum-to-visitors-please-stop-catching-pokemon-here/.

82. Frank Rose, *The Art of Immersion: How the Digital Generation Is Remaking Hollywood, Madison Avenue, and the Way We Tell Stories* (New York: W. W. Norton, 2011), 3.

83. Ibid., 166.

84. Freyermuth speaks of hyperrealistic audiovisuals and categorizes the realistic representation in VR in "virtual creation" (in the tradition of analog animation), "hybrid creation" (live-action footage blended with computer-generated images), and "procedural creation" (in the tradition of fully computer-generated experiences). See Gundolf S. Freyermuth, "From Analog to Digital Image Space: Toward a Historical Theory of Immersion," in Liptay and Dogramaci, *Immersion in the Visual Arts and Media*, 172.

85. Similar to the popular sandbox game *Minecraft*, *Fortnite* allows players to build their own worlds and battle arenas in the game's creative mode. Due to its success with younger audiences, particularly with those under 18, *Fortnite* has become a vehicle for cross-platform music marketing with artists Marshmallow and Weezer performing live in-game concerts and promoting new records. See also Tim Ingham, "Why Marshmellow's Fortnite Show Will Prove 'Revolutionary' for the Music Industry," *Rollingstone.com*, accessed March 30, 2020, https://www.rollingstone.com/music/music-features/marshmello-fortnite-show-will-prove-revolutionary-for-the-music-industry-797399/.

86. Amy M. Green, *Storytelling in Video Games: The Art of the Digital Narrative* (Jefferson, NC: McFarland, 2018).

87. Nick Yee, "The Labor of Fun," *Games and Culture* 1 (2006): 69.

88. John Seely Brown and Douglas Thomas, "You Play World of Warcraft? You're Hired!," *Wired.com*, accessed March 30, 2020, https://www.wired.com/2006/04/learn/.

**Notes**

89. Stephanie Rosenbloom, "It's Love at First Kill," *NYTimes.com*, accessed March 30, 2020, https://www.nytimes.com/2011/04/24/fashion/24avatar.html.

90. PlayStation Blog, "Escape, Explore, and Relax in Perfect, Out Tomorrow on PS VR," *Blog.US.PlayStation.com*, accessed March 30, 2020, https://blog.us.playstation.com/2016/12/12/escape-explore-and-relax-in-perfect-out-tomorrow-on-ps-vr/.

91. Treehugger VR, "Treehugger," *TreehuggerVR.com*, accessed March 30, 2020, http://www.treehuggervr.com/.

92. Marie-Laure Ryan, "From Narrative Games to Playable Stories: Toward a Poetics of Interactive Narrative," *Storyworlds: A Journal of Narrative Studies* 1 (2009): 45.

93. See Marcus Schulzke, "Translation between Forms of Interactivity: How to Build the Better Adaption," in *Game on, Hollywood! Essays on the Intersection of Video Games and Cinema*, ed. Gretchen Papazian and Joseph Michael Sommers (Jefferson, NC: McFarland, 2013), 70–85.

94. See also John Mateer, "Directing for Cinematic Virtual Reality: How the Traditional Film Director's Craft Applies to Immersive Environments and Notions of Presence," *Journal of Media Practice* 18, no. 1 (2017): 14–25.

95. Kent Bye, "The Yang and the Yin of Immersive Storytelling with Oculus' Yelena Rachitsky," *VoicesofVR.com*, accessed April 30, 2020, http://voicesofvr.com/637-the-yang-and-the-yin-of-immersive-storytelling-with-oculus-yelena-rachitsky/. This is a similar view to John Bucher's work, which sees VR storytelling as an opportunity to reflect upon our lives. See John Bucher, *Storytelling for Virtual Reality: Methods and Principles for Crafting Immersive Narratives* (New York: Routledge, 2018).

96. Ibid.

97. Paul Moody, "An 'Amuse-Bouche at Best': 360 Degree VR Storytelling in Full Perspective," *International Journal of E-Politics* 8, no. 3 (2017): 44.

98. USA Network, "Mr. Robot Virtual Reality Experience," *USANetwork.com*, accessed March 30, 2020, http://www.usanetwork.com/mrrobot/vr.

99. Christian Stiegler, "The Politics of Immersive Storytelling: Virtual Reality and the Logics of Digital Ecosystems," *International Journal of E-Politics* 8, no. 3 (2017): 1–15.

100. Internet Movie Database, "Dinner Party," *Imdb.com*, accessed March 30, 2020, https://www.imdb.com/title/tt8033802/.

101. Collisions VR, "Collisions," *CollisionsVR.com*, accessed March 30, 2020, http://www.collisionsvr.com/.

102. Virtual Reality—Notes on Blindness, "Notes on Blindness—A VR Journey into a World beyond Sight," *NotesonBlindness.co.uk*, accessed March 30, 2020, http://www.notesonblindness.co.uk/vr/.

103. See Felan Parker, "Millions of Voices: Star Wars, Digital Games, Fictional Worlds and Franchise Canon," in Papazian and Sommers, *Game on, Hollywood*, 156–168.

104. Halcyon, "The Scene Is Virtual: The Crime Is Real," *HalcyonVR.com*, accessed March 30, 2020, http://halcyonvr.com/.

105. Oculus, "Ghost in the Shell VR," *Oculus.com*, accessed March 30, 2020, https://www.oculus.com/experiences/rift/1592039100813771/.

106. Simon Hattenstone, "'After, I Feel Ecstatic and Emotional': Could Virtual Reality Replace Therapy?," *TheGuardian.com*, accessed March 30, 2020, https://www.theguardian.com/technology/2017/oct/07/virtual-reality-acrophobia-paranoia-fear-of-flying-ptsd-depression-mental-health.

107. Draw Me Close, "Draw Me Close," *Drawmeclo.se*, accessed March 30, 2020, https://www.drawmeclo.se/.

108. Internet Movie Database, "The Last Goodbye," *Imdb.com*, accessed March 30, 2020, https://www.imdb.com/title/tt6075970/.

109. Devindra Hardawar, "'The Last Goodbye' Is the VR Holocaust Memorial We Need Today," *Engadget.com*, accessed March 30, 2020, https://www.engadget.com/2017/04/22/the-last-goodbye-vr/.

110. Ibid.

## Chapter 4

1. Masahiro Mori, "The Uncanny Valley," trans. Karl F. MacDorman and Norri Kageki, *IEEE Robotics and Automation* 19, no. 2 (2012): 98–100.

2. Sigmund Freud, "The Uncanny" (1919), in *The Standard Edition of the Complete Psychological Works of Sigmund Freud*, vol. 17, ed. James Strachey and Anna Freud (London: Hogarth, 1971), 217–256.

3. Oculus, "Welcome to Facebook Horizon," *Oculus.com*, accessed February 15, 2020, https://www.oculus.com/facebookhorizon.

4. Scott Hayden, "'Facebook Spaces' Overhauls Avatars to Be More 'Fluid and Natural,'" *RoadtoVR.com*, accessed May 27, 2020, https://www.roadtovr.com/facebook-spaces-overhauls-avatars-fluid-natural/.

5. Scott Hayden, "Former HTC CEO unveils 5G-enabled VR headset & Social VR platform," *RoadtoVR.com*, accessed May 27, 2020, https://www.roadtovr.com/former-htc-ceo-xrspace-5g-headset-social-vr-platform/.

6. Scott Hayden, "'vTime' Releases New Avatar Customization Tool," *RoadtoVR.com*, accessed April 3, 2020, https://www.roadtovr.com/vtime-releases-new-avatar-customization-tool/.

7. Christian Stiegler, "Medienrealität(en): Zur Konstruktion medialer Wirklichkeiten," in *New Media Culture: Mediale Phänomene der Netzkultur*, ed. Christian Stiegler, Patrick Breitenbach, and Thomas Zorbach (Bielefeld: Transcript, 2015), 181–194.

8. E. Tory Higgins, "Self-Discrepancy: A Theory Relating Self and Affect," *Psychological Review* 94, no. 3 (1987): 320–321.

9. Joel Stein, "Millennials: The Me Me Me Generation," *Time.com*, accessed May 27, 2020, http://time.com/247/millennials-the-me-me-me-generation.

10. Ibid.

11. Christian Stiegler, "Selfies und Selfie Sticks: Automedialität des digitalen Selbstmanagements," in Stiegler, Breitenbach, and Zorbach, *New Media Culture*, 67–82.

12. David L. Jacobs, "Domestic Snapshots: Toward a Grammar of Motives," *Journal of American Culture* 4, no. 1 (1981): 104.

13. Mehita Iqani, "Spectacles or Publics? Billboards, Magazine Covers, and 'Selfies' as Spaces of Appearance," *Wiser.wits.ac.za*, accessed May 27, 2020, http://wiser.wits.ac.za/system/files/seminar/Iqani2013.pdf.

14. Brenda Laurel, *Computers as Theatre*, rev. ed. (Reading, MA: Addison-Wesley, 1993).

15. Erving Goffman, *The Presentation of Self in Everyday Life* (New York: Anchor Books, 1959).

16. In 1823, French author and critic Henri Beyle (Stendhal) told a similar anecdote about the conceptions of illusion in his comparison of the differences between the theater of Racine and that of Shakespeare, including a racist remark: "It is a warm summer night in August 1822. At the local theatre of Baltimore, Maryland, they are playing Shakespeare's tragedy of *Othello*. The production has already reached Act Five. All of a sudden, a soldier charged with keeping the order in the interior of the building is seen rushing onto the stage, shooting his weapon, breaking the arm of the actor playing the part of Othello and shouting: 'I won't have a damn nigger kill a white lady in my presence.'" Translation in Andreas Mahler, "Aesthetic Illusion in Theatre and Drama: An Attempt at Application," in *Immersion and Distance: Aesthetic Illusion in Literature and Other Media*, ed. Werner Wolf, Walter Bernhart, and Andreas Mahler (Amsterdam: Rodopi, 2013), 151.

17. James Cameron's *Avatar* (2009) draws from the same idea of being detached from reality. The main character has suffered from a combat injury and lost his legs. By transferring his consciousness into an avatar body, he fully becomes the avatar in essence, leaving his physical body behind.

18. José van Dijck, "'You Have One Identity': Performing the Self on Facebook and LinkedIn," *Media, Culture and Society* 35, no. 2 (2013): 208.

19. Hans-Thies Lehmann, *Postdramatic Theatre* (London: Routledge, 1999), 77.

20. Sarah Mannavis, "How Instagram's Plastic Surgery Filters Are Warping How We See Our Faces," *Newstatesman.com*, accessed May 27, 2020, https://www.newstatesman.com/science-tech/social-media/2019/10/how-instagram-plastic-surgery-filter-ban-are-destroying-how-we-see-our-faces.

21. Dominic Brennan, "Snapchat Launches Lens Studio for Making AR 'Lenses,'" *RoadtoVR.com*, accessed May 27, 2020, https://www.roadtovr.com/snapchat-launches-lens-studio-making-ar-lenses./ In addition to that, Snap launched Spectacles in 2016, a pair of video-recording sunglasses that allowed users to record what they see and post it directly on Snapchat. Snap distributed the glasses for a limited time only through special Snapbot vending machines, but was not able convince a larger audience due to limited content portability and the resistance to camera glasses.

22. Lev Manovich, *The Language of New Media* (Cambridge, MA: MIT Press, 2001), 180.

23. "Avatar" is often used in relation to social media and games. However, I will mostly use the term "mediated self" to describe the self in mediated experiences, not just in games or social media settings. While the term "profile" seems to be too broad to apply to specific mediated experiences, "mediated self" also includes dimensions of interactivity, sociality, and appearance.

24. Howard Rheingold, "A Slice of Life in My Virtual Community," in *Global Networks: Computers and International Communication*, ed. Linda M. Harasim (Cambridge, MA: MIT Press, 1993), 58.

25. Michael P. McCreery, Kathleen S. Krach, S., P. G. Schrader, and Randy Boone, "Defining the Virtual Self: Personality, Behavior, and the Psychology of Embodiment," *Computers in Human Behavior* 28 (2012): 977.

26. Sabina Misoch, "Avatare: Spiel(er)figuren in virtuellen Welten," in *Digitale Jugendkulturen*, ed. Kai-Uwe Hugger (Wiesbaden: Springer, 2010), 169–185.

27. Daniel Kromand, "Avatar Categorization," in *Proceedings of DiGRA 2007 Conference: Situated Play*, ed. Akira Baba, Tokyo, September 24–28, 2007, 400–406.

28. Ibid., 403.

29. Ibid.

30. Kwan Min Lee, "Presence, Explicated," *Communication Theory* 14, no. 1 (2004): 17.

31. Nick Yee and Jeremy Bailenson, "The Proteus Effect: The Effect of Transformed Self-Representation on Behavior," *Human Communication Research* 33, no. 3 (2007): 271–290.

32. Mike Brown, "Is Tinder a Match for Millennials?," *Lendedu.com*, accessed February 2, 2020, https://lendedu.com/blog/tinder-match-millennials/.

33. Jamie Feltham, "Facebook Teases Photorealistic Avatars for Social VR," *Uploadvr.com*, accessed May 3, 2020, https://uploadvr.com/facebook-teases-photorealistic-avatars-social-vr/.

34. Seth Stephens-Davidowitz, "The Songs That Bind," *NYTimes.com*, accessed February 10, 2020, https://www.nytimes.com/2018/02/10/opinion/sunday/favorite-songs.html.

35. Dana Kotter-Grühn, Maja Wiest, Peter Paul Zurek, and Susanne Scheibe, "What Is It We Are Longing For? Psychological and Demographic Factors Influencing the Contents of Sehnsucht (Life Longings)," *Journal of Research in Personality* 43, no. 3 (2009): 428–437.

36. Nintendo Entertainment System, "NES Classic Edition," *Nintendo.com*, accessed February 10, 2020, https://www.nintendo.com/nes-classic/.

37. Timothy Havens and Amanda Lotz, *Understanding Media Industries* (Oxford: Oxford University Press, 2012).

38. Josh Wigler, "'Stranger Things': All the Pop Culture References in Season 2," *Billboard.com*, accessed February 10, 2020, http://www.billboard.com/articles/news/television/8021964/stranger-things-pop-culture-references-season-2.

39. Fred Davis, *Yearning for Yesterday: A Sociology of Nostalgia* (New York: Free Press, 1979).

40. Kathrin Natterer, *Nostalgie als Zukunftsstrategie für Unterhaltungsmedien: Empirische Studien zu persönlicher und historischer Nostalgie in Medien* (Wiesbaden: Springer, 2017), 37.

41. David Harvey, *The Condition of Postmodernity: An Enquiry into the Origins of Cultural Change* (Hoboken, NJ: Wiley-Blackwell, 1991).

42. Susie Poppick, "10 Back to the Future Predictions That Came True," *Money.com*, accessed February 10, 2020, https://money.com/back-to-the-future-day-predictions-accuracy/.

43. Constantine Sedikides, Tim Wildschut, Clay Routledge, and Jamie Arndt, "Nostalgia Counteracts Self-Discontinuity and Restores Self-Continuity," *European Journal of Social Psychology* 45 (2015): 59–60.

44. Constantine Sedikides, Tim Wildschut, Jamie Arndt, and Clay Routledge, "Nostalgia. Past, Present, and Future," *Current Directions in Psychological Science* 17, no. 5 (2008): 304.

45. Kevin Kelly, "1,000 True Fans," *TheTechnium*, accessed February 10, 2020, http://kk.org/thetechnium/1000-true-fans/.

46. ABC News, "'Star Wars' Actors, Fans React to 'The Force Awakens' Trailer," *ABC7NY.com*, accessed February 10, 2020, http://abc7ny.com/entertainment/star-wars-actors-fans-react-to-the-force-awakens-trailer/1041841/.

47. Carolyn Giardina, "Ridley Scott Reveals How Kevin Spacey Was Erased from 'All the Money in the World,'" *Hollywoodreporter.com*, accessed February 10, 2020, https://www.hollywoodreporter.com/behind-screen/ridley-scott-reveals-how-kevin-spacey-was-erased-all-money-world-1068755.

48. Due to the allegations against him, Kevin Spacey mostly disappeared from the spotlight. However, as part of an interesting twist, he decided to publish cryptic Christmas videos on YouTube in 2018 and 2019, one of them being named "Let Me Be Frank." In both videos he seems to impersonate his role as *House of Cards*' Frank Underwood. Speaking in Underwood's voice, he seeks revenge against those who did him wrong. Even though the intention behind the videos remains unclear, they seem to suggest that even the actor Spacey does not distinguish anymore between himself and his roles.

49. Donald Horton and R. Richard Wohl, "Mass Communication and Para-Social Interaction: Observations on Intimacy at a Distance," *Psychiatry* 19 (1956): 215.

50. Shalin Hai-Jew, "Exploring the Immersive Parasocial: Is It You or the Thought of You?," *Merlot Journal of Online Learning and Teaching* 5, no. 3 (2009): 550–561.

51. Gayle S. Stever and Kevin Lawson, "Twitter as a Way for Celebrities to Communicate with Fans: Implications for the Study of Parasocial Interaction," *North American Journal of Psychology* 15, no. 2 (2013): 339–354.

52. Robert S. Weiss, "Attachment in Adult Life," in *The Place of Attachment in Human Behaviour*, ed. Colin Murray Parkes and Joan Stevenson-Hinde (New York: Basic Books, 1982), 171–184, and Robert S. Weiss, "The Attachment Bond in Childhood and Adulthood," in *Attachment across the Life Cycle*, ed. Colin Murray Parkes, Joan Stevenson-Hinde, and Peter Marris (New York: Routledge, 1991), 66–76.

53. Tim Cole and Laura Leets, "Attachment Styles and Intimate Television Viewing: Insecurely Forming Relationships in a Parasocial Way," *Journal of Social and Personal Relationships* 16, no. 4 (1999): 459–511.

54. J. Reid Meloy, Kris Mohandie, and Mila Green, "A Forensic Investigation of Those Who Stalk Celebrities," in *Stalking, Threatening, and Attacking Public Figures: A Psychological and Behavioral Analysis*, ed. J. Reid Meloy, Lorraine Sheridan, and Jens Hoffmann (Oxford: Oxford University Press, 2008), 37–54.

55. Brian H. Spitzberg and William R. Cupach, "Fanning the Flames of Fandom: Celebrity Worship, Parasocial Interaction, and Stalking," in Meloy, Sheridan, and Hoffmann, *Stalking, Threatening, and Attacking Public Figures*, 289.

56. Nielsen, "Binge Bunch: Two-thirds of Global VOD Viewers Say They Watch Multiple Episodes in a Single Sitting," *Nielsen.com*, accessed February 7, 2020, http://www.nielsen.com/us/en/insights/news/2016/binge-bunch-two-thirds-of-global-vod-viewers-say-they-watch-multiple-episodes-per-sitting.html.

57. Netflix, "Do You Remember Your First Time . . . Bingeing on Netflix," *Publicnow.com*, accessed February 7, 2020, http://www.publicnow.com/view/7A0A51071E2F95B73BFAF0E87B5BC16CECA62032.

58. Casey J. McCormick, "'Forward Is the Battle Cry': Binge-Viewing Netflix's *House of Cards*," in *The Netflix Effect: Technology and Entertainment in the 21st Century*, ed. Kevin McDonald and Daniel Smith-Rowsey (New York: Bloomsbury, 2016), 101.

59. While the term "binge activity" mainly refers to eating disorders and alcohol addiction, "media binge activities" focus solely on excessive media usage.

60. Silicon Valley Dictionary, "Tinder Binge," *Svdictionary.com*, accessed February 7, 2020, https://svdictionary.com/words/tinder-binge.

61. Cédric Courtois and Elisabeth Timmermans, "Cracking the Tinder Code: An Experience Sampling Approach to the Dynamics and Impact of Platform Governing Algorithms," *Journal of Computer-Mediated Communication* 23 (2018): 1.

62. Mihály Csíkszentmihályi and Jeanne Nakamura, "The Concept of Flow," in *Oxford Handbook of Positive Psychology*, ed. Charles R. Snyder and Shane L. Lopez (Oxford: Oxford University Press, 2001), 89.

63. Ibid., 92.

64. Ibid., 103.

65. Christian Stiegler, "Invading Europe: Netflix's Expansion to the European Market and the Example of Germany," in McDonald and Smith-Rowsey, *The Netflix Effect*, 235.

66. Amid the coronavirus pandemic, Netflix added 15.77 million new subscribers globally, more than double what it had initially expected, as people sought ways to entertain themselves during the lockdowns. See also Julia Alexander, "Netflix Adds 15 Million Subscribers as people Stream More Than Ever but Warns about Tough Road Ahead," *TheVerge.com*, accessed April 22, 2020, https://www.theverge.com/2020/4/21/21229587/netflix-earnings-coronavirus-pandemic-streaming-entertainment.

67. See Janet McCabe and Kim Akass, *Quality TV* (London: I. B. Tauris, 2007); Mark Jancovich and James Lyons, *Quality Popular Television* (London: BFI, 2008).

68. Brian Stelter, "New Way to Deliver a Drama: All 13 Episodes in One Sitting," *NYTimes.com*, accessed April 15, 2020, http://www.nytimes.com/2013/02/01/business/media/netflix-to-deliver-all-13-episodes-of-house-of-cards-on-one-day.html.

69. Yoon Hi Sung, Eun Yeon Kang, and Wei-Na Lee, "A Bad Habit for Your Health? An Exploration of Psychological Factors for Binge-Watching Behavior," 65th Annual International Communication Association Conference, San Juan, Puerto Rico, May, 21–25, 2015.

70. Peter Vorderer, "Interactive Entertainment and Beyond," in *Media Entertainment: The Psychology of Its Appeal*, ed. Dolf Zillman and Peter Vorderer (Mahwah, NJ: Lawrence Erlbaum Associates, 2000), 21–36.

71. In fact, excessive media usage linked to movement might even lead to health improvements. For instance, VR gamers have lost weight by standing up and moving around while playing. See David Jagneux, "The Elder Scrolls V: Skyrim V Is This Man's Weight Loss Tool," *Venturebeat.com,* accessed April 15, 2020, https://venturebeat.com/2018/04/15/the-elder-scrolls-v-skyrim-vr-is-this-mans-weight-loss-tool/, and Joe Durbin, "Man Loses More than 50 Pounds Playing a VR Game," *UploadVR.com*, accessed April 15, 2020, https://uploadvr.com/man-loses-50-pounds-playing-soundboxing/.

72. Jack Bernhardt, "Please No More Brilliant TV. I'm at Breaking Point," *TheGuardian.com*, accessed April 15, 2020, https://www.theguardian.com/commentisfree/2017/aug/18/no-more-brilliant-tv-too-many-great-shows-apple-streaming-frenzy.

73. It is worth noting that the default countdown for auto-playing episodes originally lasted 20 seconds but in recent years has been significantly reduced to a quarter of that. Only recently Netflix decided to allow users to disable the auto-play feature entirely. See Netflix, "How Can I Prevent Netflix from Auto-playing Episodes," *Help.netflix.com*, accessed March 30, 2020, https://help.netflix.com/en/node/2102.

74. Three, "Go Binge," *Three.co.uk*, accessed March 30, 2020, http://www.three.co.uk/go-binge.

75. Urban Dictionary, "Netflix," *Urbandictionary.com*, accessed March 30, 2020, http://www.urbandictionary.com/define.php?term=netflix.

76. *Netflix VR*, Netflix's VR app, invites users to a virtual log cabin up in the snowy mountains for an intimate, cozy binge experience and to watch shows on a red leather couch in front of an impressive movie screen.

77. Aaron Brown, "If You Share a Netflix Account with Someone, You NEED to Read This," *Express.co.uk*, accessed March 30, 2020, http://www.express.co.uk/life-style/science-technology/632874/Share-Netflix-Account-Policy-UK-Price-Plans.

78. Urban Dictionary, "Netflix and Chill," *Urbandictionary.com*, accessed March 30, 2020, http://www.urbandictionary.com/define.php?term=netflix and chill.

79. Oscar Rickett, "How 'Netflix and Chill' Became Code for Casual Sex," *TheGuardian.com*, accessed March 30, 2020, https://www.theguardian.com/media/shortcuts/2015/sep/29/how-netflix-and-chill-became-code-for-casual-sex.

80. Netflix, "Do You Remember Your First Time . . . Bingeing on Netflix."

81. Netflix, "Netflix Cheating Is on the Rise Globally and Shows No Signs of Stopping," *PRNewswire.com*, accessed March 30, 2020, https://www.prnewswire.com/news-releases/netflix-cheating-is-on-the-rise-globally-and-shows-no-signs-of-stopping-300406051.html.

**Notes**

82. Netflix, "A Better Netflix Experience," *Devices.netflix.com*, accessed March 30, 2020, https://devices.netflix.com/en/recommendedtv/2019/.

83. Netflix, "Projects," *Makeit.netflix.com*, accessed March 30, 2020, http://makeit.netflix.com/projects.

84. See Umberto Eco, *From the Tree to the Labyrinth: Historical Studies on the Sign and Interpretation*, trans. Anthony Oldcorn (Cambridge, MA: Harvard University Press, 2014).

85. Emerson Rosenthal, "Why Does Every Scene in 'House of Cards' Look the Same?," *Creators.Vice.com*, accessed March 30, 2020, https://creators.vice.com/en_uk/article/8qv8xp/why-does-every-scene-in-house-of-cards-look-the-same.

86. McCormick, "'Forward Is the Battle Cry,'" 105.

87. Ibid.

88. Netflix, "Interactive Storytelling on Netflix: Choose What Happens Next," *Media.netflix.com*, accessed April 5, 2020, https://media.netflix.com/en/company-blog/interactive-storytelling-on-netflix-choose-what-happens-next.

89. Ashok Chandrashekar, Fernando Amat, Justin Basilico, and Tony Jebara, "Artwork Personalization at Netflix," *Medium.com*, accessed April 5, 2020, https://medium.com/netflix-techblog/artwork-personalization-c589f074ad76.

90. See Henry E. Allison, *Kant's Theory of Taste: A Reading of the Critique of Aesthetic Judgment* (Cambridge: Cambridge University Press, 2001).

91. Alexis Madrigal, "Netflix Built Its Microgenres by Staring into the American Soul," *NPR.org*, accessed April 5, 2020, https://www.npr.org/sections/alltechconsidered/2014/01/02/259128268/netflix-built-its-microgenres-by-staring-into-the-american-soul?t=1583985594943.

92. Antoinette Rouvroy, "The End(s) of Critique: Data-Behaviourism vs. Due-Process," in *Privacy, Due Process and the Computational Turn: The Philosophy of Law Meets the Philosophy of Technology*, ed. Mireille Hildebrandt and Ekatarina De Vries (New York: Routledge, 2013), 143.

93. Ancestry, "If You Could Listen to Your DNA, What Would It Sound Like?," *Ancestry.com*, accessed April 5, 2020, https://www.ancestry.com/cs/spotify.

94. Madhumita Murgia, "The Insidious Threat of Biometrics," *Ft.com*, accessed April 5, 2020, https://www.ft.com/content/cdf0d52a-c2de-11e9-a8e9-296ca66511c9.

95. Courtois and Timmermans, "Cracking the Tinder Code," 1.

96. Jane Seidel, "The Game of Tinder: Gamification of Online Dating in the 21st Century," *Medium.com*, accessed April 5, 2020, https://medium.com/@jane_seidel/the-game-of-tinder-3c3ad575623f.

97. Courtois and Timmermans, "Cracking the Tinder Code," 7.

98. Donna Haraway, "A Cyborg Manifesto: Science, Technology and Socialist-Feminism in the Late Twentieth Century," in *Simians, Cyborgs and Women: The Reinvention of Nature* (New York: Routledge, 1991).

99. Stelarc, "Towards the Post-Human: From Psycho-body to Cybersystem," *Architectural Design* 118 (1995): 91.

100. Marwan M. Kraidy, *Hybridity, or the Cultural Logic of Globalization* (Philadelphia: Temple University Press, 2005), vi.

101. Several strategies to battle smartphone addiction have been established over the years, from imitating haptic feedback with substitute phones (therapeutic phonelike objects) to apps such as *Forest: Stay Focused* and *Hold* that support being offline with real and digital gratifications.

102. See also Adam Alter, *Irresistible: The Rise of Addictive Technology and the Business of Keeping Us Hooked* (London: Penguin Press, 2017).

103. Marina Milyavskaya, Mark Saffran, Nora Hoppe, and Richard Koestner, "Fear of Missing Out: Prevalence, Dynamics and Consequences of Experiencing FOMO," *Motivation and Emotion* 42 (2018): 725–737.

104. Apple, "About Face ID Advanced Technology," *Support.apple.com*, accessed April 25, 2020, https://support.apple.com/en-us/HT208108.

105. Apple, "How to Use Animoji on Your iPhone and iPad Pro," *Support.apple.com*, accessed April 25, 2020, https://support.apple.com/en-gb/HT208190.

106. Madeline Buxton, "Apple's Upcoming Software Update Includes Brand New Animoji," *Refinery29.uk*, accessed January 30, 2020, http://www.refinery29.uk/2017/10/179177/animoji-iphone-x-how-to-use.

107. Ben Lang, "7 Companies Aiming to Cut the Cord on High-End VR Headsets," *RoadtoVR.com*, accessed January 30, 2020, https://www.roadtovr.com/5-wireless-vr-companies-htc-vive-oculus-rift-technology-overview/.

108. Ibid.

109. Dieter Bohn, "Up Close with Pixel Buds, Google's Answer to AirPods," *TheVerge.com*, accessed January 30, 2020, https://www.theverge.com/2017/10/4/16405202/google-pixel-buds-wireless-headphones-photos-video-hands-on.

110. Judith Butler, *Bodies that Matter: On the Discursive Limits of "Sex"* (New York: Routledge, 1993).

111. Donna Haraway, "A Game of Cat's Cradle: Science Studies, Feminist Theory, Cultural Studies," *Configurations* 2, no. 1 (1994): 59–71.

112. Karen Barad, "Posthumanist Performativity: Toward an Understanding of How Matter Comes to Matter," *Signs: Journal of Women in Culture and Society* 28, no. 3 (2003): 812.

113. Kashif Saleem, Basit Shahzad, Mehmet A. Orgun, Jalal Al-Muhtadi, Joel J. P. C. Rodrigues, and Mohammed Zakariah, "Design and Deployment Challenges in Immersive and Wearable Technologies," *Behaviour and Information Technology* 36, no. 7 (2017): 688.

114. See also Argyro P. Karanasiou and Sharanjit Kang, "My Quantified Self, My FitBit and I. The Polymorphic Concept of Health Data and the Sharer's Dilemma," *Digital Culture & Society* 2, no. 1 (2016): 123–142.

115. Gina Neff and Dawn Nafus, *The Quantified Self* (Cambridge, MA: MIT Press, 2016).

116. The Medical Futurist, "How to Improve Your Mental Health with Technology," *Medicalfuturist.com*, accessed April 5, 2020, https://medicalfuturist.com/improve-mental-health-with-technology/.

117. Barad, "Posthumanist Performativity," 816.

118. Karen Barad, "Getting Real: Technoscientific Practices and the Materialization of Reality," *Differences* 10, no. 2 (1998): 87–128.

119. Bruno Latour, *Reassembling the Social: An Introduction to Actor-Network-Theory* (Oxford: Oxford University Press, 2005).

120. Barad, "Posthumanist Performativity," 816–817.

121. Ibid., 817.

122. Ibid.

123. Google AI, "Advancing AI for Everyone," *Ai.google*, accessed April 5, 2020, https://ai.google/.

124. Lev Manovich, "Can We Think without Categories?," *Digital Culture and Society* 4, no. 1 (2018): 20.

125. Christoph Kucklick, *Die granulare Gesellschaft. Wie das Digitale unsere Wirklichkeit auflöst* (Berlin: Ullstein, 2010).

126. In 2017, Apple published its first research paper, see Ashish Shrivastava, Tomas Pfister, Oncel Tuzel, Josh Susskind, Wenda Wang, and Russ Web, "Learning from Simulated and Unsupervised Images through Adversarial Training," *Arxiv.org*, accessed April 5, 2020, https://arxiv.org/pdf/1612.07828.pdf.

127. Etienne Wenger, *Artificial Intelligence and Tutoring Systems: Computational and Cognitive Approaches to the Communication of Knowledge* (Burlington, MA: Morgan Kaufmann, 2014).

128. John Johnston, *The Allure of Machinic Life: Cybernetics, Artifical Life, and the New AI* (Cambridge, MA: MIT Press, 2008), 278.

129. Toshinori Munakata, *Fundamentals of the New Artificial Intelligence: Neural, Evolutionary, Fuzzy and More* (New York: Springer, 2008).

130. Pew Research Center, "Americans' Attitudes towards a Future in which Robots and Computers Can Do Many Human Jobs," *Pewinternet.com*, accessed April 5, 2020, http://www.pewinternet.org/2017/10/04/americans-attitudes-toward-a-future-in-which-robots-and-computers-can-do-many-human-jobs/.

131. Steve Rose, "Reality Bites: Are Virtual Actors about to Put Hollywood's Humans Out of Work?," *TheGuardian.com*, accessed May 26, 2020, https://www.theguardian.com/film/2020/may/25/are-virtual-actors-about-to-put-hollywoods-humans-out-of-work-miquela.

132. Angela Watercutter, "The Irishman Gets De-Aging Right—No Tracking Dots Necessary," *Wired.com*, accessed March 11, 2020, https://www.wired.com/story/the-irishman-netflix-ilm-de-aging/.

133. Dieter Bohn, "Amazon Says 100 Million Alexa Devices Have Been Sold—What's Next?," *TheVerge.com*, accessed March 11, 2020, https://www.theverge.com/2019/1/4/18168565/amazon-alexa-devices-how-many-sold-number-100-million-dave-limp.

134. Ben Popper, "Ray Kurzweil Is Building a Chatbot for Google," *TheVerge.com*, accessed March 11, 2020, https://www.theverge.com/2016/5/27/11801108/ray-kurzweil-building-chatbot-for-google.

135. Guardian Music, "Warner Music Signs First Ever Record Deal with an Algorithm," *TheGuardian.com*, accessed March 25, 2020, https://www.theguardian.com/music/2019/mar/22/algorithm-endel-signs-warner-music-first-ever-record-deal.

136. Rob Arcand, "The Artists Using Artificial Intelligence to Dream up the Future of Music," *Spin.com*, accessed March 25, 2020, https://www.spin.com/featured/ai-music-artificial-intelligence-feature-holly-herndon-yacht/.

137. AIArtists, "Dr. Ahmed Elgammal," *Aiartists.org*, accessed March 25, 2020, https://aiartists.org/ahmed-elgammal.

138. Chris Welch, "You Can Now Hear Samuel L. Jackson's Voice on Your Amazon Echo," *TheVerge.com*, accessed March 25, 2020, https://www.theverge.com/2019/12/12/21013145/amazon-echo-alexa-samuel-l-jackson-celebrity-voice-now-available-price.

139. Christopher D. Manning, "Computational Linguistics and Deep Learning," *Computational Linguistics* 41, no. 4 (2015): 701–707.

140. Eric Mack, "These 27 Expert Predictions about Artificial Intelligence Will Both Disturb and Excite You," *Inc.com*, accessed March 25, 2020, https://www.inc.com

/eric-mack/heres-27-expert-predictions-on-how-youll-live-with-artificial-intelligence-in-near-future.html.

141. Ray Kurzweil, "Foreword to Virtual Humans," *Kurzweilai.net*, accessed June 29, 2020, https://www.kurzweilai.net/foreword-to-virtual-humans.

142. SXSW 2017, "Dark Days: AI and the Rise of Fascism," *Schedule.sxsw.com*, accessed June 29, 2020, https://schedule.sxsw.com/2017/events/PP93821.

143. Kurzweil, "Foreword to Virtual Humans."

**Chapter 5**

1. TMZ, "Justin Timberlake First Super Bowl Sneak Peek!!! (Spoiler Alert)," *TMZ.com*, accessed March 30, 2020, http://www.tmz.com/2018/02/03/justin-timberlake-super-bowl-halftime-performance-secrets/.

2. The Music Interview Archive, "Prince 1998 Guitar World Interview," *TheMusicInterviewArchive.com*, accessed March 30, 2020, https://sites.google.com/site/themusicinterviewarchive/prince/prince-1998-guitar-world-interview.

3. John Bream, "Plan for Prince Hologram in Super Bowl Halftime Show Apparently Dropped," *StarTribune.com*, accessed March 30, 2020, http://www.startribune.com/prince-to-play-super-bowl-halftime-via-hologram/472535173/.

4. Instagram, "ABBA Official," *Instagram.com/ABBAofficial*, accessed April 30, 2020, https://www.instagram.com/p/BiEl5cslkdR/?hl=en&taken-by=abbaofficial.

5. Eamonn Forde, "Touring Goes Beyond the Grave: Behind the Business, Technology and Ethics of the Hologram Concert Revolution," *Synchtank.com*, accessed April 30, 2020, https://www.synchtank.com/blog/touring-goes-beyond-the-grave-behind-the-business-technology-and-ethics-of-the-hologram-concert-revolution/.

6. Mark Zuckerberg, "Facebook, 09 January, 2020," *Facebook.com*, accessed April 12, 2020, https://www.facebook.com/zuck/posts/10103154542263811.

7. Oculus, "Introducing 'Facebook Horizon,' a New Social VR World, Coming to Oculus Quest and the Rift Platform in 2020," *Oculus.com*, accessed April 18, 2020, https://www.oculus.com/blog/introducing-facebook-horizon-a-new-social-vr-world-coming-to-oculus-quest-and-the-rift-platform-in-2020/.

8. See also Ray Kurzweil, *How to Create a Mind: The Secret of Human Thought Revealed* (New York: Viking Books, 2012).

9. Amazon, "Your Dash Buttons," *Amazon.com*, accessed March 30, 2020, https://www.amazon.com/ddb/learn-more.

10. Raw TV, "The Sex Robots Are Coming," *Raw.co.uk*, accessed April 14, 2020, https://www.raw.co.uk/sex-robots-coming.

11. Finlay Greig, "The Sex Robots Are Coming Review: Awkward, Unsettling—and Horrifying," *iNews.co.uk,* accessed April 18, 2020, https://inews.co.uk/culture/television/sex-robots-coming-review/.

12. Tiina Männistö-Funk and Tanja Sihvonen, "Voices from the Uncanny Valley: How Robots and Artificial Intelligences Talk Back to Us," *Digital Culture and Society* 4, no. 1 (2018): 45–64.

13. John Danaher and Neil McArthur, *Robot Sex: Social and Ethical Implications* (Cambridge, MA: MIT Press, 2017).

14. See also Zabet Patterson, "Going On-line: Consuming Pornography in the Digital Era," in *Porn Studies,* ed. Linda Williams (Durham: Duke University Press 2004), 104–125.

15. Antonio Rafele, "The Concept of Immersion: Pornography and Virtual Reality," *Sociétés* 4, no. 142 (2018): 19–31.

16. Feona Attwood, "Immersion: 'Extreme' Texts, Animated Bodies and the Media," *Media, Culture and Society* 36, no. 8 (2014): 1192.

17. Nick Bilton, "Coronavirus Is Creating a Fake-News Nightmarescape," *Vanityfair.com,* accessed April 18, 2020, https://www.vanityfair.com/news/2020/03/coronavirus-is-creating-fake-news-nightmarescape-social-media.

18. Bradley E. Wiggins, "Navigating an Immersive Narratology: Factors to Explain the Reception of Fake News," *International Journal of E-Politics* 8, no. 3 (2017): 16–33.

19. Yi-Fu Tuan, *Escapism* (Baltimore: Johns Hopkins University Press, 1998), 7.

20. Attwood, "Immersion," 1192.

21. Violet Kim, "Virtual Reality, Real Grief: A South Korean Documentary Reunited a Grieving Mother with Her Dead Daughter—in VR," *Slate.com,* accessed May 28, 2020, https://slate.com/technology/2020/05/meeting-you-virtual-reality-documentary-mbc.html.

22. Robert Nozick, *Anarchy, State, and Utopia* (New York: Basic Books, 1974), 44.

23. AzanaBand, "Introducing the AzanaBand. The World's First Sensory Gaming Device," *Azanaband.com,* accessed April 18, 2020, https://www.azanaband.com/.

24. Channel 4, "Kiss Me First," *Channel4.com,* accessed April 20, 2020, http://www.channel4.com/programmes/kiss-me-first.

25. Warner Bros. Pictures, "Ready Player One—Official Trailer 1 (HD)," *YouTube.com,* accessed April 20, 2020, https://www.youtube.com/watch?v=cSp1dM2Vj48.

# References

Adorno, Theodor W. *Ästhetische Theorie*. Frankfurt am Main: Suhrkamp, 1973.

Alsever, Jennifer. "Is Virtual Reality the Ultimate Empathy Machine?" *Wired.com*. Accessed June 29, 2020. https://www.wired.com/brandlab/2015/11/is-virtual-reality-the-ultimate-empathy-machine/.

Alter, Adam. *Irresistible: The Rise of Addictive Technology and the Business of Keeping Us Hooked*. London: Penguin Press, 2017.

Altheide, David and Robert Snow. *Media Logic*. London: Sage, 1979.

Andrejevic, Mark. "Facebook als neue Produktionsweise," In *Generation Facebook: Über das Leben im Social Net*, edited by Oliver Leistert and Theo Röhle, 31–49. Bielefeld: Transcript, 2011.

———. "The Webcam Subculture and the Digital Enclosure." In *MediaSpace: Place, Scale and Culture in a Media Age*, edited by Nick Couldry and Anna McCarthy, 193–208. London: Routledge, 2004.

Andrejevic, Mark, John Banks, John Edward Campbell, Nick Couldry, Adam Fish, Alison Hearn, and Laurie Ouellette. "Participations: Dialogues on the Participatory Promise of Contemporary Culture and Politics. Part 2: Labor." *International Journal of Communication* 8 (2015): 1089–1106.

Ancestry. "If You Could Listen to Your DNA, What Would It Sound Like?" *Ancestry.com*. Accessed April 5, 2020. https://www.ancestry.com/cs/spotify.

Ang, Ien. *Desperately Seeking the Audience*. London: Routledge, 1991.

Apple. "About Face ID advanced technology." *Support.apple.com*. Accessed April 25, 2020. https://support.apple.com/en-us/HT208108.

Assmann, Aleida, and Jan Assmann. "Schrift—Kognition—Evolution. Eric A. Havelock und die Technologie kultureller Kommunikation." In *Schriftlichkeit. Das griechische Alphabet als kulturelle Revolution*, edited by Eric A. Havelock, 1–35. Weinheim: VCH, 1990.

Attwood, Feona. "Immersion: 'Extreme' Texts, Animated Bodies and the Media." *Media, Culture and Society* 36, no. 8 (2014): 1186–1195.

Balázs, Béla. *Early Film Theory: Visible Man and the Spirit of Film.* Edited by Erica Carter, translated by Rodney Livingstone. New York: Berghahn Books, 2010.

Barad, Karen. "Posthumanist Performativity: Toward an Understanding of How Matter Comes to Matter." *Signs: Journal of Women in Culture and Society* 28, no. 3 (2003): 801–831.

———. "Getting Real: Technoscientific Practices and the Materialization of Reality." *Differences* 10, no. 2 (1998): 87–128.

Barlow, John Perry. "A Declaration of the Independence of Cyberspace." *Eff.org.* Accessed February 18, 2020, https://www.eff.org/cyberspace-independence

Barthes, Roland. "Leaving the Movie Theater." In *The Art of the Personal Essay: An Anthology from the Classical Era to the Present.* Edited by Phillip Lopate. Translated by Richard Howard, 418–421. New York: Anchor Books, 1995.

———. "The Death of the Author." In *Image / Music / Text.* Translated by Stephen Heath, 142–147. New York: Hill and Wang, 1977.

———. "L'effet de réel." *Communications* 11 (1968): 84–89.

Baudrillard, Jean. *Simulacra and Simulation.* Translated by Sheila Faria Glaser. Ann Arbor: University of Michigan Press, 1994.

———. *Selected Writings.* Edited by Mark Poster. Stanford: Stanford University Press, 1988.

———. *Simulations.* Translated by Paul Foss, Paul Patton, and Philip Beitchman. New York: Semiotext[e], 1983.

Bauman, Zygmunt. *Liquid Modernity.* Cambridge: Polity Press, 2000.

Bazin, André. "The Myth of Total Cinema." In *What Is Cinema*, vol. 1, edited and translated by Hugh Gray, 17–22. 1967; Berkeley: University of California Press, 2005.

Bell, Alice, Astrid Ensslin, Isabelle van der Bom, and Jen Smith. "Immersion in Digital Fiction." *International Journal of Literary Linguistics* 7, no. 1 (2018): 1–22.

Belting, Hans. *Bild und Kult. Eine Geschichte des Bildes vor dem Zeitalter der Kunst.* 6th edition. Munich: C. H. Beck, 2004.

Benjamin, Walter. *The Arcades Project.* Translated by Howard Eiland and Kevin McLaughlin. Cambridge, MA: Belknap Press of Harvard University Press, 1999.

Bennett, Tony. "A Thousand and One Troubles: Blackpool Pleasure Beach." In *Formations of Pleasure.* Edited by Fredric Jameson, Terry Eagleton, and Cora Kaplan, 138–155. London: Routledge and Kegan Paul, 1983.

Bente, Gary, and Ansgar Feist. "Affect-Talk and Its Kin." In *Media Entertainment: The Psychology of Its Appeal*, edited by Dolf Zillmann and Peter Vorderer, 21–36. Mahwah, NJ: Lawrence Erlbaum Associates, 2000.

Bentham, Jeremy. *The Panopticon Writings*. New York: Verso, 2010.

Bernhardt, Jack. "Please no more brilliant TV. I'm at breaking point." *TheGuardian.com*. Accessed April 15, 2020. https://www.theguardian.com/commentisfree/2017/aug/18/no-more-brilliant-tv-too-many-great-shows-apple-streaming-frenzy.

Bieger, Laura. *Ästhetik der Immersion. Raum-Erleben zwischen Welt und Bild. Las Vegas, Washington und die White City*. Bielefeld: Transcript, 2007.

Biggin, Rose. *Immersive Theatre and Audience Experience: Space, Game and Story in the Work of Punchdrunk*. London: Palgrave Macmillan, 2017.

Bilton, Nick. "Coronavirus is creating a fake-news nightmarescape." *Vanityfair.com*. Accessed April 18, 2020. https://www.vanityfair.com/news/2020/03/coronavirus-is-creating-fake-news-nightmarescape-social-media.

Biocca, Frank. "Virtual Reality Technology: A Tutorial." *Journal of Communication* 42, no. 4 (1992): 23–72.

Biocca, Frank, and Ben Delaney. "Immersive Virtual Reality Technology." In *Communication in the Age of Virtual Reality*, edited by Frank Biocca and Mark R. Levy, 57–124. Hillsdale, NJ: Lawrence Erlbaum, 1995.

Boehm, Gottfried. *Wie Bilder Sinn erzeugen. Die Macht des Zeigens*. Berlin: Berlin University Press, 2007.

Bogatyrev, Petr. "Les signes du théatre." *Poétique* 8 (1971): 517–530.

Böhme, Hartmut. *Natur und Subjekt*. Frankfurt am Main: Suhrkamp, 1988.

Bolter, Jay David, and Richard Grusin. *Remediation. Understanding New Media*. Cambridge, MA: MIT Press, 1998.

Bortolussi, Marisa, and Peter Dixon. *Psychonarratology: Foundations for the Empirical Study of Literary Response*. Cambridge: Cambridge University Press, 2003.

Boyd, Brian. *On the Origins of Stories: Evolution, Cognition, and Fiction*. Cambridge, MA: Harvard University Press, 2010.

Bracken, Cheryl Campanella, and Renée A. Botta. "Telepresence and Television." In *Immersed in Media: Telepresence in Everyday Life*, edited by Cheryl Campanella Bracken and Paul Skalski, 39–62. London: Routledge, 2009.

Brecht, Bertolt. "On Chinese Acting." *The Tulane Drama Review* 6, no. 1 (1961): 130–136.

Bricken, Meredith. "Virtual Worlds: No Interface to Design." In *Cyberspace: First Steps*, edited by Meredith Bricken, 363–382. Cambridge, MA: MIT Press, 1991.

Brown, Emily, and Paul Cairns. "A Grounded Investigation of Game Immersion." In *CHI 2004: Extended Abstracts on Human Factors in Computing Systems*, 1297–1300. Vienna, April 24–29, 2004.

Brown, Mike. "Is Tinder a Match for Millennials?" *Lendedu.com*. Accessed February 2, 2020. https://lendedu.com/blog/tinder-match-millennials/.

Bucher, John. *Storytelling for Virtual Reality. Methods and Principles for Crafting Immersive Narratives*. New York: Routledge, 2018.

Butler, Judith. *Bodies that Matter: On the Discursive Limits of "Sex."* New York: Routledge, 1993.

———. *Gender Trouble: Feminism and the Subversion of Identity*. New York: Routledge, 1990.

Butsch, Richard. *Screen Culture: A Global History*. Cambridge: Polity, 2019.

Caillois, Roger. *Les jeux et les hommes: Le masque et le vertige*. Rev. ed. 1958; Paris: Gallimard, 1991.

Calleja, Gordon. "Immersion in Virtual Worlds." In *The Oxford Handbook on Virtuality*, edited by Mark Grimshaw, 222–236. Oxford: Oxford University Press, 2015.

———. *In-Game: From Immersion to Incorporation*. Cambridge, MA: MIT Press, 2011.

———. "Digital Games and Escapism." *Games and Culture* 5, no. 4 (2010): 335–353.

Camille, Michael. "Simulacrum." In *Critical Terms for Art History*, edited by Robert S. Nelson and Richard Shiff, 35–50. Chicago: University of Chicago Press, 2003.

Carey, James W. "A Cultural Approach to Communication." In *Communications as Culture: Essays on Media and Society*, rev. ed., edited by James W. Carey, 11–28. London: Routledge, 2009.

Carpenter, Jordan M., and Melanie C. Green. "Flying with Icarus: Narrative Transportation and the Persuasiveness of Entertainment." In *The Psychology of Entertainment Media: Blurring the Lines between Entertainment and Persuasion*, edited by L. J. Shrum, 169–194. New York: Routledge, 2012.

Castronova, Edward. *Exodus to the Virtual World: How Online Fun Is Changing Reality*. London: Palgrave Macmillan, 2007.

———. *Synthetic Worlds: The Business and Culture of Online Games*. Chicago: University of Chicago Press, 2005.

Chan, Melanie. *Virtual Reality. Representations in Contemporary Media*. New York: Bloomsbury, 2014.

Chandrashekar, Ashok, Fernando Amat, Justin Basilico, and Tony Jebara. "Artwork Personalization at Netflix." *Medium.com*. Accessed April 5, 2020. https://medium.com/netflix-techblog/artwork-personalization-c589f074ad76.

# References

Cohen, Annabel J. "Music as a Source of Emotion in Film." In *Music and Emotion: Theory and Research*, edited by Patrik N. Juslin and John A. Sloboda, 249–271. Oxford: Oxford University Press, 2001.

Cole, Tim, and Laura Leets. "Attachment Styles and Intimate Television Viewing: Insecurely Forming Relationships in a Parasocial Way." *Journal of Social and Personal Relationships* 16, no. 4 (1999): 459–511.

Coplan, Amy. "Empathic Engagement with Narrative Fictions." *Journal of Aesthetics and Art Criticism* 62, no. 2 (2004): 141–152.

Courtois, Cédric, and Elisabeth Timmermans. "Cracking the Tinder Code: An Experience Sampling Approach to the Dynamics and Impact of Platform Governing Algorithms." *Journal of Computer-Mediated Communication* 23 (2018): 1–16.

Csíkszentmihályi, Mihály, and Jeanne Nakamura. "The Concept of Flow." In *Oxford Handbook of Positive Psychology*, edited by Charles R. Snyder and Shane L. Lopez, 89–105. Oxford: Oxford University Press, 2001).

———. *Flow: The Psychology of Optimal Experience*. New York: Harper Perennial, 1990.

Curtis, Robin. "Immersion and Abstraction as Measures of Materiality." In *Immersion in the Visual Arts and Media*, edited by Fabienne Liptay and Burcu Dogramaci, 41–66. Leiden: Brill Rodopi, 2016.

———. "'Einfühlung' and Abstraction in the Moving Image." *Science in Context* 25, no. 3 (2012): 425–446.

Danaher, John, and Neil McArthur. *Robot Sex: Social and Ethical Implications*. Cambridge, MA: MIT Press, 2017.

Daraiseh, Isra, and M. Keith Booker. "Unreal City: Nostalgia, Authenticity, and Posthumanity in 'San Junipero.'" In *Through the Black Mirror: Deconstructing the Side Effects of the Digital Age*, edited by Terence McSweeney and Stuart Joy, 151–164. London: Palgrave Macmillan, 2019.

Davis, Fred. *Yearning for Yesterday: A Sociology of Nostalgia*. New York: Free Press, 1979.

Debord, Guy-Ernest. *La société du spectacle*. Paris: Buchet-Chastel, 1967.

Deleuze, Gilles. "Plato and the Simulacrum." In *The Logic of Sense*, translated by Mark Lester and Charles Stivale, edited by Constantin V. Boundas, 253–266. New York: Columbia University Press, 1990.

Dijck, José van. "'You Have One Identity': Performing the Self on Facebook and LinkedIn." *Media, Culture and Society* 35, no. 2 (2013): 199–215.

———. "Facebook and the engineering of connectivity: A multi-layered approach to social media platforms." *Convergence* 19, no. 2 (2012): 141–155.

———. "Facebook as a Tool for Producing Sociality and Connectivity." *Television and Media* 13, no. 2 (2012): 160–176.

Dijck, José van, and Thomas Poell. "Understanding Social Media Logic." *Media and Communication* 1, no. 1 (2013): 2–14.

Ditlea, Steve. "Inside Virtual Reality." *PC/Computing* (1998): 91–101.

Dixon, Steve. *Digital Performance: A History of New Media in Theater, Dance, Performance Art, and Installation* (Cambridge, MA: MIT Press, 2007).

Dogramaci, Burcu, and Fabienne Liptay. "Immersion in the Visual Arts and Media." In *Immersion in the Visual Arts and Media*, edited by Fabienne Liptay and Burcu Dorgramaci, 1–17. Leiden: Brill Rodopi, 2016.

Dyson, Frances. *Sounding New Media: Immersion and Embodiment in the Arts and Culture*. Berkeley: University of California Press, 2009.

Eco, Umberto. *From the Tree to the Labyrinth: Historical Studies on the Sign and Interpretation*. Translated by Anthony Oldcorn. Cambridge, MA: Harvard University Press, 2014.

Ehrmann, Jacques. "Homo Ludens Revisited." In *Game, Play, Literature*, edited by Jacques Ehrmann, 31–57. Boston: Beacon Press, 1971.

Elsaesser, Thomas. "The 'Return' of 3-D: On Some Logics and Genealogies of the Image in the Twenty-First Century." *Critical Inquiry* 39, no. 2 (2013): 217–246.

———. "The Mind-Game Film." In *Puzzle Films: Complex Storytelling in Contemporary Cinema*, edited by Warren Buckland, 13–41. London: Wiley-Blackwell, 2009.

Emri, Laura, and Frans Mäyrä. "Fundamental Components of the Gameplay Experience: Analysing Immersion." In *Worlds in Play: International Perspectives on Digital Games Research*, edited by Suzanne de Castell and Jennifer Jenson, 15–27. New York: Peter Lang, 2005.

Evans, Andrew. *This Virtual Life: Escapism and Simulation in Our Media World*. London: Vision, 2001.

Faccio, Mara, and John J. McConnell. "Death by Pokémon Go: The Economic and Human Cost of using apps while driving." *SSRN.com*. Accessed February 20, 2020. https://papers.ssrn.com/sol3/Papers.cfm?abstract_id=3073723.

Facebook. "Reports fourth quarter and full year 2019 results." *Facebook.com*. Accessed February 15, 2020. https://investor.fb.com/investor-news/press-release-details/2020/Facebook-Reports-Fourth-Quarter-and-Full-Year-2019-Results/default.aspx.

———. "Facebook Spaces." *Facebook.com*. Accessed June 30, 2020. https://www.facebook.com/spaces.

Felski, Rita. *Uses of Literature*. Malden, MA: Blackwell, 2008.

Fermoso, Jose. "Facebook Invests $250m More in VR as Zuckerberg Shows Off Wireless Oculus." *TheGuardian.com*. Accessed July 20, 2020. https://www.theguardian

.com/technology/2016/oct/06/zuckerberg-facebook-virtual-reality-wireless-oculus-connect-3.

Fichte, Johann Gottlieb. *The Science of Knowledge*. Edited and translated by Peter Heath and John Lachs. Cambridge: Cambridge University Press, 1982.

Flusser, Vilém. "The Codified World." In *Writings*, edited by Andreas Ströhl, translated by Erik Eisel, 35–41. Minneapolis: University of Minnesota Press, 2002.

———. *Kommunikologie. Schriften 4*. Edited by Vera Eckstein and Stefan Bollmann. Mannheim: Bollmann, 1996.

Ford, Samuel Earl. "As the World Turns in Convergence Culture." MIT Comparative Media Studies thesis, 2007. Accessed July 20, 2020. https://cmsw.mit.edu/as-the-world-turns-in-a-convergence-culture/.

Forde, Eamonn. "Touring goes beyond the grave: Behind the business, technology and ethics of the hologram concert revolution." *Synchtank.com*. Accessed April 30, 2020. https://www.synchtank.com/blog/touring-goes-beyond-the-grave-behind-the-business-technology-and-ethics-of-the-hologram-concert-revolution/.

Foucault, Michel. "Of Other Spaces." *Diacritics* 16, no. 1 (1986): 22–27.

———. "The Confessions of the Flesh." In *Power/Knowledge: Selected Interviews and Other Writings 1972–1977*, edited by Colin Gordon, 194–228. New York: Harvester Press, 1980.

———. *Discipline and Punish: The Birth of the Prison*. New York: Random House, 1975.

Freud, Sigmund. "The Uncanny." In *The Standard Edition of the Complete Psychological Works of Sigmund Freud*, edited by James Strachey and Anna Freud, vol. 17, 217–256. London: Hogarth, 1971.

———. *Group Psychology and the Analysis of the Ego*. Translated by James Strachey. New York: Boni and Liveright, 1922.

Freyermuth, Gundolf S., "From Analog to Digital Image Space: Toward a Historical Theory of Immersion," In *Immersion in the Visual Arts and Media*, edited by Fabienne Liptay and Burcu Dogramaci, 165–203. Leiden: Brill Rodopi, 2016.

Fuchs, Philippe. *Virtual Reality Headsets: A Theoretical and Pragmatic Approach*. London, New York: CRC Press, 2017.

Gadamer, Hans-Georg. *Truth and Method*. 2nd ed. Translated by Joel Weinsheimer and Donald G. Marshall. London: Continuum, 2004.

Gallagher, Shaun. *How the Body Shapes the Mind*. Oxford: Clarendon Press, 2005.

Gehlen, Arnold. *Zeitbilder*. Frankfurt am Main: Klostermann, 1986.

Genette, Gérard. *Narrative Discourse. An Essay on Method*. Ithaca: Cornell University Press, 1972.

Genosko, Gary. *Baudrillard and Signs: Signification Ablaze*. London: Routledge, 1994.

Gerlitz, Carolin, and Anne Helmond. "The Like Economy: Social Buttons and the Data-Intensive Web." *New Media and Society* 15, no. 8 (2013): 1348–1365.

Gerrig, Richard J. *Experiencing Narrative Worlds: On the Psychological Activities of Reading*. New Haven: Yale University Press, 1993.

Giannachi, Gabriella, and Nick Kaye. *Performing Presence: Between the Life and the Simulated*. Manchester: Manchester University Press, 2011.

Gibbs, Samuel. "Google Glass review: useful—but overpriced and socially awkward." *TheGuardian.com*. Accessed July 6, 2020. https://www.theguardian.com/technology/2014/dec/03/google-glass-review-curiously-useful-overpriced-socially-awkward.

Gilbert, Ben. "Facebook is changing its logo to make sure users know it owns Instagram and WhatsApp." *Inc.com*. Accessed February 12, 2020. https://www.inc.com/business-insider/facebook-new-logo-instagram-whatsapp-parent-company.html.

Goffman, Erving. *The Presentation of Self in Everyday Life*. New York: Anchor Books, 1959.

Google AI. "Advancing AI for everyone." *Ai.google*. Accessed April 5, 2020. https://ai.google/

Grant, Stuart. "Performing an Aesthetics of Atmospheres." *Literature and Aesthetics* 23, no. 1 (2014): 12–32.

Grau, Oliver. "Immersion and Interaction: From Circular Frescoes to Interactive Image Spaces." Translated by Gloria Custance. *Media Art Net*. Accessed July 20, 2020. http://medienkunstnetz.de/themes/overview_of_media_art/immersion/.

———. *Virtual Art: From Illusion to Immersion*. Cambridge, MA: MIT Press, 2003.

Green, Amy M. *Storytelling in Video Games. The Art of the Digital Narrative*. Jefferson, NC: McFarland, 2018.

Griffiths, Alison. *Shivers Down Your Spine: Cinemas, Museums and the Immersive View*. New York: Columbia University Press, 2008.

Groys, Boris. "Die Dauer der Bilder." In *Beat Streuli: City*, edited by Boris Groys and Rupert Pfab, 13–19. Ostfildern: Hatje Cantz, 1999.

Gumbrecht, Hans Ulrich. *Production of Presence: What Meaning Cannot Convey*. Stanford: Stanford University Press, 2004.

Gunning, Tom. "The Cinema of Attractions: Early Film, Its Spectator and the Avant-Garde." In *Early Cinema: Space Frame Narrative*, edited by Thomas Elsaesser, 56–62. London: British Film Institute, 1990.

———. "An Aesthetic of Astonishment: Early Film and the (In)Credulous Spectator." *Art and Text* 34 (1989): 31–45.

# References

Hai-Jew, Shalin. "Exploring the Immersive Parasocial: Is it You or the Thought of You?" *Merlot Journal of Online Learning and Teaching* 5, no. 8 (2009): 550–561.

Haraway, Donna. "A Game of Cat's Cradle: Science Studies, Feminist Theory, Cultural Studies." *Configurations* 2, no. 1 (1994): 59–71.

———. "A Cyborg Manifesto: Science, Technology and Socialist-Feminism in the Late Twentieth Century." In *Simians, Cyborgs and Women: The Reinvention of Nature*, 149–181. New York: Routledge, 1991.

Harvey, David. *The Condition of Postmodernity: An Enquiry into the Origins of Cultural Change*. Hoboken, NJ: Wiley-Blackwell, 1991.

Havens, Timothy, and Amanda Lotz. *Understanding Media Industries*. Oxford: Oxford University Press, 2012.

Hayles, Katherine N. *How We Became Posthuman: Virtual Bodies in Cybernetics, Literature, and Informatics*. Chicago and London: The University of Chicago Press, 1999.

Heeter, Carrie. "The Subjective Experience of Presence." *Presence: Teleoperators and Virtual Environments* 1, no. 2 (1992): 262–271.

Heidegger, Martin. *The Question Concerning Technology and other Essays*. Translated by William Lovitt. New York: Harper and Row, 1993.

Heim, Michael, R. "The Paradox of Virtuality." In *The Oxford Handbook on Virtuality*, edited by Mark Grimshaw, 111–125. Oxford: Oxford University Press, 2015.

Hesmondhalgh, David. *The Cultural Industries*. 3rd ed. London: Sage, 2013.

Higgins, Tory E. "Self-Discrepancy: A Theory Relating Self and Affect." *Psychological Review* 94, no. 3 (1987): 319–340.

Hills-Duty, Rebecca. "Apple team up with IKEA to create AR shopping app." *VR Focus*. Accessed July 6, 2020. https://www.vrfocus.com/2017/06/apple-team-up-with-ikea-to-create-ar-shopping-app/.

Horton, Donald, and R. Richard Wohl. "Mass Communication and Para-Social Interaction: Observations on Intimacy at a Distance." *Psychiatry* 19 (1956): 215–229.

Ibrahim, Yasmin. *Production of the "Self" in the Digital Age*. London: Palgrave Macmillan, 2018.

———. "Self-Production through the Banal and the Fictive: Self and the Relationship with the Screen." *International Journal of E-Politics* 7, no. 2 (2016): 51–61.

Ijsselstein, Wijnand, and Giuseppe Riva. "Being There: The Experience of Presence in Mediated Environments." In *Being There: Concepts, Effects and Measurements of User Presence Synthetic Environments*, edited by Wijnand Ijsselstein and Giuseppe Riva, 3–16. Amsterdam: Ios Press, 2003.

Ingham, Tim. "Why Marshmellow's Fortnite Show Will Prove 'Revolutionary' for the Music Industry." *Rollingstone.com*. Accessed March 30, 2020. https://www.rollingstone.com/music/music-features/marshmello-fortnite-show-will-prove-revolutionary-for-the-music-industry-797399/.

International Society for Presence Research. "Presence defined." *Ispr.info*. Accessed July 8, 2020. https://ispr.info/about-presence-2/about-presence/.

Iqani, Mehita. "Spectacles or Publics? Billboards, magazine covers, and 'selfies' as spaces of appearance." *Wiser.wits.ac.za*. Accessed May 27, 2020. http://wiser.wits.ac.za/system/files/seminar/Iqani2013.pdf.

Iser, Wolfgang. *The Act of Reading: A Theory of Aesthetic Response* (Baltimore: Johns Hopkins University Press, 1978).

Jacobs, David L. "Domestic Snapshots: Toward a Grammar of Motives." *Journal of American Culture* 4, no. 1 (1981): 93–105.

Jancovich, Mark, and James Lyons. *Quality Popular Television*. London: BFI, 2008.

Jenkins, Henry. "Transmedia Storytelling 101." *Henryjenkins.org*. Accessed July 20, 2020. http://henryjenkins.org/blog/2007/03/transmedia_storytelling_101.html.

———. *Convergence Culture: Where Old and New Media Collide*. New York: New York University Press, 2006.

Johnston, John. *The Allure of Machinic Life: Cybernetics, Artifical Life, and the New AI*. Cambridge, MA: MIT Press, 2008.

Juul, Jesper. *Half-Real: Video Games between Real Rules and Fictional Worlds*. Cambridge, MA: MIT Press, 2005.

Kant, Immanuel. *Kritik der praktischen Vernunft*. Stuttgart: Reclam, 1973.

Karanasiou, Argyro P., and Sharanjit Kang. "My Quantified Self, My FitBit and I. The Polymorphic Concept of Health Data and the Sharer's Dilemma." *Digital Culture and Society* 2, no. 1 (2016): 123–142.

Kari, Tuomas. "Pokémon Go 2016: Exploring Situational Contexts of Critical Incidents in Augmented Reality." *Journal of Virtual Worlds Research* 9, no. 3 (2016): 1–12.

Kelly, Kevin. "AR will spark the next big tech platform—Call it Mirrorworld." *Wired.com*. Accessed February 2, 2020. https://www.wired.com/story/mirrorworld-ar-next-big-tech-platform/.

———. "1,000 True Fans." *The Technium*. Accessed February 10, 2020. http://kk.org/thetechnium/1000-true-fans/.

Kim, Violet. "Virtual Reality, Real Grief. A South Korean Documentary Reunited a Grieving Mother with Her Dead Daughter—in VR." *Slate.com*. Accessed May 28, 2020.

# References

https://slate.com/technology/2020/05/meeting-you-virtual-reality-documentary-mbc.html.

Koestlé-Cate, Jonathan. *Art and the Church: A Fractious Embrace. Ecclesastical Encounters with Contemporary Art*. London: Routledge, 2016.

Kokkinara, Elena, and Rachel McDonnell. "Animation Realism Affects Perceived Character Appeal of a Self-Virtual Face." In *Proceedings of the 8th ACM SIGGRAPH Conference on Motion in Games*, 221–226. New York: AMC, 2015.

Kotter-Grühn, Dana, Maja Wiest, Peter Paul Zurek, and Susanne Scheibe. "What Is It We Are Longing For? Psychological and Demographic Factors Influencing the Contents of Sehnsucht (Life Longings)." *Journal of Research in Personality* 43, no. 3 (2009): 428–437.

Koumarianos, Myrto, and Cassandra Silver. "Dashing at a Nightmare: Haunting Macbeth in *Sleep No More*." *The Drama Review* 57, no. 1 (2013): 167–175.

Kraidy, Marwan M. *Hybridity, or the Cultural Logic of Globalization*. Philadelphia: Temple University Press, 2005.

Kromand, Daniel. "Avatar Categorization." In *Proceedings of DiGRA 2007 Conference: Situated Play*, edited by Akira Baba, 400–406. Tokyo, September 24–28, 2007.

Kucklick Christoph. *Die granulare Gesellschaft. Wie das Digitale unsere Wirklichkeit auflöst*. Berlin: Ullstein, 2010.

Kuo, Andrew, Richard J. Lutz, and Jacob L. Hiller. "Brave New World of Warcraft: A Conceptual Framework for Active Escapism." *Journal of Consumer Marketing* 33, no. 7 (2016): 498–506.

Kurzweil, Ray. *How to Create a Mind: The Secret of Human Thought Revealed*. New York: Viking Books, 2012.

———. "Foreword to Virtual Humans," *Kurzweilai.net*, accessed February 2, 2020, https://www.kurzweilai.net/foreword-to-virtual-humans.

———. "By 2030 We'll Have Full-Immersion, Shared, Virtual-Reality Environments." *Kurzweilai.net*. Accessed June 29, 2020. https://www.kurzweilai.net/ray-kurzweil-by-2030-full-immersion-vr.

Kwastek, Katja. "Immersed in Reflection? The Aesthetic Experience of Interactive Media Art." In *Immersion in the Visual Arts and Media*, edited by Fabienne Liptay and Burcu Dogramaci, 67–86. Leiden: Brill Rodopi, 2016.

Lacan, Jacques "The Mirror Stage." In *Identity: A Reader*, edited by Paul du Gay, Jessica Evans, and Peter Redman, 44–50. London: Sage, 2000.

Lanier, Jaron. *Dawn of the New Everything: A Journey through Virtual Reality*. London: Bodley Head, 2017.

———. "Virtual Reality: The Promise of the Future." *Interactive Learning International* 8, no. 4 (1992): 275–279.

Lasswell, Harold D. "The Structure and Function of Communication in Society." In *The Communication of Ideas: A Series of Addresses*, edited by Lyman Bryson, 32–52. New York: Institute for Religious and Social Studies, 1948.

Latour, Bruno. *Reassembling the Social: An Introduction to Actor-Network-Theory*. Oxford: Oxford University Press, 2005.

Laurel, Brenda. *Computers as Theatre*. Rev. ed. Reading, MA: Addison-Wesley, 1993.

Leach, Robert. *Theatre Studies: The Basics*. London: Routledge, 2008.

Lee, Kwan Min. "Presence, Explicated." *Communication Theory* 14, no. 1 (2004): 27–50.

Lévy, Pierre. *Collective Intelligence*. Translated by Robert Bononno. New York: Perseus Books, 1999.

Liptay, Fabienne. "Neither Here nor There: The Paradoxes of Immersion." In *Immersion in the Visual Arts and Media*, edited by Fabienne Liptay and Burcu Dogramaci, 87–110. Leiden: Brill Rodopi, 2016.

Loiperdinger, Martin. "Lumière's Arrival of the Train: Cinema's Founding Myth." *Moving Image* 4, no. 1 (2004): 89–118.

Lombard, Matthew, and Theresa Ditton. "At the Heart of It All: The Concept of Presence." *Journal of Computer-Mediated Communication* 3, no. 2 (1997): unpaginated.

Lotman, Juri. "On the Semiosphere." Translated by Wilma Clark. *Sign System Studies* 33, no. 1 (2005): 205–229.

MacDonald, Scott. *The Garden in the Machine: A Field Guide to Independent Films about Place*. Berkeley: University of California Press, 2001.

Machon, Josephine. *Immersive Theatres: Intimacy and Immediacy in Contemporary Performance*. London: Palgrave Macmillan, 2013.

Madrigal, Alexis. "Netflix built its microgenres by staring into the American soul." *NPR.org*. Accessed April 5, 2020. https://www.npr.org/sections/alltechconsidered/2014/01/02/259128268/netflix-built-its-microgenres-by-staring-into-the-american-soul?t=1583985594943.

Mahler, Andreas. "Aesthetic Illusion in Theatre and Drama: An Attempt at Application." In *Immersion and Distance: Aesthetic Illusion in Literature and Other Media*, edited by Werner Wolf, Walter Bernhart, and Andreas Mahler, 151–182. Amsterdam: Rodopi, 2013.

———. "Glauben, Nicht-Glauben, Anders-Sagen: Wege des Fingierens in Englands früher Neuzeit." In *Fiktionen des Faktischen in der Renaissance*, edited by Ulrike Schneider and Anita Traninger, 23–44. Stuttgart: Steiner, 2010.

———. "Presented Representation: Intermedial Go-Betweens on the Shakespearean Stage." *Shakespeare Jahrbuch* 143 (2007): 147–158.

Mann, Steve, and Woodrow Barfield. "Introduction to Mediated Realities." *International Journal of Human-Computer Interaction* 15, no. 2 (2003): 205–208.

Mannavis, Sarah. "How Instagram's Plastic Surgery Filters Are Warping the We See Our Faces." *Newstatesman.com*. Accessed May 27, 2020. https://www.newstatesman.com/science-tech/social-media/2019/10/how-instagram-plastic-surgery-filter-ban-are-destroying-how-we-see-our-faces.

Manning, Christopher D. "Computational linguistics and deep learning." *Computational Linguistics* 41, no. 4 (2015): 701–707.

Männistö-Funk, Tiina, and Tanja Sihvonen. "Voices from the Uncanny Valley. How Robots and Artificial Intelligences Talk Back to Us." *Digital Culture and Society* 4, no. 1 (2018): 45–64.

Manovich, Lev. "Can We Think without Categories?" *Digital Culture and Society* 4, no. 1 (2018): 17–27.

———. *The Language of New Media*. Cambridge, MA: MIT Press, 2001.

———. "To Lie and to Act: Potemkin's Village, Cinema and Telepresence." In *The Robot in the Garden: Telerobotics and Telepistemology in the Age of the Internet*, edited by Ken Goldberg, 164–179. Cambridge, MA: MIT Press, 2000.

Mateer, John. "Directing for Cinematic Virtual Reality: How the Traditional Film Director's Craft Applies to Immersive Environments and Notions of Presence." *Journal of Media Practice* 18, no. 1 (2017): 14–25.

Mathwick, Charla, and Edward Rigdon. "Play, Flow, and the Online Search Experience." *Journal of Consumer Research* 31, no. 2 (2004): 324–332.

Mayer-Schönberger, Viktor, and Kenneth Cukier. *Big Data: A Revolution That Will Transform How We Live, Work and Think*. London: John Murray, 2013.

McCabe, Janet, and Kim Akass. *Quality TV*. London: I. B. Tauris, 2007.

McCormick, Casey J. "'Forward Is the Battle Cry': Binge-Viewing Netflix's *House of Cards*." In *The Netflix Effect: Technology and Entertainment in the 21st Century*, edited by Kevin McDonald and Daniel Smith-Rowsey, 101–116. New York: Bloomsbury, 2016.

McCreery, Michael P, Kathleen S. Krach, S., P. G. Schrader, and Randy Boone. "Defining the Virtual Self: Personality, Behavior, and the Psychology of Embodiment." *Computers in Human Behavior* 28 (2012): 976–983.

McLuhan, Marshall. *Understanding Media*. New York: McGraw-Hill, 1964.

McMahan, Alison. "Immersion, Engagement, and Presence: A Method for Analyzing 3-D Video Games." In *The Video Game Theory Reader*, edited by Mark J. P. Wolf and Bernard Perron, 67–86. London: Routledge, 2003.

McQuail, Denis. *McQuail's Mass Communication Theory*. London: Sage, 2010.

Meloy, J. Reid, Kris Mohandie, and Mila Green. "A Forensic Investigation of Those Who Stalk Celebrities." In *Stalking, Threatening, and Attacking Public Figures: A Psychological and Behavioral Analysis*, edited by J. Reid Meloy, Lorraine Sheridan, and Jens Hoffmann, 37–54. Oxford: Oxford University Press, 2008.

Merrin, William. "To Play with Phantoms: Jean Baudrillard and the Evil Demon of the Simulacrum." *Economy and Society* 30, no. 1 (2001): 85–111.

Metzinger, Thomas. "The Myth of Cognitive Agency: Subpersonal Thinking as a Cyclically Recurring Loss of Mental Autonomy." *Frontiers in Psychology* 4. Accessed January 30, 2020. https://www.frontiersin.org/articles/10.3389/fpsyg.2013.00931/full.

———. *The Ego Tunnel. The Science of the Mind and the Myth of the Self*. New York: Basic Books, 2009.

———. *Being No One. The Self-Model Theory of Subjectivity*. Cambridge, MA: MIT Press, 2001.

Microsoft. "Introducing Windows Mixed Reality." *Microsoft.com*. Accessed January 30, 2020. https://www.microsoft.com/en-gb/windows/windows-mixed-reality.

Milgram, Paul, and Fumio Kishino. "A Taxonomy of Mixed Reality Visual Displays." *IEICE TRANSACTIONS on Information and Systems* E77-D, no. 12 (1994): 1321–1329.

Milyavskaya, Marina, Mark Saffran, Nora Hoppe, and Richard Koestner. "Fear of Missing Out: Prevalence, Dynamics and Consequences of Experiencing FOMO." *Motivation and Emotion* 42 (2018): 725–737.

Misoch, Sabina. "Avatare: Spiel(er)figuren in virtuellen Welten." In *Digitale Jugendkulturen*, edited by Kai-Uwe Hugger, 169–185. Wiesbaden: Springer, 2010.

Moody, Paul. "An 'Amuse-Bouche at Best': 360 Degree VR Storytelling in Full Perspective," *International Journal of E-Politics* 8, no. 3 (2017): 42–50.

Mori, Masahiro. "The Uncanny Valley." Translated by Karl F. MacDorman and Norri Kageki. *IEEE Robotics and Automation* 19, no. 2 (2012): 98–100.

Mulvey, Laura. "Visual Pleasure and Narrative Cinema." In *Film Theory and Criticism: Introductory Readings*, edited by Leo Braudy and Marshall Cohen, 833–844. New York: Oxford University Press, 1999.

Munakata, Toshinori. *Fundamentals of the New Artificial Intelligence. Neural, Evolutionary, Fuzzy and More*. New York: Springer, 2008.

Murgia, Madhumita. "The Insidious Threat of Biometrics." *Ft.com*. Accessed April 5, 2020. https://www.ft.com/content/cdf0d52a-c2de-11e9-a8e9-296ca66511c9.

Murray, Janet. *Hamlet on the Holodeck: The Future of Narrative in Cyberspace*. New York: Free Press, 1997.

Nassar, Dalia. *The Relevance of Romanticism: Essays on German Romantic Philosophy.* Oxford: Oxford University Press, 2014.

Natterer, Kathrin. *Nostalgie als Zukunftsstrategie für Unterhaltungsmedien: Empirische Studien zu persönlicher und historischer Nostalgie in Medien.* Wiesbaden: Springer, 2017.

Neff, Gina, and Dawn Nafus. *The Quantified Self.* Cambridge, MA: MIT Press, 2016.

Negroponte, Nicholas. *Being Digital.* New York: Knopf, 1995.

Nell, Victor. *Lost in a Book: The Psychology of Reading for Pleasure* (New Haven: Yale University Press, 1988.

Netflix. "Projects." *Makeit.netflix.com.* Accessed March 30, 2020. http://makeit.netflix.com/projects.

———. "Netflix Cheating Is on the Rise Globally and Shows No Signs of Stopping." *PRNewswire.com.* Accessed March 30, 2020. https://www.prnewswire.com/news-releases/netflix-cheating-is-on-the-rise-globally-and-shows-no-signs-of-stopping-300406051.html.

Nielsen. "Binge Bunch: Two-thirds of Global VOD viewers Say They Watch Multiple Episodes in a Single Sitting." *Nielsen.com.* Accessed February 7, 2020. http://www.nielsen.com/us/en/insights/news/2016/binge-bunch-two-thirds-of-global-vod-viewers-say-they-watch-multiple-episodes-per-sitting.html.

Nilsson, Niels Christian, and Rolf Nordahl. "Immersion Revisited: A Review of Existing Definitions of Immersion and Their Relation to Different Theories of Presence." *Human Technology* 12, no. 2 (2016): 108–134.

Novalis (Friedrich von Hardenberg). *Henry of Ofterdingen.* Translated by Palmer Hiltey. New York: Ungar, 1964.

Nozick, Robert. *Anarchy, State, and Utopia.* New York: Basic Books, 1974.

O'Brian, Daniel Paul. "The Pervasive and the Digital: Immersive Worlds in Blast Theory's 'A Machine to See With' and Dennis Del Favero's 'Scenario.'" *International Journal of E-Politics* 8, no. 3 (2017): 30–41.

Oculus. "Introducing 'Facebook Horizon,' a New Social VR World, Coming to Oculus Quest and the Rift Platform in 2020." *Oculus.com.* Accessed April 18, 2020. https://www.oculus.com/blog/introducing-facebook-horizon-a-new-social-vr-world-coming-to-oculus-quest-and-the-rift-platform-in-2020/.

———. "Welcome to Facebook Horizon." *Oculus.com.* Accessed February 15, 2020. https://www.oculus.com/facebookhorizon/.

Öztürk, Seyda. "Simulation Reloaded." *Cinetext.philo.at.* Accessed June 29, 2020. http://cinetext.philo.at/magazine/ozturk/seyda_ozturk-simulation_reloaded.pdf.

Papenburg, Bettina. "Touching the Screen, Striding through the Mirror: The Haptic in Film." In *What Does a Chameleon Look Like? Topographies of Immersion*, edited by Stefanie Kiwi Menrath and Alexander Schwinghammer, 113–136. Cologne: Halem, 2011.

Parker, Felan. "Millions of Voices: Star Wars, Digital Games, Fictional Worlds and Franchise Canon." In *Game on, Hollywood! Essays on the Intersection of Video Games and Cinema*, edited by Gretchen Papazian and Joseph Michael Sommers, 156–168. Jefferson, NC: McFarland, 2013.

Patterson, Zabet. "Going On-line: Consuming Pornography in the Digital Era." In *Porn Studies*, edited by Linda Williams, 104–125. Durham: Duke University Press 2004.

Pertaub, David-Paul, Mel Slater, and Chris Barker. "An Experiment on Public Speaking Anxiety in Response to Three Types of Virtual Audience." *Presence: Teleoperators and Virtual Environments* 11, no. 1 (2002): 68–78.

Peterson, Andrea. "Holocaust Museum to Visitors: Please Stop Catching Pokémon Here." *Washington Post*. Accessed January 15, 2020. https://www.washingtonpost.com/news/the-switch/wp/2016/07/12/holocaust-museum-to-visitors-please-stop-catching-pokemon-here/.

Politico. "Coronavirus will change the world permanently. Here's how." *Politico.com*. Accessed March 23, 2020. https://www.politico.com/news/magazine/2020/03/19/coronavirus-effect-economy-life-society-analysis-covid-135579.

Postman, Neil. *Technopoly: The Surrender of Culture to Technology*. New York: Vintage Books, 1993.

———. *Amusing Ourselves to Death: Public Discourse in the Age of Show Business*. New York: Penguin, 1986.

Price, Hanna O. *Internet Addiction*. Hauppauge, NY: Nova Science Publishers, 2001.

Prümm, Karl. "From the Unchained to the Ubiquitous Motion-Picture Camera: Camera Innovations and Immersive Effects." In *Immersion in the Visual Arts and Media*, edited by Fabienne Liptay and Burcu Dogramaci, 139–164. Leiden: Brill Rodopi, 2016.

PwC. "Perspectives from The Global Entertainment and Media Outlook 2017–2021." *Pwc.com*. Accessed February 12, 2020. https://www.pwc.com/gx/en/entertainment-media/pdf/outlook-2017-curtain-up.pdf.

Rabinovitz, Lauren. *Electric Dreamland: Amusement Parks, Movies, and American Modernity*. New York: Columbia University Press, 2012.

Rafele, Antonio. "The Concept of Immersion: Pornography and Virtual Reality." *Sociétés* 4, no. 142 (2018): 19–31.

Reed, Stephen K. *Cognition: Theory and Applications*. 6th edition. Belmont, CA: Wadsworth, 2004.

ResearchAndMarkets. "Virtual Reality—Global Market Outlook (2017–2026)." *ResearchAndMarkets.com*. Accessed February 12, 2020. https://www.researchandmarkets.com/reports/5017503/virtual-reality-global-market-outlook-2018?w=4.

Rheingold, Howard. "A Slice of Life in My Virtual Community." In *Global Networks: Computers and International Communication*, edited by Linda M. Harasim, 57–80. Cambridge, MA: MIT Press, 1993.

———. *Virtual Reality*. New York: Summit Books/Simon & Schuster, 1991.

Rickett, Oscar. "How 'Netflix and Chill' Became Code for Casual Sex." *TheGuardian.com*. Accessed March 30, 2020. https://www.theguardian.com/media/shortcuts/2015/sep/29/how-netflix-and-chill-became-code-for-casual-sex.

Riva, Giuseppe, and John A. Waterworth. "Being Present in a Virtual World." In *The Oxford Handbook on Virtuality*, edited by Mark Grimshaw, 205–221. Oxford: Oxford University Press, 2015.

Rose, Frank. *The Art of Immersion: How the Digital Generation Is Remaking Hollywood, Madison Avenue, and the Way We Tell Stories*. New York: W. W. Norton, 2011.

Rosenthal, Emerson. "Why Does Every Scene in 'House of Cards' Look the Same?" *Creators.Vice.com*. Accessed October 12, 2017. https://creators.vice.com/en_uk/article/8qv8xp/why-does-every-scene-in-house-of-cards-look-the-same.

Rouvroy, Antoinette. "The End(s) of Critique: Data-Behaviourism vs. Due-Process." In *Privacy, Due Process and the Computational Turn: The Philosophy of Law meets the Philosophy of Technology*, edited by Mireille Hildebrandt and Ekatarina De Vries, 143–168. New York: Routledge, 2013.

Ryan, Marie-Laure. *Narrative as Virtual Reality 2: Revisiting Immersion and Interactivity in Literature and Electronic Media*. Baltimore: Johns Hopkins University Press, 2015.

———. "Impossible Worlds and Aesthetic Illusion." In *Immersion and Distance: Aesthetic Illusion in Literature and Other Media*, edited by Werner Wolf, Walter Bernhart, and Andreas Mahler, 131–148. Amsterdam: Rodopi, 2013.

———. "From Narrative Games to Playable Stories: Toward a Poetics of Interactive Narrative." *Storyworlds: A Journal of Narrative Studies* 1 (2009): 43–59.

———. *Avatars of Story*. Minneapolis: University of Minneapolis Press, 2006.

———. *Narrative as Virtual Reality. Immersion and Interactivity in Literature and Electronic Media*. Baltimore: Johns Hopkins University Press, 2001.

———. "Immersion vs. Interactivity: Virtual Reality and Literary Theory." *SubStance* 28 (1999): 110–137.

———. *Possible Worlds, Artificial Intelligence and Literary Theory*. Bloomington, IN: Indiana University Press, 1991.

Saleem, Karshif, Basit Shahzad, Mehmet A. Orgun, Jalal Al-Muhtadi, Joel J.P.C. Rodrigues, and Mohammed Zakariah. "Design and Deployment Challenges in Immersive and Wearable Technologies." *Behaviour and Information Technology* 36, no. 7 (2017): 687–698.

Saussure, Ferdinand de. *Course in General Linguistics*. New York: McGraw-Hill, 1959.

Schmerheim, Philipp. *Skepticism Films: Knowing and Doubting the World in Contemporary Cinema*. New York: Bloomsbury, 2015.

Schulzke, Marcus. "Translation between Forms of Interactivity: How to Build the Better Adaption." In *Game on, Hollywood! Essays on the Intersection of Video Games and Cinema*, edited by Gretchen Papazian and Joseph Michael Sommers, 70–85. Jefferson, NC: McFarland, 2013.

Sedikides, Constantine, Tim Wildschut, Clay Routledge, and Jamie Arndt. "Nostalgia Counteracts Self-Discontinuity and Restores Self-Continuity." *European Journal of Social Psychology* 45 (2015): 52–61.

———. "Nostalgia. Past, Present, and Future." *Current Directions in Psychological Science* 17, no. 5 (2008): 304–307.

Seer, Martin. *Aesthetics of Appearing*. Translated by John Farell. Stanford: Stanford University Press, 2005.

Seidel, Jane. "The Game of Tinder: Gamification of Online Dating in the 21st Century." *Medium.com*. Accessed April 5, 2018. https://medium.com/@jane_seidel/the-game-of-tinder-3c3ad575623f.

Sherman, William R., and Alan B. Craig. *Understanding Virtual Reality: Interface, Application, and Design*. San Francisco: Morgan Kaufmann, 2003.

Shrivastava, Ashish, Tomas Pfister, Oncel Tuzel, Josh Susskind, Wenda Wang, and Russ Web. "Learning from simulated and unsupervised images through adversarial training." *Arxiv.org*. Accessed April 5, 2020. https://arxiv.org/pdf/1612.07828.pdf.

Simanowski, Roberto. *Digital Art and Meaning: Reading Kinetic Poetry, Text Machines, Mapping Art, and Interactive Installations*. Minneapolis: University of Minnesota Press, 2011.

Simeon, Daphne, and Jeffrey Abugel. *Feeling Unreal: Depersonalization Disorder and the Loss of the Self*. Oxford: Oxford University Press, 2006.

Simmel, Georg. "The Picture Frame: An Aesthetic Study." *Theory, Culture and Society* 11, no. 1 (1994): 11–17.

Slater, Mel. "A Note on Presence Terminology." *Cs.ucl.ac.uk*. Accessed July 7, 2020. http://www0.cs.ucl.ac.uk/research/vr/Projects/Presencia/ConsortiumPublications/ucl_cs_papers/presence-terminology.htm.

Slater, Mel, and Sylvia Wilbur. "A Framework for Immersive Virtual Environments (Five): Speculations on the Role of Presence in Virtual Environments." *Presence: Teleoperators and Virtual Environments* 6, no. 6 (1997): 603–616.

Sloterdijk, Peter. "Architektur als Immersionskunst." *Arch+* 178 (2006): 58–61.

Solimini, Angelo G. "Are There Side Effects to Watching 3D Movies? A Prospective Crossover Observational Study on Visually Induced Motion Sickness," *PloS ONE* 8, no. 2 (2009): 1–8.

Sontag, Susan. *Against Interpretation and Other Essays*. London: Picador, 1966.

Spitzberg, Brian H., and William R. Cupach. "Fanning the Flames of Fandom: Celebrity Worship, Parasocial Interaction, and Stalking." In *Stalking, Threatening, and Attacking Public Figures: A Psychological and Behavioral Analysis*, edited by J. Reid Meloy, Lorraine Sheridan, and Jens Hoffmann, 287–324. Oxford: Oxford University Press, 2008.

Stein, Joel. "Millennials: The Me Me Me Generation." *Time.com*. Accessed May 27, 2020. http://time.com/247/millennials-the-me-me-me-generation/.

Steinicke, Frank, and Gerd Bruder. "A Self-Experimentation Report about Long-Term Use of Fully-Immersive Technology." In *The 2nd ACM Symposium*, edited by Andy Wilson, Frank Steinicke, Evan Suma, and Wolfgang Stürzlinger, 66–69. Honolulu: AMC, 2014.

Stelarc, "Towards the Post-Human: From Psycho-body to Cybersystem." *Architectural Design* 118 (1995): 91–96.

Stelter, Brian. "New Way to Deliver a Drama: All 13 Episodes in One Sitting." *NYTimes.com*. Accessed April 15, 2020. http://www.nytimes.com/2013/02/01/business/media/netflix-to-deliver-all-13-episodes-of-house-of-cards-on-one-day.html.

Stephens-Davidowitz, Seth. "The Songs That Bind." *NYTimes.com*. Accessed February 10, 2020. https://www.nytimes.com/2018/02/10/opinion/sunday/favorite-songs.html.

Stever, Gayle S., and Kevin Lawson. "Twitter as a Way for Celebrities to Communicate with Fans: Implications for the Study of Parasocial Interaction." *North American Journal of Psychology* 15, no. 2. (2013): 339–354.

Stiegler, Christian. "The Politics of Immersive Storytelling: Virtual Reality and the Logics of Digital Ecosystems." *International Journal of E-Politics* 8, no. 3 (2017): 1–15.

———. "Invading Europe: Netflix's Expansion to the European Market and the Example of Germany." In *The Netflix Effect: Technology and Entertainment in the 21st Century*, edited by Kevin McDonald and Daniel Smith-Rowsey, 235–246. New York: Bloomsbury, 2016.

———. "Medienrealität(en): Zur Konstruktion medialer Wirklichkeiten." In *New Media Culture: Mediale Phänomene der Netzkultur*, edited by Christian Stiegler, Patrick Breitenbach, and Thomas Zorbach, 181–194. Bielefeld: Transcript, 2015.

———. "Selfies und Selfie Sticks: Automedialität des digitalen Selbstmanagements." In *New Media Culture: Mediale Phänomene der Netzkultur*, edited by Christian Stiegler, Patrick Breitenbach, and Thomas Zorbach, 67–82. Bielefeld: Transcript, 2015.

Sung, Yoon Hi, Eun Yeon Kang, and Wei-Na Lee. "A Bad Habit for Your Health? An Exploration of Psychological Factors for Binge-Watching Behavior." 65th Annual International Communication Association Conference, San Juan, Puerto Rico, May 21–25, 2015.

Sutherland, Ivan. "The Ultimate Display." In *Information Processing 1965: Proceedings of International Federation for Information Processing Congress 65*. Edited by Wayne A. Kalenich, 506–508. Washington, DC: Spartan Books, 1965.

Thon, Jan-Noël. "Immersion Revisited: On the Value of a Contested Concept." In *Extended Experiences: Structure, Analysis and Design of Computer Game Player Experience*, edited by Amyris Fernandez, Olli Leino, and Hanna Wirman, 29–43. Rovaniemi: Lapland University Press, 2008.

Thrift, Nigel. "Afterwords." *Environment and Planning D: Society and Space* 18 (2000): 213–255.

Toro, Fernando de, and Carole Hubbard. *Theatre Semiotics: Text and Staging in Modern Theatre*. Toronto: University of Toronto Press, 1995.

Torop, Peeter. "Semiosphere and/as the Research Object of Semiotics of Culture." *Sign System Studies* 33, no. 1 (2005): 159–173.

Tuan, Yi-Fu. *Escapism*. Baltimore: Johns Hopkins University Press, 1998.

Urban Dictionary, "Netflix and chill." *Urbandictionary.com*, Accessed March 30, 2020. http://www.urbandictionary.com/define.php?term=netflix and chill.

Vorderer, Peter. "Interactive Entertainment and Beyond." In *Media Entertainment: The Psychology of Its Appeal*, edited by Dolf Zillman and Peter Vorderer, 21–36. Mahwah, NJ: Lawrence Erlbaum Associates, 2000.

Wagner, Richard. "Outlines of the Artwork of the Future." Reprinted in *Multimedia: From Wagner to Virtual Reality*, edited by Randall Packer and Ken Jordan, 3–9. New York: W. W. Norton, 2001.

Walton, Kendall L. *Mimesis as Make-Believe*. Cambridge, MA: Harvard University Press, 1993.

Warnke, Martin. "On the Spot. The Double Immersion of Virtual Reality." In *Immersion in the Visual Arts and Media*, edited by Fabienne Liptay and Burcu Dogramaci, 205–213. Leiden: Brill Rodopi, 2016.

Weber, Stefan. "Media and the Construction of Reality." *Mediamanual.at*. Accessed July 6, 2020. https://www.mediamanual.at/en/pdf/Weber_etrans.pdf.

## References

Weinbaum, Stanley Grauman. "Pygmalion's Spectacles." *Gutenberg.org*. Accessed July 6, 2020. https://www.gutenberg.org/files/22893/22893-h/22893-h.htm.

Weiss, Robert S. "The Attachment Bond in Childhood and Adulthood." In *Attachment across the Life Cycle*, edited by Colin Murray Parkes, Joan Stevenson-Hinde, and Peter Marris, 66–76. New York: Routledge, 1991.

———. "Attachment in Adult Life." In *The Place of Attachment in Human Behaviour*, edited by Colin Murray Parkes and Joan Stevenson-Hinde, 171–184. New York: Basic Books, 1982.

Wenger, Etienne. *Artificial Intelligence and Tutoring Systems: Computational and Cognitive Approaches to the Communication of Knowledge*. Burlington, Mass.: Morgan Kaufmann, 2014.

Werry, Margaret, and Bryan Schmidt. "Immersion and the Spectator." *Theatre Journal* 66 (2014): 467–479.

Whitehead, Alfred North. *Process and Reality*. London: Macmillan, 1967.

Wiggins, Bradley E. "Navigating an Immersive Narratology: Factors to Explain the Reception of Fake News." *International Journal of E-Politics* 8, no. 3 (2017): 16–33.

Williams, Raymond. *Television: Technology and Cultural Form*. London: Fontana, 1974.

Wilson, Edward O. *Consilience: The Unity of Knowledge*. New York: Vintage Books, 1990.

Winner, Langdon. *Autonomous Technology: Technics-Out-of-Control as a Theme in Political Thought*. Cambridge, MA: MIT Press, 1977.

Witmer, Bob G., and Michael J. Singer. "Measuring Presence in Virtual Environments: A Presence Questionnaire." *Presence: Teleoperators and Virtual Environments* 7, no. 3 (1998): 225–240.

Wolf, Werner. "Aesthetic Illusion." In *Immersion and Distance: Aesthetic Illusion in Literature and Other Media*, edited by Werner Wolf, Walter Bernhart, and Andreas Mahler, 1–67. Amsterdam: Rodopi, 2013.

———. "Aesthetic Illusion as an Effect of Lyric Poetry?" In *Immersion and Distance: Aesthetic Illusion in Literature and Other Media*, edited by Werner Wolf, Walter Bernhart, and Andreas Mahler, 183–233. Amsterdam: Rodopi, 2013.

———. "Introduction: Frames, Framings and Framing Borders in Literature and Other Media." In *Framing Borders in Literature and Other Media*, edited by Werner Wolf and Walter Bernhart, 1–40. Amsterdam: Rodopi, 2006.

———. *Ästhetische Illusion und Illusionsdurchbrechung in der Erzählkunst. Theorie und Geschichte mit Schwerpunkt auf englischem illusionsstörenden Erzählen*. Tübingen: Max Niemeyer Verlag, 1993.

Yee, Nick. "The Labor of Fun." *Games and Culture* 1 (2006): 68–71.

Yee, Nick, and Jeremy Bailenson. "The Proteus Effect: The Effect of Transformed Self-Representation on Behavior." *Human Communication Research* 33, no. 3 (2007): 271–290.

Young, Kimberly S. "Internet Addiction: The Emergence of a New Clinical Disorder." *CyberPsychology and Behavior* 1, no. 3 (1998): 237–244.

Žižek, Slavoj. "The Reality of the Virtual." *Openculture.com*. Accessed July 2, 2020. http://www.openculture.com/2014/07/the-reality-of-the-virtual-zizek.html.

# Index

360° cameras, 148–149
360° gaze framework, 9–10, 42, 65–70, 78, 80, 85, 87, 95, 101, 109, 113, 122, 146, 174, 179 183, 207, 226, 231
360° installations, 103, 127
360° video, 15, 16, 20, 27, 30, 69, 76, 128, 135, 138, 146, 149–151, 153–154
3D technology, 140
5G, 214

ABBA, 222
Abrams, J. J., 96, 98
Actor-network theory (ANT), 212
Adorno, Theodor W., 60
Algorithms, 14, 100, 174, 189–192, 200–202, 214–216, 224
  algorithmic flow, 200–201
*Alice's Adventures in Wonderland*, 6, 7, 16, 25, 26, 63, 107, 121–122
All-surrounding experiences, viii, 5, 9, 21, 53, 65, 83, 120, 217
Alphabet Inc., 223. *See also* Google
Alternative realities, 8, 14, 56, 63–64, 74, 104, 203
AltspaceVR, 94, 175, 176
Amacher, Maryann, 126
Amazon, 26, 224
  Alexa, 216
Amenábar, Alejandro, 3, 139
Andrejevic, Mark, 101
Apple, 19, 26, 209, 224
  iPhone, 32, 206–207
  Siri, 215, 216
Aristotle, 111
Artaud, Antonin, 120
ARTE, 11
Artificial intelligence (AI), 5, 10, 105, 202, 209, 212–219, 220
  AI systems, 213, 215–216, 217–218
Assmann, Aleida and Jan, 60
Atkins, Ed, 124, 163, 217–218, 219–220
Attwood, Feona, 226
Augmented reality (AR), 5, 19, 22, 32, 35
Automediality, 167–168
Avatars, digital, 4, 14, 15, 52, 67, 86, 88, 139, 161–165, 171–176, 205, 219, 223, 230
  Facebook and, 76–77, 84, 92, 161, 163–164, 172
  in gaming, 82, 172
  immersion and, 61
AzanaBand, 231

Baker, Frederick, 125
Balázs, Béla, 58–59, 81, 136, 138
Barad, Karen, 210, 211–212
Barker, Robert, 103, 104
Barthes, Roland, 54, 111, 116

Baudrillard, Jean, 25, 40–41, 42–43, 63, 80
Bauman, Zygmunt, 57, 59
Bazin, André, 137
*Beat Saber*, 21, 29
Beck, 128
Bell, Alice, 112
Belting, Hans, 71
Benayoun, Maurice, 124
Benjamin, Walter, 58
Bennett, Tony, 80
Bente, Gary, 92
Bentham, Jeremy, 70
Berliner Festspiele, 11
Bieger, Laura, 80
Binge-watching, 10, 30, 65, 75, 91, 93, 99, 131, 185, 189–202, 224
Björk, 29, 128–129, 130, 132, 133
*Black Mirror*, 1–4, 5, 6, 24, 26
Blast Theory, 124
Boehm, Gottfried, 71
Böhme, Hartmut, 60
Bolter, Jay David, 59, 87
Bon Iver, 127
Booth, Michael, 77
Borders. *See* Thresholds
Bortolussi, Marisa, 110
Boyd, Brian, 86
*Breaking Bad*, 98
Brecht, Bertolt, 118–119
Brewster, David, 136
Bricken, Meredith, 34
Broadcasting, 33, 142
Brooker, Charlie, 1
Brown, Emily, 45
Butler, Judith, 210

Cairns, Paul, 45
Calleja, Gordon, 39, 45, 48, 51–52
Carey, James W., 71, 74
Carroll, Lewis, 6, 16. *See also Alice's Adventures in Wonderland*
Castronova, Edward, 8, 86

Cinema, 38, 45, 51, 54, 58, 74, 90, 127, 136–140, 149, 155, 158, 198
 360°, 59, 150
 cinematic VR, 15, 20–21, 59, 141, 146, 150, 222
 immersive, 123–124
 virtual, 16
CinemaScope, 141
Cinerama, 141
Cline, Ernest, 6
Cole, Tim, 188
*Collisions*, 151, 157–159
Cope, David, 215
Courtois, Cédric, 201
COVID-19 pandemic, 7, 226
Crowe, Cameron, 3
Csíkszentmihályi, Mihály, 87, 190–191
Cukier, Kenneth, 100
Cultural industries, 26, 42, 123–124, 179
 and digital ecosystems, 9, 10, 26, 66–68, 93–101, 179, 226
Cunningham, Chris, 128
Cupach, William R., 188
Curtis, Robin, 80, 91–92

Danielewski, Mark Z., 114
Data, 213–219, 223–224
 datafication, 99–101
 data mining, 99, 200, 213
 data protection, 229
Davis, Charlotte, 124
De Brigard, Felipe, 229
Deep learning, 210, 213, 215
Del Favero, Dennis, 124
Delight Games, 112
Depersonalization, 91, 138. *See also* Personalization
De Toro, Fernando, 116
Digital ecosystem, 2, 9, 17–18, 26, 66–68, 93, 99–101, 146, 166, 179, 185, 216, 223, 224, 226
Digital games, 11, 44–45, 47–48, 51, 53, 67, 104, 146, 148, 150

# Index

Digital masks, 156, 170
Digital self, 37, 67, 142, 143, 165–168, 171, 174, 175. *See also* Selfhood
Dinner Party, 150–151
Disney, 96–97
Disneyland, 42
Dixon, Peter, 110
Dogramaci, Burcu, 57
Dolby Atmos, 127
Doyle, Arthur Conan, 38
*Draw Me Close*, 152–153, 154–156
Dünne, Jörg, 167

Eaton, Scott, 215
Ehrmann, Jacques, 115
Elgammal, Ahmed, 216
Elsaesser, Thomas, 77–78
Emerging technologies, 1, 4, 9, 76, 157, 231
Ende, Michael, 109
Endel, 215
Enfants Terribles, Les, 81, 121–122, 130–131
Engage, 18
Eno, Brian, 127
Ensslin, Astrid, 112
Ermi, Laura, 52
*eXistenZ*, 26, 203
Extended reality (XR), 5, 8–9, 19, 22, 31, 34–35, 174–175, 231

Facebook, 1–2, 4, 8, 15–18, 26, 76–77, 84, 94, 222–223. *See also* Oculus
360° Spatial Workstation, 15
DeepFace, 51
Facebook Live, 16, 150
Horizon, 17, 18, 28, 76, 84, 92, 163–164, 223
Spaces, 15–16, 17, 62, 76, 84
Venues, 17, 28
Fandom, 29, 63, 65, 183–189
immersive, 187
Father John Misty, 55, 103, 161, 221

Feist, Ansgar, 92
Fforde, Jasper, 109
Fichte, Johann Gottlieb, 36, 105
Field, Syd, 111
Film noir, 138–139
Fleischmann, Monika, 124
Flow, 190–191, 193, 198
algorithmic flow, 200–201
Flusser, Vilém, 44, 73
Foer, Jonathan Safran, 113–114
Ford, Sam, 112
Formation of reality, 35, 43, 73, 138, 161
Foucault, Michel, 14, 69, 79–80, 117
Fove VR, 33
Frames, 51, 58–61, 63, 108, 151, 189, 208
dissolution of, 60, 61, 73, 83, 86, 104, 119, 168
Freud, Sigmund, 70, 162
Freyermuth, Gundolf S., 58, 75
Friedrich, Caspar David, 104–105, 106, 107
Frost, Ben, 127
Fuchs, Philippe, 22, 27

Gadamer, Hans-Georg, 86
Gale, Bob, 124
Gehlen, Arnold, 60
Generative adversarial networks, 215
Gerrig, Richard J., 83
Giannachi, Gabriella, 50, 79
Goethe, Johann Wolfgang von, 109
Goffman, Erving, 168
Gómez-Peña, Guillermo, 204
Gondry, Michel, 128
González Iñárritu, Alejandro, 21
Google, 26, 33, 223–224
AI, 213
Cardboard, 30, 31
Daydream View, 30, 84
Glass, 19, 20, 209
*Inside Music*, 127
Street View, 41, 84

Gorillaz, 128, 222
Gorlitz, Carolin, 1
Grau, Oliver, 36, 57, 59, 73–74, 82, 103, 104
Griffiths, Alison, 141–142
Groys, Boris, 56
Gruber, William, 136
Grusin, Richard, 59, 87
Gunning, Tom, 136

Hai-Jew, Shalin, 187
*Halcyon*, 152
*Half-Life*, 29
Haraway, Donna J., 2, 210
*Hardcore Henry*, 139, 140
*Harry Potter* (franchise), 21, 38–40, 41, 63–64, 110, 120, 186
Harvey, David, 181
Havens, Timothy, 97, 179
Headsets, 5, 7, 13–21, 23, 26–34, 48, 58, 61, 84, 90, 125, 135, 152, 155–156, 208, 220, 231
Heilig, Morton, 140
Helmond, Anne, 1
Herndon, Holly, 127, 215
Hesmondhalgh, David, 98
Higgins, E. Tory, 165
Hoffmann, E. T. A., 106, 107, 162
Holograms, 20, 32, 221–222, 228, 230
Horton, Donald, 186
*House of Cards*, 192–193, 198–199
Howells, Adrian, 156
HP Reverb 2, 31, 32
HTC, 28–29, 31, 32–33, 208
  Lighthouse, 29
  Vive, 28, 29, 31, 33, 84, 152
Hubbard, Carole, 116
Humanities and social sciences perspective, 9, 11, 45, 53, 65, 66
Human-machine hybridization, 10, 84, 202–203, 207, 229
Human-machine interaction, 5

Huxley, Aldous, 13
Hyperreality, 40–43, 80, 141, 146, 181, 219

Ibrahim, Yasmin, 142
IJsselsteijn, Wijnand, 46
IKEA, 19
Illusion, 5, 20, 23, 34, 42, 51, 111–114, 117–120
  illusion-forming, 86, 93, 187
  of immediacy, 139
ILMxLAB, 21, 34
IMAX, 59, 141–142
Immediacy, 5, 15, 16, 34, 50, 59–61, 139, 141
Immersion, 3–11, 51–54, 61, 65–66, 70
  and addiction, 193–194
  and literature, 104–115
  and liquid space, 55–58
  performativity, 78–85
  and presence, 45–54
  psychology of, 85–93
  semiotics of, 76–78
Immersive experiences, 1, 2
Immersive media, 1, 8–9, 26, 53, 66, 69, 71, 76, 103–104, 145, 148, 154, 158, 227–231
Immersive storytelling, 146, 149–150, 154
Instagram, 8, 16, 17, 76
Institute of Sound and Music, Berlin, 126
International Society for Presence Research, 48
Iqani, Mehita, 167–168

Jacobs, David L., 167
Jenkins, Henry, 26–27, 112
Johnston, John, 214
Jonze, Spike, 216
Juul, Jesper, 82

Kac, Eduardo, 204
Kant, Immanuel, 36, 71, 200

# Index

Kaprow, Allan, 121
Kardashian, Kim, 167
Kaufman, Charlie, 115
Kaye, Nick, 50, 79
Kelly, Katie, 164–165, 175–177
Kelly, Kevin, 22, 184
Kiesler, Friedrich, 118
King, Stephen, 97–98
Klingemann, Mario, 215
Koestlé-Cate, Jonathan, 74
Kokkinara, Elena, 92
Koumarianos, Myrto, 122
Kraidy, Marwan M., 204
Kristeva, Julia, 110
Kromand, Daniel, 172
Kucklick, Christoph, 213
Kuo, Andrew, 90
Kurzweil, Ray, 22, 61, 215, 217

Lacan, Jacques, 70
L-Acoustics, 127
*Lady in the Lake*, 138–139
Lanier, Jaron, 24–25, 27
Lasswell, Harold, 72
*Last Goodbye, The*, 153–154
Latour, Bruno, 212
Laurel, Brenda, 168, 169
*Lawnmower Man, The*, 25
Lee, Ang, 215
Lee, Kwan Min, 172
Leets, Laura, 188
Lehmann, Hans-Thies, 170
Lemmerz, Christian, 125
LinkedIn, 170
Liptay, Fabienne, 57
Liquidity, 8, 57–60, 63, 65, 80, 86
Liquid modernity, 57
Liquid spaces, 7, 8, 9, 53, 55–65, 68, 79, 86–87, 104, 145, 181, 187, 202, 232
Lotz, Amanda, 97, 179
Luckey, Palmer, 94
Lumière brothers, 87, 136, 155

MacDonald, Scott, 137
Machine learning, 190, 202, 213–215, 217, 224
Machon, Josephine, 119
Madame Tussauds, 34, 87–89, 120, 186
Magic Leap, 32, 128–129
  One, 20
Magritte, René, 38
Mahler, Andreas, 116–117
Manovich, Lev, 47, 74, 171, 213
Mass media, 4, 9, 11, 18, 35, 37, 51, 56, 65, 92, 100, 179, 232
*Matrix* series, 25, 131
Mayer-Schönberger, Viktor, 100
Mäyrä, Frans, 52
McCormick, Casey J., 135, 191, 198–199
McDonnell, Rachel, 92
McGregor, Wayne, 215
McLuhan, Marshall, 22, 46
McQuail, Denis, 72
McTiernan, John, 137
Media convergence, 16
Media literacy, 49, 230
Mediated representation, 37, 38, 43, 53, 59–60, 66–69, 73, 86, 101, 107, 142, 167, 169–170, 184–188
Mediation, 4, 11, 36–44, 49–54, 57–73, 78–79, 83, 87–90, 104–108, 113, 136, 144, 151, 165–168, 181–190, 210–213, 216, 225–229
Metzinger, Thomas, 67, 88
Microsoft, 94
  HoloLens, 20, 32, 129
  Microsoft Mixed Reality, 94
Minsky, Marvin, 45, 47, 213
Misoch, Sabina, 172
Mixed reality (MR), 2, 5, 19, 22, 32, 35, 94, 108, 128, 146
MMORPG (massively multiplayer online role-playing games), 24, 146–147
Mobile screens, 125, 142, 143

Mobile World Congress, 13, 14
Moody, Paul, 150
Mori, Masahiro, 162
Moser, Christian, 167
*Mr. Robot VR*, 21, 150
Multisensory experiences, 140
Murray, Janet, 80
Muse, 128
Muybridge, Eadweard, 136

Narrative gaps, 110–111, 115
National, The (band), 125, 126, 132
Natterer, Kathrin, 181
Nell, Victor, 93
Netflix, 26, 30, 99, 186, 190, 191–201
Neuhaus, Max, 126
NextVR, 17
Nintendo, 26, 143
Nostalgia, 3–4, 6, 98, 178–183, 201
*Notes on Blindness*, 151–152
Novalis (Friedrich von Hardenberg), 107–108
Nozick, Robert, 229

Oculus, 16, 18, 28, 76, 208. *See also* Facebook
  Go, 16, 33
  Parties, 16, 17, 28
  Quest, 16, 28, 33
  Quest 2, 16, 28, 27, 31, 33
  Rift, 15, 28, 152
  Rooms, 16, 17, 76, 84
  Touch, 16, 28
Oculus Connect, 15, 17
Orwell, George, 13

Panoramas, 70, 79, 95, 103–104, 136, 141–142
Parasocial, 52, 82, 184–187
  immersive, 89, 187–189
Pearl Jam, 19
Penny, Simon, 124
Perfume Genius, 127

Personalization, 92, 99–100, 110, 130, 154, 172, 177, 190, 200, 215, 224.
  *See also* Depersonalization
Phoenix, 127
Physical reality and mediated reality, 8, 35–39, 43, 52–53, 58–61, 73, 109, 115, 193, 226–232
Pimax, 33
Poell, Thomas, 100
*Pokémon Go*, 19, 35, 143–144
Posthuman, 204, 210, 212, 214, 217–219
  posthuman hybridity, 204–205, 208, 212
Postman, Neil, 13, 56
Postmodernism, 4, 25, 35, 43–44, 58, 65, 110–114, 118–119, 124, 139, 178–182, 212, 225
  postmodern culture, viii, 6–11, 22, 51, 57, 64, 123, 128, 170, 178, 182, 191, 205, 228
Presence, 44
Prince, 221
Prümm, Karl, 94–95, 138
Psychology and reception, 9, 66–70, 85–93
Punchdrunk, 81–82, 121

Rabbit holes, 6–8, 16, 24, 39, 57, 69, 75, 78, 96, 103, 108, 113–114, 134, 142, 146, 181, 225, 227
Rabinovitz, Lauren, 123
Rachitsky, Yelena, 149
Radiohead, vii, 127, 130
Ragnar Kjartansson, 125, 126, 132–133
*Ready Player One*, 6, 7, 26, 131, 147, 232
R.E.M., vii–viii, 127, 179, 183
Representation, 5, 36–42, 66, 71, 79, 137, 165
Rheingold, Howard, 171–172
Rimini Protokoll, 121
Riva, Giuseppe, 46, 49
Robotics, 9, 162, 213–214, 225, 231

# Index

Romanticism, 104–109
Rose, Frank, 145
Rouvroy, Antoinette, 200
Rowling, J. K., 110. *See also Harry Potter*
Ryan, Marie-Laure, 17, 51, 104, 111, 112, 173, 187

Saleem, Kashif, 210
Samsung, 20–21
 Galaxy, 21, 30, 134
 Gear, 20, 21, 30
 Wall, 133–134, 135
Sarin, Helena, 215
Schechner, Richard, 121
Schubert, Franz, 126, 132
Schubert, Gotthilf Heinrich von, 105
Screen cultures, 10, 58, 142, 144
Seager, James, 121, 122, 130–131
Secret Cinema, 123–124
Sedikides, Constantine, 183
Sega, 26
Selfhood, 37, 49, 52, 70, 85, 91–92, 97, 104, 105, 120, 130, 163, 166–168, 171–172, 184, 189, 191, 219. *See also* Digital self
 mediated, 66–67, 120, 142, 158, 165, 172, 181
Semiotic codes, 10, 35, 42, 44, 63–64, 70, 74, 76, 83, 96, 101, 116, 207, 226
 and social activities, 9, 66–70, 75, 78, 80, 97, 146, 174, 180, 183
Sensorama, 140
Sex robots, 225–227
Shakespeare, William, 116
Shaw, Jeffrey, 124
Shelley, Mary Wollstonecraft, 105
Sigur Rós, 32, 128–129
Silver, Cassandra, 122
Simanowski, Roberto, 112
Simmel, Georg, 59
*Sims, The*, 81, 172
Singer, Michael J., 45
Slater, Mel, 45, 46–47

Sloterdijk, Peter, 58
Smart technologies, 203, 211
Smith, Patti, 13
Snapchat, 19, 171
Snell, Ben, 215
Social interaction, 4, 66, 88, 162–163, 184, 193, 197, 230
 parasocial interactions, 88, 185–186
Social VR, 15–18, 28, 61, 70, 76–77, 83–84, 94, 159, 161–164, 172, 174–176, 218, 223
Song Exploder, 127
Sontag, Susan, 72
Sony, 29
 *Perfect*, 148
 PlayStation VR, 29, 31
Spacey, Kevin, 185–186
Spatial audio, 15, 16, 127
Spielberg, Steven, 6
Spitzberg, Brian H., 188
Spotify, 134, 201
 Time Capsule, 177
Squarepusher, 128
*Star Wars*, 21, 96–97, 184
Stelarc, 204
Stephens-Davidowitz, Seth, 177
Stevenson, Robert Louis, 107
Stockhausen, Karlheinz, 126, 132
Story worlds, 21, 97–98, 109–114, 146, 222
*Stranger Things*, 179–180
Strauss, Wolfgang, 124
Sutherland, Ivan, 22–24

Tannahill, Jordan, 152–153, 154–156
Tannen, Deborah, 8
Telepresence, 45–47, 124, 213
Theater
 immersive, 10, 50, 63, 68, 81–82, 87, 100–101, 115–124, 130–131, 146, 147, 148, 152, 155
 performance, 87, 116–117, 120, 168
 postmodern, 118–119

Théâtre du Grand-Guignol, 118
Thon, Jan-Noël, 52
Three-dimensional environments, 46, 117, 124, 127
Thresholds, 7, 39, 57–58, 61, 86, 134, 137, 143, 202–203
TikTok, 19, 214
Timberlake, Justin, 221
Timmermans, Elisabeth, 201
Tinder, 201–202
*Treehugger: Wawona*, 148
Tuan, Yi-Fu, 89–90, 227
Twitch, 8

U2, 127
Uncanny valley, 162, 165, 175

Valve, 29, 31, 32
Van Dijck, José, 100
*Vanilla Sky*, 3
Varèse, Edgard, 126
Vesna, Victoria, 124
VirBela, 18
Virtual reality (VR), 5–7, 13–34, 35–36, 61, 228–231
   Björk and, 133
   Prince and, 221
   technologies of, 13–16, 23–24, 26–34
   virtual reality sickness, 90
   VR experiences, 3, 16, 20–21, 27, 28, 32, 38–39, 48, 61, 96, 128, 148–151, 154, 174–176, 223
   VR games, 21, 29, 84, 147
Void, The, 33–34

Wagner, Richard, 118
Wallworth, Lynette, 151, 157–159
Waterworth, John A., 49
Wearable technologies, 85, 182, 202, 210
Weber, Stefan, 36–37
Weeknd, The, 128
Weinbaum, Stanley G., 23

Whats App, 16, 17, 18, 76
Wheatstone, Charles, 136
Whitehead, Alfred North, 74
Widescreen technologies, 141, 143
Wilbur, Sylvia, 45
Wilson, Edward O., 67
Winner, Langdon, 43
Witmer, Bob G., 45
Wohl, R. Richard, 186
World-building, 10, 74, 104, 112–113, 139, 146, 164, 174
*World of Warcraft*, 24, 73, 146, 147

Xiaomi, 134, 135

Yorke, Thom, 127
You Me Bum Bum Train, 121
YouTube, 8

Zemeckis, Robert, 182
Zimmerman, Tom, 25
Žižek, Slavoj, 36
Zuckerberg, Mark, 13–15, 17, 76, 98, 159, 161, 222